Mining Law and Governance in Africa

This book explores the various issues that characterise the African mining sector, drawing examples from different African countries and regional organisations. Although there is a massive literature on the subject, some issues have been neglected, including the crucial role of digitalisation and technological advancement in resolving the environmental and social challenges faced in Artisanal and Small-Scale Mining (ASM), deep-sea mining, mining contract negotiations and modernising mining laws to reflect the increasing role of critical minerals, to mention but a few. Therefore, the book unpacks the critical issues associated with the mining sector, explicitly reflecting on the practical solutions needed to address the challenges in the African mining sector. This book uniquely analyses and adds flavour to international mining's fundamental concepts by describing a simulated annealing-based approach appropriate for complex mining projects in Africa. Book contributors comprise of academics from different universities including professors, practitioners, government policymakers, NGO executives and a variety of different experts. This multidisciplinary book will be of interest to African policymakers, governments, academics, industry professionals, energy and mining institutions, international organisations, universities across the globe and companies.

Victoria R. Nalule is a lawyer and an energy and mining expert, with extensive experience working on various projects in different parts of the world. She is a holder of a PhD in International Energy Law and Policy (Dundee, UK). She is an author of five books. Victoria is the founder of a charity NGO, the African Energy and Minerals Management Initiative. She is the CEO and lead consultant at Nalule Energy and Minerals Consultancy (NEM Energy). She is also a lecturer in Energy and Natural Resources Law at the University of Bradford, UK. Victoria is enthusiastic about energy arbitration and dispute resolution. As such, she is an Energy Arbitrator at the Energy Disputes Arbitration Center (EDAC), Turkey. She is a committee member at the International Law Association (ILA) and at Access for Women in Energy.

Mining Law and Governance in Africa

Transformation and Innovation for a Sustainable Mining Sector

Edited by
Victoria R. Nalule

Routledge
Taylor & Francis Group
LONDON AND NEW YORK

First published 2023
by Routledge
4 Park Square, Milton Park, Abingdon, Oxon OX14 4RN

and by Routledge
605 Third Avenue, New York, NY 10158

Routledge is an imprint of the Taylor & Francis Group, an informa business

British Library Cataloguing-in-Publication Data
A catalogue record for this book is available from the British Library

ISBN: 978-1-032-25659-7 (hbk)
ISBN: 978-1-032-25660-3 (pbk)
ISBN: 978-1-003-28443-7 (ebk)

DOI: 10.4324/9781003284437

Typeset in Times New Roman
by Taylor & Francis Books

To the African Youth, for your continued interest in the sector.

Contents

Illustrations

Figures

Tables

Box

Acknowledgments

I would like to extend my sincere gratitude to the book contributors, for their hard work and patience during the publication of this book. I would also like to thank my colleagues at Nalule Energy and Minerals Consultancy (NEM Energy), for their support and encouragement.

Foreword 1

The mining sector is vital to the global economy. There is a need for further research in this area and this book contributes to solving this issue. The text focuses on specific African countries and on issues facing the continent generally. There are some vital topics covered in the analysis that will be well worth reading for mining law scholars.

The book focuses on mining from a perspective that includes beyond just oil and gas extraction. The geographical scope is on Africa, and this is a highlight as more legal and associated interdisciplinary research on this region is needed. That this book has this focus signifies its originality, and for me that makes it an enjoyable and impressive read. I am sure many will find this an interesting, insightful and well-written text.

Particular and notable research analysis of value is contained within the book, which I am sure will prove itself in the longevity of the lessons from the book. The coverage of many of these important issues, and to do so with a legal lens, is impressive. Overall, for me, this research offers originality and emphasises pathways of how countries can proceed in furthering economic development whilst avoiding the past pitfalls of the mining sector.

All the researchers who have contributed have opened the door to many possible avenues of future research. I think a whole range of legal stakeholders will be very interested in the discussions and opinions contained within the text. The main editor should be commended for her time, effort and perseverance which has enabled her to introduce us to this compelling account of what is happening in the mining sector in Africa from a legal perspective at this time.

Raphael J. Heffron
France, 14 October 2022

Foreword 2

It is my greatest pleasure and honour to write a foreword for this important book edited by Dr Victoria Nalule. The book exhaustively examines regulatory developments and legal innovations in the African mining sector, drawing examples from different African countries and regional organisations.

The African mining sector has the potential to contribute to the economic development of the countries which possess the minerals. Whereas the sector can contribute to the achievement of the United Nations Sustainable Development Goal (UN SDG) 1 on poverty eradication, it equally has the same potential in achieving the UN SDG 13 on climate action. This is so, given the increasing role of critical minerals in transitioning to a low-carbon economy. However, there are pertinent challenges in the sector that ought to be addressed to ensure that the African people benefit from their minerals. In light of these facts, this book unpacks some of the key issues in the African mining sector. It starts by drawing the nexus between mining, energy transitions and climate change (with a focus on the role of critical minerals).

With a brief historical background, the book reflects on the various generation of mining laws and proposes a new generation that attempts to ensure a special legislation for critical minerals, aimed at ensuring that the African continent maximises the revenues from these minerals.

The book also contributes to the existing literature regarding the social, environmental and governance issues associated with Artisanal and Small-Scale Mining (ASM). It adds flavour to the discussion by spotlighting the role of digitalisation in tackling the challenges associated with ASM.

Notable and diverse research analyses are contained in the book, drawing the readers to the impact of Chinese investments in the African mining sector; issues that ought to be addressed concerning deep-sea mining and the blue economy; the fiscal regime as reflected in the African mining sector and the increasing role of the African Mining Vision.

All the researchers who have contributed have brought first-hand experiences from different African countries. This research offers originality and draws the attention of policymakers on the pathway to take to ensure that they maximise the benefits that accrue from the mining sector. The contributors and the main editor should therefore be highly commended for their hard work in highlighting these pertinent issues in the African mining sector.

The book is timely, and it will be a great asset to professionals, government institutions, universities and a whole range of legal stakeholders who are interested in understanding the mining sector from an African perspective. Readers will feel well informed about mining law and governance in Africa after reading such an impressive work.

Damilola S. Olawuyi
SAN, FCIArb Professor and UNESCO Chair on Environmental Law and Sustainable Development, Hamad Bin Khalifa University, Doha, Qatar
Independent Expert, United Nations Working Group on Business and Human Rights

Preface

The mining sector's importance has been emphasised in the twenty-first century, given the increasing demand and role of critical minerals such as cobalt which are essential in the global transition to a low-carbon economy. The African continent is home to enormous reserves of oil, gas, energy and solid minerals. These resources are crucial for these countries' economic development. Nevertheless, in recent years, while we have witnessed initiatives to address climate change, the 2022 global energy crisis spotlighted the continued role of extractives (both fossil fuels and minerals) in our economies. There is therefore a need to strike a balance in managing the extractive industries in order to adequately achieve the net zero targets and other climate change goals, as envisaged in the 2015 Paris Agreement and Goal 13 of the United Nations Sustainable Development Goals (SDGs). Scholars have advocated for a just energy transition, while others, such as Victoria Nalule, have insisted on the 'progressive' way of managing the extractive industries and climate change. Consequently, the present and future laws, legislations and policies of the African extractive industries including oil, gas and minerals should take into consideration these environmental and sustainable global efforts.

Moving forward, in terms of solid minerals production – including gold, cobalt, industrial diamond, bauxite, phosphate, vermiculite and zirconium – African countries stand out as major producers, including South Africa, Botswana, Ghana, Zimbabwe, the Democratic Republic of the Congo and Equatorial Guinea.

We cannot ignore the importance of the mining sector to the general economic development of the continent. This is because various minerals are used as raw materials in different sectors including construction and transport. Most importantly, mineral materials also play an important role in industrialisation and urbanisation, which are now rapidly escalating on the African continent. The unique challenges and opportunities experienced in the African mining sectors have necessitated the publication of this book which focuses on the African continent. This is because common challenges do require common solutions which can easily be achieved through regionalism under the auspices of the various African regional organisations including the African Union, East African Community (EAC), Southern African Development Community (SADC), Economic Community of West African States (ECOWAS) and specialised mining regional initiatives such as the African Mining Vision (AMV). Whereas the mining sector presents opportunities such as employment, revenue generation, infrastructural development and many others, the common challenges in the sector ought also to be recognised. These challenges include the social and environmental impacts of Artisanal and Small-Scale Mining (ASM), ineffective mining legislation, poor negotiation skills for mining agreements, lack of advanced technology and poor governance, to mention just a few.

There is a growing body of scholarly articles and national, regional and international initiatives to ensure that these countries benefit from their resources. Law, policies and regulations have been at the centre of these initiatives. For countries endowed with mineral resources, the mining sector contributes to these countries' economic development as it can finance infrastructural development. In Botswana, for instance, approximately 70–80% of export earnings are attributed to diamond mining. Indeed, if well managed, revenues from the mining sector can be used to tackle the poverty challenges facing the continent. However, some countries have fallen prey to the resource curse, due to poor governance, corruption and mismanagement.

Against this backdrop, this book explores the various issues characterising the African mining sector, drawing examples from different African countries and regional organisations. Although there is a massive body of literature on the subject, some issues have been neglected, including the crucial role of digitalisation and technological advancement in resolving the social and environmental challenges faced in artisanal and small-scale mining, deep-sea mining, mining contract negotiations and the modernisation of mining laws to reflect the crucial role of critical minerals, to mention but a few. Therefore, this book unpacks the key issues associated with the mining sector, explicitly reflecting on the practical solutions needed to address the challenges in the African mining sector. The book uniquely analyses and adds flavour to international mining's fundamental concepts by describing a simulated annealing-based approach appropriate for complex mining projects in Africa. The book explores the African mining sector using a multidisciplinary approach, which has spotlighted not only the legal and social issues, but also the fiscal aspects of the mining sector. The book authoritatively illustrates the nexus between energy, mining and climate change. Book contributors comprise of academics from different universities including professors, practitioners, government policymakers, NGO executives and a variety of different experts. This multidisciplinary book will be of interest to African policymakers, governments, academics, industry professionals, energy and mining institutions, international organisations, universities across the globe and companies.

Victoria R. Nalule
Bingley, United Kingdom
12 October 2022

List of Contributors

Dr Victoria R. Nalule
Theophilus Acheampong
Bridget Okyerebea Menyeh
Peter Nuhu
Ishmael Ackah
Dramani Bukari
Cynthia Andiabutswa Opakas
Antoine Floribert Mulaba-Bafubiandi
Michael Uche Ukponu
Austin Ukpata Odo
Nirdesh Singh
Beejo Blessing Chauke
David Yibakuo Amakiri
Solomon Ugochukwu Ezike
Kangwa-Musole George Chisanga
Mwaba Chileya Mulenga
Chileshe Mwaba
Solomon N. Obulor
Eddy L. Wifa
Selina Kudzai Zhuwarara
Dr Abubakar I. Umar
Dr Muhammad Bello
King Carl Tornam Duho
Leezola Zongwe
Obvious Terrence Mutambara
Pablo Ferrara

List of Abbreviations

AETR	average effective tax rate
AfCFTA	African Continental Free Trade Agreement
ASGM	Artisanal and Small-Scale Gold Mining
ASM	Artisanal and Small-Scale Mining
ADIA	Accelerated Industrial Development of Africa
AU	African Union
AfCFTA	African Continental Free Trade Area
AfDB	African Development Bank
AMDC	African Minerals Development Centre
AMV	Africa Mining Vision
APRM	African Peer Review Mechanism
BE	Blue Economy
BIT	bilateral investment treaty
BRI	Belt and Road Initiative
CAJAC	China–Africa Joint Arbitration Center
CCDI	climate change disclosure index
CDA	community development agreement
CDSB	Climate Disclosure Standards Board
CDM	clean development mechanism
CERD	Convention on Elimination of Racial Discrimination
CFU	Carbon Finance Unit
CIT	corporate income tax
CMV	country mining vision
CONMESA	Common Market for Eastern and Southern Africa
COP	Conference of the Parties
CPA	China Policy on Africa
CSO	civil society organisation
CSR	corporate social responsibility
DFI	development finance institution
DRC	Democratic Republic of Congo
DSM	dispute settlement mechanism
EBIT	earnings before interest and taxes
EEZ	Exclusive Economic Zones
EIA	environmental impact assessment

EITI	Extractive Industries Transparency Initiative
ESG	environmental, social, and governance
ET	emission trading
EU	European Union
EV	electric vehicles
FARI	Fiscal Analysis of Resources Industries
FDI	foreign direct investment
FiTs	feed-in tariffs
FOCAC	Forum for China–Africa Cooperation
FPIC	free, prior, and informed consent
GDP	gross domestic product
GDPG	gross domestic product growth
GHG	greenhouse gases
GIS	geographic information systems
GRI	Global Reporting Initiative
HDI	Human Development Index
HSE	health, safety, and environment
IBA	impact and benefit agreement
ICCPR	International Covenant on Civil and Political Rights
ICJ	International Court of Justice
ICSID	International Centre for the Settlement of Investment Disputes
ICT	information communication technologies
IEA	International Energy Agency
IFC	International Finance Corporation
IFRS	International Financial Reporting Standards
IGF	Intergovernmental Forum on Mining, Minerals, Metals and Sustainable Development
IGY	International Geophysical Year
IIRC	International Integrated Reporting Council
ILC	International Law Commission
ILC	International Labour Organization
INDC	Intended Nationally Determined Contribution
IRENA	International Renewal Energy Agency
ISDS	investor–state dispute settlement
ISA	International Seabed Authority
ISA	International Space Agreement
ISO	International Organization for Standardization
IWOSS	industries without smokestacks
JSE	Johannesburg Stock Exchange
LCOE	levelised cost of energy
LDC	least developed countries
LDCF	Least Developed Country Fund
LSGM	large-scale gold mining
LSM	large-scale mining
MIC	Multilateral Investment Court
MIGA	Multilateral Investment Guarantee Agency

MOP	Meeting of the Parties
MOST	Maltese Ocean Space Treaty
MOU	Memorandum of Understanding
MRT	mineral royalty tax
NDCs	nationally determined contributions
NGO	non-governmenal organisation
NO_x	nitrogen oxides
NYSE	New York Stock Exchange
ODS	ozone-depleting substances
OFDI	outward foreign direct investment
PRI	Principles of Responsible Investment
PSC	production sharing contract
R&D	research and development
RDI	research, development, and innovation
RE	renewable energy
ROS	renewable obligations standard
SA	Republic of South Africa
SADC	Southern Africa Development Community
SASB	Sustainability Accounting Standards Board
SDGs	Sustainable Development Goals
SDR	sustainable development report
SLO	social licence to operate
SO_x	sulphur oxides
SOE	state-owned enterprise
SSA	Sub-Saharan Africa
SWF	sovereign wealth fund
SWX	Swiss Exchange
TCFD	Task Force on Climate-related Financial Disclosures
UN	United Nations
UNEP	United Nations Environment Programme
UNFCC	United Nations Framework Convention on Climate Change
UNCITRAL	United Nations Commission for International Trade Law
UNCLOS	United Nations Convention on the Law of the Sea
UNCTAD	United Nations Conference on Trade and Development
UNECA	United Nations Economic Commission for Africa
UNEP	United Nations Environment Programme
UNGC	United Nations Global Compact
US GAAP	United States Generally Accepted Accounting Principles
VAT	value added tax
VIF	variance inflation factor
WAGP	West African Gas Pipeline
WGI	Worldwide Governance Index
WHO	World Trade Organisation
WSSD	World Summit on Sustainable Development

Part I

Introductory Context on the African Mining Sector

1 Modernisation of the Mining Laws and Key Issues for Consideration in Africa

Victoria R. Nalule

Therefore, the purpose of this book is to explore the various issues characterising the African mining sector, drawing examples from different countries and regional organisations. Although there is a massive body of literature on the subject, some issues have been neglected, including the crucial role of digitalisation in resolving the challenges faced in Artisanal and Small-Scale Mining, deep-sea mining, the increasing importance of the mining sector in ensuring the energy transition, to mention but a few. This introductory chapter will highlight the key issues in the African mining sector, with a focus on the modernisation of the mining laws.

1 Introduction: The Mining Sector in Light of the Energy Crisis

1.1 The African Extractive Industries

The twenty-first century has been characterised by the twin challenge of addressing climate change and managing the impact of the extractive industries. The African mining sector, especially the presence of critical minerals, has a significant role to play in not only meeting the economic aspirations of the countries concerned but also in ensuring an effective energy transition.

Although the African continent is home to massive energy and mining resources, it is also home to some of the poorest countries. The resources have been exploited on the continent for centuries; however, most of the local people have not fully benefited. There are various reasons for this state of affairs, including poor governance, corruption, and lack of advanced technology.[1] However, we cannot underestimate the historical influences that have left most Africans poor. In particular, African governments have long negotiated petroleum and mining agreements to the detriment of their citizens. This was in the past associated with a lack of adequate knowledge, expertise, and skills to negotiate complex energy and mining contracts. However, this has changed, as the twenty-first century is now characterised by elite Africans. Despite the

1 Nalule VR. *Land Law and the Extractive Industries: Challenges and Opportunities in Africa*. London: Bloomsbury Publishing; 2021.

DOI: 10.4324/9781003284437-2

various improvements, African countries are still prone to unfavourable extractive contracts. This is in addition to the lack of necessary infrastructure such as refineries to ensure value addition for the minerals and energy resources. Consequently, we have witnessed a situation where some African countries mostly suffer negative impacts associated with petroleum and mining projects.[2]

Regarding extractive contract negotiations, research to date on computational models of negotiation has focused almost exclusively on defining contracts consisting of one or a few independent issues and tractable contract spaces. Many real-world contracts, by contrast, are much more complex, consisting of multiple interdependent issues and intractably large contract spaces and clauses.[3] This book, therefore, describes a simulated annealing-based approach appropriate for negotiating such complex mining contracts that achieves near-optimal social welfare for negotiations with binary issue dependencies. Additionally, the book gives an analytical outlook on the various practical issues associated with the African extractives industry, with a focus on mining.

Important to note, however, is that the extractives industry does not only imply extraction of oil, gas, and other minerals from the earth's surface. Behind these activities are people, corporations, governments, and financial institutions that support the extraction of the resources. Besides contract negotiations and land access issues including ownership and compensation, the finite nature of the extractives industry implies that countries must employ robust measures to protect the environment by, for example, ensuring mining closures and decommissioning at the end of the project life cycle.[4] The entire process of the extractives industry has an impact on the social, political, and economic sphere of the country.[5] In essence, the extractives industry comprises two main subsectors: petroleum and solid minerals. The processes in these sectors – for instance, oil and gas exploration, solid minerals mining, dredging, and quarrying – are associated with environmental concerns that directly affect the local communities and nearby landowners. Nevertheless, these activities are also responsible for the economic empowerment of various African countries including Nigeria, Botswana, Namibia, South Africa which heavily rely on revenues from the petroleum and mining sectors.[6]

2 Nalule VR. Social and Environmental Impacts of Mining. In *Mining and the Law in Africa: Exploring the Social and Environmental Impacts* (pp. 51–81). Cham: Palgrave Pivot; 2020.
3 An example of the complex clauses found in petroleum and mining agreements is stabilisation clauses. See Nalule VR. What is the Problem with Stabilization Clauses in Petroleum Agreements? *Journal of Sustainable Development Law and Policy*. 2022; 13(1): 85–102.
4 Cameron PD, Stanley MC. *Oil, Gas, and Mining: A Sourcebook for Understanding the Extractive Industries*. Washington, DC: World Bank Publications; 2017.
5 Olawuyi DS. *Extractives Industry Law in Africa*. Cham: Springer International Publishing; 2018.
6 Nalule VR. *Mining and the Law in Africa: Exploring the Social and Environmental Impacts*. Cham: Palgrave Pivot; 2020.

For instance, with respect to the mining sector, which is the main focus of this book, we note that Africa is home to 30% of the global mineral reserves, and these contribute significantly to exports and tax revenues for most countries on the continent.[7] Besides agriculture, mining is one of the oldest industries to be found on the African continent.[8] The region is endowed with various minerals including gold, diamonds, vermiculite, manganese, cobalt, zirconium, salt, and phosphate rock. These minerals are also geographically distributed across the continent, with Namibia, Zimbabwe, South Africa, Botswana, and the DRC representing some of the African historical giants in the mining industry. Namibia, for instance, is historically known for diamond mining. The country is also the world's fourth-largest producer of uranium oxide. Mining is considered the country's leading economic sector, accounting for roughly 10% of Namibia's annual GDP. In Ghana, the mining sector accounts for about 5% of the national GDP, with gold accounting for over 90% of total mineral exports. Mozambique is known for minerals such as aluminium; Zambia is a major producer of copper; Niger is responsible for around 44% of Africa's uranium supply; and Guinea is responsible for more than 95% of Africa's bauxite production. These are just a few examples of the different African countries that possess mineral wealth.[9]

The importance of various minerals including metallic, industrial, and construction minerals in today's economy cannot be overstated in both developed and developing countries. Industrialisation and urbanisation are escalating in developing countries such as those in Africa and Asia, making minerals essential for the economic growth of these countries. According to the BP Energy Outlook, the global boom in urbanisation is projected to increase, as almost 2 billion people are likely to live in urban centres by 2040, and Africa is projected to contribute one-third of this increase in urbanisation.[10] Productivity levels are also expected to increase, and it is estimated that 2.5 billion people will be lifted from the low-income bracket.[11]

All the above developments will necessitate an increase in the mining of different minerals across the African continent. Additionally, the developed world is heavily industrialised and, as such, it relies mostly on minerals in the form of raw materials, which are an essential primary input for industry. Consequently, the mining sector has become key in tackling the development of the other sectors of the economy.

7 Nalule VR. Introduction to Mining in Africa. In *Mining and the Law in Africa* (pp. 1–17). Cham: Palgrave Pivot; 2020.
8 Ibid.
9 See Nalule VR. *Mining and the Law in* Africa. Cham: Palgrave Pivot; 2020.
10 Nalule VR, Babajide NA. Protecting Natural Gas Investments: Trends in Investment Treaties and Investors Projections in Africa. In *The Palgrave Handbook of Natural Gas and Global Energy Transitions* (pp. 461–481). Cham: Palgrave Macmillan; 2022.
11 Ibid.

1.2 Nexus between the Mining Sector and the Energy Sector: Critical Minerals

It is crucial to emphasise the nexus between the mining sector and the energy sector, especially in this era of climate change. The importance of the mining sector in this energy transition era is reflected in the increasing demand for critical minerals, which are crucial in low-carbon technologies. Before a discussion on critical minerals, it is imperative to note the 2022 energy transition challenges, especially considering efforts to address the global energy crisis. Some countries in the Global North, which had initially made progress to achieve the net-zero targets, have without doubt witnessed a 'backward energy transition'.[12] The geopolitics in Europe (Russia–Ukraine war), coupled with ineffective energy policies and the divestment in fossil fuels,[13] partly contributed to the global energy crisis. As a result of the hiked oil and gas prices, many developed countries opted for fossil fuels. Ideally, with energy transitions, countries should be able to shift from traditional energy to modern energy; from fossil fuels to renewables.[14] However, the energy crisis led developed countries such as the United Kingdom to embrace firewood,[15] and Germany reopened its coal-fired power plants. In essence, this is what is referred to as the 'backward energy transition'.[16]

With respect to the African energy sector, we should not underestimate the crucial role of energy in ensuring the economic development of different African countries. Energy is used in everyday life for lighting, heating, cooking, and transport, to mention but a few. However, it is estimated that approximately 1.2 billion people globally do not have access to modern energy such as electricity, and nearly 3 billion people rely on traditional biomass such as wood and charcoal for cooking and heating.[17] The African continent is endowed with massive but untapped energy resources, including both conventional and unconventional resources. Despite these resources, the region faces various energy challenges, including low electrification rates and a heavy reliance on inefficient energy sources such as traditional biomass.

12 Dr Victoria R Nalule, 'Backward Energy Transition', www.youtube.com/watch?v=F15UCgIQiTw&t=141s. Last accessed on 10 October 2022.

13 Dr Victoria R Nalule, 'Ineffective Energy Policies', https://youtu.be/3k5E09ZkaGk. Last accessed on 10 October 2022.

14 For a discussion on modern energy and traditional energy, see Nalule VR. *Energy Poverty and Access Challenges in Sub-Saharan Africa: The Role of Regionalism*. Cham: Springer Nature; 2018.

15 Vaughan A. 'UK energy crisis sparks rush for firewood despite air pollution fears', *New Scientist*, 2 September 2022. www.newscientist.com/article/2336109-uk-energy-crisis-sparks-rush-for-firewood-despite-air-pollution-fears. Last accessed on 10 October 2022.

16 Dr Victoria R Nalule, 'Backward Energy Transition', www.youtube.com/watch?v=F15UCgIQiTw&t=82s. Last accessed on 4 October 2022.

17 United Nations Foundation. Achieving Universal Energy Access, 2019. www.unfoundation.org/what-we-do/issues/energy-and-climate/clean-energy-development.html

Whereas the challenge of energy access is rampant on the African continent, we must also worry about the challenge of climate change, which has increased the global debate on how countries can justly transition their energy systems from conventional fossil-fuel sources to zero-carbon by 2050.[18] Indeed, global discussions about energy transition must consider the social and economic discrepancies between different countries and the different energy challenges they face.[19] The African region, for instance, has a natural gas potential of approximately 558 trillion cubic feet (Tcf).[20] This makes Africa the fourth-largest gas reserves holder in the world after North America.[21] The region also has an estimated 65 billion barrels of proven oil reserves, equivalent to about 5% of the world total,[22] and yet most African countries are dependent on biomass and waste as their primary energy supply source. Nevertheless, the role of natural gas as a transitional energy has been emphasised by different scholars.[23]

Moving forward, we note that the mining sector, specifically critical minerals, remains crucial in achieving an effective energy transition. Minerals considered critical or strategic are those scarce metals and non-metals that are crucial for the economy.[24] They are characterised by geological scarcity, hence limiting their supply. Geography and availability of domestic supply often defines which minerals are deemed 'critical' for any particular region or country.[25] The importance of critical minerals has increased in recent years given the global move to transition to a low-carbon economy. Basically, energy transitions are characterised by the deployment of renewable energy and energy efficiency technologies. Among the minerals that are considered to be critical are those metals and semi-metals used in the manufacture of wind turbines, electric cars, solar panels, and other high-tech applications which are crucial for shifting to a low-carbon economy.[26] For instance, copper is

18 Heffron RJ. Applying energy justice into the energy transition. *Renewable and Sustainable Energy Reviews*. 2022; 156: 111936.
19 Bridge G, Bouzarovski S, Bradshaw M, Eyre N. Geographies of energy transition: Space, place and the low-carbon economy. *Energy Policy*. 2013; 53: 331–340.
20 OPEC, World proven natural gas reserves by country. 2019. Can be accessed at https://asb.opec.org/data/ASB_Data.php. Last visited on 20 March 2020.
21 Ibid.
22 International Energy Agency (IEA). Special Report: World Energy Investment Outlook, 2014. www.iea.org/publications/freepublications/publication/WEIO2014.pdf.
23 Nalule VR, Babajide NA. Protecting Natural Gas Investments: Trends in Investment Treaties and Investors Projections in Africa. In *The Palgrave Handbook of Natural Gas and Global Energy Transitions* (pp. 461–481). Cham: Palgrave Macmillan; 2022.
24 Dr Victoria Nalule, 'Critical Minerals Concerns', www.youtube.com/watch?v=svMx0m4sjAo&t=104s. Last accessed on 10 October 2022
25 Acheampong T. *The Energy Transition and Critical Minerals in Ghana: Diversification Opportunities and Governance Challenges.* Ghana Extractive Industries Transparency Initiative – GHEITI, Accra, Ghana; 2022.
26 Nalule VR. *Land Law and the Extractive Industries.* London: Bloomsbury Publishing; 2021.

essential for the use of electricity throughout the energy system, and platinum is a key material in many clean energy technologies and emissions control devices. Lithium, cobalt, and nickel provide batteries with greater charging performance and higher energy density.[27] Some rare earth elements such as neodymium make powerful magnets that are vital for wind turbines and electric vehicles.

The above therefore highlights the vital role of critical/strategic minerals in this era of energy transition. As such, African countries should position themselves to ensure that they benefit from critical minerals. This will necessitate not only the establishment of value addition infrastructure such as refineries but also the amendment of the relevant mining laws and regulations.

2 Key Issues of Concern in the African Mining Sector

2.1 Categorisation of Minerals and Its Impact on Policies

Categorisation of minerals and having a clear understanding of what mining entails are key in determining the necessary mining policies relevant to the exploration and development of the industry. There is an economic and social risk in excluding essential materials such as sand as part of the mining sector. This can lead not only to loss of revenue but also to the mismanagement of the resources – especially if there is a lacuna in the legal and regulatory framework. Additionally, the health and environmental impacts associated with mining are exacerbated in unregulated minerals/materials, which is the case with sand mining and salt mining in some countries.[28]

So, one might ask, what does mining actually entail? In simple terms, mining refers to the extraction of coal and other substances from the earth. A mineral refers to a naturally occurring metallic or non-metallic substance formed through biogeochemical processes.[29] There is no agreed universal definition of the term 'mineral', as different legislations confer a slightly different meaning, depending on the jurisdiction concerned. In this respect, the legal connotation of the term can include all solid, gaseous or liquid natural resources, or, in some jurisdictions, be limited solely to solid metals. For instance, the Kenyan Mining Act 2016 defines 'mineral' as:

> a geological substance whether in solid, liquid or gaseous form occurring naturally in or on the earth, in or under water, in mine waste or tailing and includes the minerals specified in the First Schedule but does not include petroleum, hydrocarbon gases or groundwater.[30]

27 Ibid.
28 Dr Victoria R Nalule, 'Sand Mining in Africa', www.youtube.com/watch?v=TfEU7w7GoZ0&t=1s. Last accessed on 1 September 2022.
29 Nalule VR. 2019. *Mining and the Law in Africa: Exploring the Social and Environmental Impacts.* Cham: Springer Nature; 2019.
30 Section 4 of the Kenyan Mining Act, 2016.

One issue that has raised a lot of concern with respect to the definition of 'mineral' is the aspect of sand and gravel. Previously, most mining legislation throughout the world did not recognise sand as a mineral. For instance, Article 244 (3) of the Ugandan Constitution states:

> For the purpose of this article, 'mineral' does not include clay, murram, sand or any stone commonly used for building or similar purposes.[31]

Other jurisdictions also exclude sand and gravel in their definition of a mineral, although this exclusion is subject to ownership. A case in point is the Alberta Mines and Minerals Act Chapter M-17, in Alberta, Canada. Section 1 of the Act defines 'mineral' as illustrated below. The summary below also provides definitions from South Africa.

Definition of a mineral

Alberta definition[32]

'[M]inerals' means all naturally occurring minerals, and without restricting the generality of the foregoing, includes (i) gold, silver, uranium, platinum, pitchblende, radium, precious stones, copper, iron, tin, zinc, asbestos, salts, sulphur, petroleum, oil, asphalt, bituminous sands, oil sands, natural gas, coal, anhydrite, barite, bauxite, bentonite, diatomite, dolomite, epsomite, granite, gypsum, limestone, marble, mica, mirabilite, potash, quartz rock, rock phosphate, sandstone, serpentine, shale, slate, talc, thenardite, trona, volcanic ash, sand, gravel, clay and marl, but (ii) does not include (A) sand and gravel that belong to the owner of the surface of land under section 58 of the *Law of Property Act*, (B) clay and marl that belong to the owner of the surface of land under section 57 of the *Law of Property Act*, or (C) peat on the surface of land and peat obtained by stripping off the overburden, excavating from the surface, or otherwise recovered by surface operations…

South African definition[33]

'[M]ineral' means any substance, whether in solid, liquid or gaseous form, occurring naturally in or on the earth or in or under water and which was formed by or subjected to a geological process, and includes sand, stone, rock, gravel, clay, soil and any mineral occurring in residue stockpiles or in residue deposits, but excludes – (a) water, other than water taken from land or sea for the extraction of any mineral from such water, (b) petroleum; or (c) peat…

31 Article 244 (3) of the Constitution of Uganda, 1995 as amended.
32 The Alberta Mines and Minerals Act Chapter M-17.
33 South African Mineral and Petroleum Resources Development Act 28 of 2002.

The above definitions are elaborative and include all the various types of minerals, including precious minerals, critical minerals, and construction minerals. As illustrated above, the ownership of construction minerals is an issue of concern in different countries. In Kenya, unlike other minerals which are expressly vested in the ownership of the state, section 7 of the Mining Act makes an exception for sand and other construction minerals. As such, these can still be owned or mined by the local communities, as part of their customary practices. The section states that:

> (1) Nothing in this Act shall prevent any person from taking, subject to such conditions as may be prescribed from time to time by the Cabinet Secretary, soil, clay iron, salt or soda from any land, except land within the area of a mineral right, from which it has been the custom of the member of the community to which that person belongs to take the same. (2) The Cabinet Secretary may, by notice in the Gazette, and with the advice of the Mineral Rights Board, prescribe materials of customary usage.[34]

The above provision considers the customary practices and attachments with respect to construction minerals. However, with the increasing role of sand in the twenty-first century, countries have been persuaded to legislate sand mining, and hence to amend mining laws to consider sand as a mineral.

2.1.1 Impact of Categorisation of Minerals

In the previous section, we spotlighted the increasing role of critical minerals such as cobalt in this energy transition era. Besides critical minerals, we have noted the impact of construction minerals such as sand, for urbanisation, and precious minerals such as gold and diamonds. In essence, it is imperative to identify the impact and demand of different minerals – and, consequently, formulate tailored laws and policies aimed at responding to their unique characteristics. Most mining laws and policies apply uniform rules and regulations to all the different types of minerals. Although this is an accepted practice in many African countries, it has made it difficult for the same countries to position themselves in the global mining markets. For instance, given the crucial role of critical minerals, the mining laws and regulations should be formulated in such a way that they make it mandatory for beneficiation of these minerals to take place in the African countries concerned, such as the DRC. This will, however, require investors to cooperate with the governments in the establishment of the necessary infrastructures, such as refineries. Consequently, the categorisation of minerals should be the basis for the various mining policies in a country. Additionally, the categorisation of minerals also plays a role in understanding the economic and financial impact of such minerals. For instance, construction minerals are different from

34 Section 7 of the Mining Act of Kenya, 2016.

precious minerals, and, as such, the technical and financial requirements to develop these resources may differ greatly. Are these differences taken into consideration by the policymakers when including provisions relating to construction minerals in the mining laws? What about the rehabilitation and mine closure requirements? Are the companies concerned able to offer financial guarantees to ensure that the land is rehabilitated at the end of their operations? All these are questions that need to be addressed when legislating the mining sector.

The above notwithstanding, the main characteristics of the mining industry are that mining activities are capital-intensive, and mineral deposits are location-bound and exhaustible.[35] The main unique feature of the mining sector is its dependence on geological surveys to map the exact location of minerals. In this respect, then, the geological potential for the target mineral is a key factor to consider before investing in the prospection, exploration, and exploitation of mineral raw materials. Other relevant factors include the security of tenure, measure of profitability, legal and regulatory framework, investment protection, and the taxation regime, to mention but a few.[36]

2.2 Digitalisation and Automation: Artisanal and Small-Scale Mining as a Barrier

Digitalisation and automation have been a marked feature of the twenty-first century, especially in the various sectors of the economy. The Oxford Dictionary defines digitalisation as the process of changing data into a digital form that can be easily read and processed by a computer.[37] The Cambridge Dictionary, on the other hand, defines automation as the use of machines and computers that can operate without needing human control.[38]

This new technological era has reshaped the way industries work. In the mining sector, these are visible and have been embraced by large-scale mining companies, especially in the developed world. However, the sector continues to be one of the most hazardous industries. Although there are technological improvements in the large-scale mining industries, especially for productive agendas, we must devise means of ensuring that digitalisation and technological advancement can be used to address the social and environmental impacts associated with the mining sector.[39]

35 Nalule VR. *Mining and the Law in Africa: Exploring the Social and Environmental Impacts.* Cham: Springer Nature; 2019.
36 Ibid.
37 Oxford Learner's Dictionary. Can be accessed at www.oxfordlearnersdictionaries.com/definition/english/digitalization
38 Cambridge Dictionary. Can be accessed at https://dictionary.cambridge.org/dictionary/english/automation
39 Duarte J, Fernanda Rodrigues M, Baptista JS. Data Digitalisation in the Open-Pit Mining: Preliminary Results. In Occupational and Environmental Safety and Health II (pp. 715–723). Springer, Cham; 2020.

Additionally, an important question that remains to be addressed is how the new technologies can be incorporated in the Artisanal and Small-Scale Mining (ASM) sector. There is literature aimed at spotlighting how the mining industry is addressing the issues of automation, digitalisation, and interoperability.[40] However, little attention is paid to the impact of digitalisation in ASM. Basically, ASM involves miners who carry out mining activities using their own resources and tools. These miners work independently from mining companies. Whereas artisanal mining focuses on substance miners, small-scale mining includes enterprises or individuals that employ workers for mining. The main characteristic of ASM is the use of hand tools to carry out mining activities.[41] So a question arises as to how digitalisation and automation can be embraced in the ASM sector.

To achieve effective digitalisation and automation in the mining sector, all the processes of the sector, including mineral winning, transport, and preparation, must be automated. Additionally, special software, intelligent sensors, and actuators should be utilised to improve the mining networking. This is essential to ensure that all deployed plant and equipment are capable of communicating with one another and with the central data processing unit.[42] In essence, it is next to impossible to incorporate digitalisation and automation in ASM because of the rudimentary methods used by the miners.

With respect to large-scale mining, however, there are several advantages of digitalisation and automation, including increased productivity and reduced personnel costs. Additionally, digitalisation has minimised the health and social impacts of mining. For instance, several modern mining companies use self-propelled machines and mine robots to carry out potentially hazardous tasks associated with mineral winning and transport.[43]

From the African perspective, however, there are various barriers to digitalisation in the mining sector. First, the presence of ASM, as discussed above, poses a major barrier to digitalisation and automation in the mining sector, especially the use of rudimentary methods which make it impossible to introduce digitalisation. Second is the issue of energy access, whereby some regions don't have access to modern energy such as electricity which is key in digitalisation and automation.[44] The other challenge relates to the lack of technological transfer and finances required to deploy the automated machinery.

40 Lehmann J. Digitalisation, Automation, Interoperability – Standardisation at International Level. *Mining Report*. 2021; 157(5).

41 Nalule VR. Social and Environmental Impacts of Mining. In *Mining and the Law in Africa* (pp. 51–81). Cham: Palgrave Pivot; 2020.

42 Füllenbach C. Digitisation in the Mining Industry: Networked Machines Making for Greater Efficiency. Mining Report. 2018; 154(4).

43 Ibid.

44 Nalule VR. *Energy Poverty and Access Challenges in Sub-Saharan Africa*. Cham: Springer Nature; 2018.

Moving forward, it is worth noting that when it comes to digitalisation, there is a big discrepancy between developed countries and developing countries. Nevertheless, even for developed countries such as Germany, it has been observed that whereas digitalisation has moved at a steady pace in other industries, this has been delayed in the mining sector. This is due to the unique challenges experienced in the mining industry, including flagging productivity, high energy costs, increasingly strict environmental and safety regulations, and longer transport paths. There is therefore a need to address these challenges in both the Global North and Global South. For the African mining industry, the formalisation and legislation of the ASM sector is crucial not only in addressing the social and environmental impacts of ASM but also in ensuring effective digitalisation and automation.[45]

2.3 How to Benefit from Linkages in the African Mining Sector

Linkages in the mining sector are the main driver of industrial development in the regions concerned. Linkage development is considered the basis for economic development and a way to avoid the resource curse through diversification. Development is described as the record of how one thing leads to another, and the linkages are that record of the developments associated with the mining sector.[46] To fully understand the benefits associated with linkages, we must be aware of the nature of the mining sector in Africa.

It has already been established that mining is carried out on both small-scale or ASM and large-scale. For ASM, there are minimal production linkages realised; however, the large-scale mining sectors in Africa are dominated by multinational companies, and, as such, there are some linkages associated with their operations in Africa. However, for countries to benefit from the linkages, there should be effective public and private sector policies. Host governments should be able to translate and implement the relevant policies and laws to ensure that they benefit from the mining sector and related linkages.[47]

The presence of large-scale mining (LSM) companies in different African mining communities is prima facie associated with increased public investment in social and economic infrastructure, including transport systems, health care, and educational facilities. In this respect, investment in the extractives sector does create economically viable linkages. To ensure that the host communities benefit from such investments, governments have embraced various strategies and policies including local content policies, corporate

45 Nalule VR. Regulation of Mining in Africa. In *Mining and the Law in Africa* (pp. 19–50). Cham: Palgrave Pivot; 2020.

46 Atienza M, Lufin M, Soto J. Mining linkages in the Chilean copper supply network and regional economic development. *Resources Policy*. 2021; 70: 101154.

47 Mjimba V. The nature and determinants of backward linkages in emerging minerals commodity sectors: A case study of gold mining in Tanzania. PhD thesis, The Open University.

social responsibility (CSR), and local processing standards.[48] Linkage development has therefore been described by some scholars as an evolutionary process of knowledge and technological capabilities accumulation among the local supplier firms.[49]

There are different types of linkages, but the common ones are economic linkages. These include fiscal linkages and production linkages. Fiscal linkages deal with payments and taxes. In the extractive industries, these are intended to enable the state to maximise economic returns. They include, among others, cost recovery provisions, mining royalties, fiscal pricing, service contracts, and bonus payments. Production linkages, on the other hand, include backward (or upstream) linkages and forward (or downstream) linkages. Backward linkages involve not only employment opportunities but also the supply of goods, services, and security. On the one hand, the enforcement of local content policies in different countries is, in principle, aimed at achieving the backward/upstream linkages. On the other hand, forward/downstream linkages are focused on adding value to the commodities extracted.

In essence, in order to benefit from the economic linkages, countries should ensure that they have a competitive fiscal regime. Local content policies should also be implemented and, most importantly, these countries should invest more in infrastructural development, especially refineries. Forward linkages are focused on beneficiation or value addition, which can lead to tax and export earnings. Some resource-rich African countries such as Botswana have benefited from both forward and backward linkages. Specifically, the establishment of Debswana enabled De Beers' sales and sorting operations to move from London to Botswana.

Besides the production linkages discussed above, there are other types of linkages including side-stream linkages which are associated with the establishment of developmental infrastructure such as roads, healthcare services, and water facilities. These can also be linked to the social licence to operate (SLO), especially where the host community demands are related to such infrastructural development. However, it is imperative to note that the SLO is unwritten and can be withdrawn at any time, as such side-stream linkages are not an excuse for disregarding the SLO. Lastly, both ASM and LSM are associated with increased earning, hence leading to the realisation of consumption linkages.

In a nutshell, it is not guaranteed that mining activities lead to economic development, and hence we have witnessed that many resource-rich countries have been victims of a resource curse. To benefit from these resources, countries should strategise on how to formulate and enforce the relevant laws and

48 Nalule VR. *Land Law and the Extractive Industries*. London: Bloomsbury Publishing; 2021.

49 Atienza M, Lufin M, Soto J. Mining linkages in the Chilean copper supply network and regional economic development. *Resources Policy*. 2021; 70 :101154.

policies. The mining laws and policies should be competitive to ensure that they advocate for value addition in the country (forward linkages). They should empower the local people to take part in these mining activities (local content). SLO expectations should be clear, and the host communities should be guided on how to benefit from their SLO rights (side-stream linkages). The fiscal regimes and the mining agreements should be well negotiated, and, if possible, detrimental clauses such as 'stabilisation clauses'[50] should be avoided (fiscal linkages). Additionally, besides large-scale mining, ASM can present economic linkages in the mining sector. However, this can only happen if countries are successful with the formalisation and legalisation of ASM. Additionally, there is a need to ensure technological transfer from LSM to ASM; this will not only improve productivity but also lead to more earnings in both sectors – all of which contribute to enhancing economic linkages in the mining industry.

3 Modernisation of the Mining Sector: Impact of the Different Generations of Mining Laws

The mining sector has been going through a progressive evolution in not only technological enhancement but also the modernisation of mining laws and regulations. To appreciate how this modernisation has impacted the industry, it is worth having an understanding of the history of mining in Africa. Therefore, before a discussion on the legal and regulatory framework, this section briefly discusses the history of mining in Africa.

3.1 What Is the Influence of History on the African Mining Sector?

When discussing the history of mining, it is important to note that many important cultural eras are associated with and identified by various minerals or their derivatives. These include the Stone Age (prior to 4000 BCE), the Bronze Age (2000–700 BCE), the Iron Age (1500 BCE to 1780 CE), the Steel Age (1780–1945), and the Nuclear Age (1945–the present).[51]

Indeed, in discussions about mining in Africa, it is worth exploring the connection between Africa's mineral wealth and colonialism. The mineral wealth of African states is one of the reasons why colonisation escalated in the nineteenth and twentieth centuries. The diamond rush in Southern Africa, for instance, was the basis for the prolongation of European rule in Africa. In 1869, an 83.5-carat rough diamond was discovered by a Griqua herdsman in Hopetown in South Africa. The first diamond rush was followed by the second in 1870 when diamonds were found in river diggings at Klip Drift (now Barkly West); diamonds were also found at Bultfontein farm on the

50 Nalule VR. What is the Problem with Stabilization Clauses in Petroleum Agreements? *Journal of Sustainable Development Law and Policy*. 2022; 13(1): 85–102.
51 Nalule VR. *Mining and the Law in Africa*. Cham: Palgrave Pivot; 2020. Page 12.

edge of modern-day Kimberley, and in December of that year, diamonds were found in the adjacent Du Toit's Pan, triggering the third diamond rush. The diamond rush in Southern Africa attracted a host of prospectors, smugglers, and speculators who were supported by British Imperialism.[52] Indeed, some well-known mining companies were established during this period. For instance, both the De Beers mine and Kimberly mines were discovered in 1871. Consequently, De Beers Consolidated Mines Limited was established on 12 March 1888, and Cecil Rhodes was named founding chairman. Besides De Beers, other prominent companies including Rio Tinto-Zinc and Consolidated Gold Fields were also established.[53]

Taking the historical perspective above, we note that the mining laws and regulations that were enacted during the colonial era focused on the exploration of the resources for the benefit of the colonial masters. In essence, the principle of permanent sovereignty over natural resources was not in existence. However, after the independence of most countries, the right of peoples to self-determination and to a means of subsistence were some of the justifications for permanent sovereignty over natural resources.

Taking stock of the above, it is clear that the challenges experienced in the African mining sector have a historical root, implying that these laws have evolved into different generations, and each generation of the mining codes has unique challenges to address, as will be discussed in the next section.

3.2 Generations of Mining Laws

Legislation of the mining sector in Africa has been going through a transformative process and different mining codes have characterised a particular stage in this process. Before a discussion on the different generations of mining laws, it is worth noting that the mining sector is governed by various legislations, primarily mining laws and regulations. Besides the mining laws, there are other laws that are relevant to the governance of the mining sector, and these include environmental laws, taxation laws, land laws, and employment laws, among others.

In many African countries, the Constitution is the supreme law of the land and, as such, the basis for all other laws, including those relating to the mining sector. In Uganda, for instance, the 1995 Constitution (as amended) under Article 244 provides for the exploitation of minerals. The article states that the exploitation of minerals should be done for the interest of the people, and should consider the interests of the individual landowners, local government, and the government. We note that the Constitution provides for sovereignty over natural resources and, in this respect, the state owns the minerals

52 Nalule VR. Introduction to Mining in Africa. In *Mining and the Law in Africa* (pp. 1–17). Cham: Palgrave Pivot; 2020.

53 For a full history of African Mining, see Nalule VR. *Mining and the Law in Africa*. Cham: Palgrave Pivot; 2020.

as a custodian for the people. When discussing ownership over natural resources, however, it is important to note the distinction between the natural resources contained in the subsoil and the extracted resources. This distinction is key due to the basic fact that a natural resource that is not yet extracted is not clearly known in terms of quality and quantity, and, as such, it is considered immovable property, and the financing requirements for both are different.[54]

Additionally, mining policies have an impact on the different generations of mining laws. A mining policy is defined as the entirety of actions of a state for influencing the supply of and demand for mineral resources on its territory and beyond that.[55] There are common objectives of mining policies which are emphasised in different African countries. These are outlined below:

- To regulate and improve artisanal mining.
- To strengthen local capacity for mineral development.
- To promote private sector participation.
- To develop and strengthen local capacity for mineral development.
- To address child labour in the mining sector.
- To promote gender equality in the mining sector.
- Value addition and revenue generation.

Moving forward, the mining policies do influence the enaction of mining codes/laws. These codes are divided into different generations including the first generation which entails mining legislation of the 1980s, the second generation which entails mining legislation of the early and mid-1990s, and the third generation which basically refers to mining legislation of the late 1990s. The new wave of reforms in the current mining laws can arguably be referred to as the fourth generation of mining codes.

3.2.1 First Generation of Mining Codes

The first generation of mining laws is characterised by the withdrawal of the state from actively being involved in the mining sector as a way of attracting investors. Ghana is a good example of this generation of laws. Generally, liberalisation of the mining sector took off in the 1980s, and this resulted in the reduction of the state's role in mining activities. Although it was against the interest of states, international monetary organisations, such as the World Bank, encouraged African countries to completely withdraw from the sector as a prerequisite for attracting much-needed investment. In this respect, with the influence of organisations such as the World Bank, developing countries formulated laws that had the effect of reducing state involvement,

54 Olawuyi DS. *Extractives Industry Law in Africa*. Cham: Springer Nature; 2018.

55 Nalule VR. Regulation of Mining in Africa. In *Mining and the Law in Africa* (pp. 19–50). Cham: Palgrave Pivot; 2020.

privatisation, increasing incentives given to foreign investors, and lowering of tax rates and royalties, to mention but a few. All these changes meant that the state's role was merely to facilitate and ensure that private investors had a conducive environment in which to carry out their activities. This also meant a relaxation in employment and labour laws, thus making it easy for international companies to hire foreigners, and it also meant relaxation on procurement requirements in the country. There are various countries whose previous mining laws fall under the first generation of mining codes, a perfect example being Ghana's 1986 Minerals and Mining Law.[56]

3.2.2 Second Generation of Mining Codes

The second generation is characterised by the regulation of the mining sector, including consideration of environmental and health protections, which, contrary to the present practice, were left to become a responsibility of the private sector – that is, non-state actors. Guinea provides a good example of second-generation mining legislation. The need to ensure the enforcement of social and environmental rules by multinational companies was the cornerstone of the second generation of mining codes which were enacted in the early 1990s. However, due to the limited powers and influence of the state in mining activities, they could hardly enforce such rules against multinational companies.[57]

3.2.3 Third Generation of Mining Codes

The late 1990s witnessed the emergence of the third generation of mining codes. Unlike the first generation, the third generation recognises and encourages the role of the state in the facilitation and regulation of the mining sector. However, the reforms made had the effect of reducing the state's role and ownership in the mining sector. Mali, Madagascar, and Tanzania provide a good example of the third-generation laws. In Mali, for instance, the 1999 Mining Code emphasised the need to attract foreign investors in the country, and many incentives were provided by the government to foreign mining corporations. The same objective was shared by the 1999 Madagascar Mining Code which aimed to reduce state engagement in mining operations and increase foreign investments in the mining sector. The disengagement of states and governments in the mining operations created a gap where social and environmental aspects were not being catered for. In response to this, several initiatives were created to address some of these issues. In this respect, the Global Mining Initiative was launched in 1998, led by a consortium of mining company leaders.[58]

56 Ibid.
57 Ibid.
58 Ibid.

3.2.4 Fourth Generation of Mining Codes

There have been various reforms in the mining sector, and these have necessitated the adoption of new mining laws, which can arguably form the fourth generation of mining codes of the 2000s. For instance, in 2015, Burkina Faso passed its new mining law; in 2016, Kenya also adopted its new mining Act; and, in 2017, Tanzania adopted new laws relating to the mining sector. The new laws basically deal with the attraction of foreign investments and, most importantly, with ensuring local participation and economic development arising from mining activities. Given global efforts to tackle climate change, the fourth generation of mining codes also emphasise the need for environmental protection.[59]

3.2.5 Proposed Fifth Generation of Mining Codes

Throughout this chapter, we have identified the main challenges of the African mining sector. Some of these challenges, such as the informality of ASM and the related environmental and social impacts, are not new, and the previous generation of mining laws have endeavoured to address them. However, there is a need to analyse the current state of affairs relating to the increasing role of critical minerals in ensuring energy transitions and tackling climate change. Consequently, the proposed fifth generation of mining codes should categorise the different minerals and regulate them accordingly. Construction minerals, precious minerals, and critical minerals should have tailored legislation aimed at balancing the interests of the African exporting countries. This new fifth generation of mining codes should also clearly stipulate how digitalisation and automation can be embraced in Africa not only in LSM but also in ASM. This would require mandatory technology transfer provisions. Additionally, value addition and beneficiation should be featured in these new codes, especially for the critical minerals, to ensure that the African countries which possess these resources effectively benefit from exporting refined products. Nevertheless, all the other aspects addressed in the fourth generation of mining laws, such as the local content policies, should be featured in the fifth generation of mining codes.

4 Aim, Scope, and Structure of the Book

There is a growing body of scholarly articles and national, regional, and international initiatives to ensure that African countries benefit from their resources. Law, policies, and regulations have been at the centre of these initiatives. For countries endowed with mineral resources, the mining sector contributes to these countries' economic development as it can finance infrastructural development. In Botswana, for instance, approximately 70–80% of

59 Ibid.

export earnings are attributed to diamond mining. Indeed, if well managed, revenues from the mining sector can be used to tackle the poverty challenges facing the continent. However, some countries have fallen prey to the resource curse, due to poor governance, corruption, and mismanagement.

Therefore, this book explores the various issues characterising the African mining sector, drawing examples from different African countries and regional organisations. Although there is a massive literature on the subject, some issues have been neglected, including the crucial role of digitalisation in resolving the challenges faced in ASM, deep-sea mining, dispute resolution in mining, and contract negotiations, to mention but a few. As such, this book uniquely analyses and adds flavour to international mining's fundamental concepts by describing a simulated annealing-based approach appropriate for complex mining projects in Africa. Book contributors comprise of academics from different universities including professors, practitioners, government policymakers, NGO executives, and a variety of different experts. This multidisciplinary book will be of interest to African policymakers, governments, academics, industry professionals, energy and mining institutions, international organisations, universities across the globe, and companies.

4.1 Structure of the Book

The book is organised into three main parts. Part I includes introductory chapters on mining in Africa. This part is made up of three chapters, exploring various key issues in the African mining sector including the modernisation of mining laws, the increasing role of critical minerals in energy transition, and how to balance the interests relating to the African mining codes. Part II of the book explores case studies from different parts of Africa. It consists of seven chapters covering ASM in Kenya, ASM in Nigeria, the fiscal regime in Zambia, regional cooperation in the mining sector with a focus on the African Mining Vision, digitalisation in South Africa's mining sector, and Chinese investments in the African mining sector. Part III concludes the book and it consists of four chapters focusing on climate change reporting in the mining sector, deep-sea mining, the blue economy, and an epilogue. Each of the chapters of the book provides a detailed and rigorous background of the legal, regulatory, and governance issues of the mining sector in the surveyed country, especially in the context of the African experience. The chapters explore the twenty-first-century issues that need to be addressed in the African mining sector.

5 Conclusion

The mining sector's importance has been emphasised in the twenty-first century, given the increasing demand for and role of critical minerals such as cobalt in the global transition to a low-carbon economy. The African continent is home to enormous reserves of oil, gas, energy, and solid minerals.

These resources are crucial for these countries' economic development. Nevertheless, in recent years we have seen increasing global attention to tackle climate change as envisaged in the 2015 Paris Agreement and Goal 13 of the United Nations Sustainable Development Goals (SDGs). Consequently, the present and future laws, legislation, and policies of the African extractive industries including oil, gas, and minerals should take into consideration these environmental and sustainable global efforts.

2 The Energy Transition, Critical Minerals and Industrialisation in Sub-Saharan Africa

Needs, Opportunities, and Strategies

Theophilus Acheampong and Bridget Okyerebea Menyeh

Highlights

- Tackling energy poverty in Sub-Sharan Africa is the foremost priority.
- Green industrialisation can co-exist with traditional resource-based industrialisation to drive structural transformation.
- Synergies between energy and industrial policy, evidenced by linkages and value chain development, are critical.
- African policymakers need to craft a pan-African transition strategy.

1 Introduction

The clean energy transition is no longer a fringe idea. Although there are various issues associated with the energy transition, a notable one is the relevance of the African mining sector in this energy transition and climate change era. Africa has an abundance of the critical minerals needed for the energy transition. Metals such as copper, lithium, nickel, cobalt, manganese, and graphite are vital to the transition for sustaining battery longevity, performance, and energy density of all-electric vehicles (EVs) motors, solar panels, and wind turbines. The energy transition, which is a pathway toward transforming the global energy sector to zero-carbon by 2050 and beyond, represents one way to mitigate the impacts of human-induced climate change.[1] Governments, companies, and individuals worldwide are under increasing pressure to do more to tackle the existential threat of climate change by switching to low and zero-carbon energy sources. Aside from climate change concerns, the transition is driven by (1) advances in technology, (2) new energy policies being promoted by global governments, especially in a post-COVID-19 pandemic context, and (3) evolving consumer preferences, especially with environmental, social, and governance

1 IPCC (2021). Sixth Assessment Report. Available at: www.ipcc.ch/report/ar6/wg1/downloads/report/IPCC_AR6_WGI_TS.pdf (Accessed: 20 September 2021).

DOI: 10.4324/9781003284437-3

(ESG) pressures.[2] In the post-COVID-19 context, several multinational energy companies have lost landmark cases or suffered shareholder revolts on their climate plans. For example, in May 2021, Royal Dutch Shell was ordered by a Netherlands court to reduce its emissions by 45% by 2030[3] compared to 2019 baseline levels.[4] At the same time, three climate-minded activist shareholders were elected to the Exxon Mobil Board following a shareholder revolt led by Engine No. 1, a small hedge fund, while Chevron investors also passed resolutions to cut carbon emissions. The above indicates the extent to which ESG pressures concerning the need for multinationals to do more to combat climate change are beginning to be felt in company boardrooms. At the government level, the U.S. Government also announced that it would use part of its fiscal stimulus to achieve a 50–52% reduction in greenhouse gases (GHGs) by 2030 from 2005 levels under the 'building back better' theme.[5] The European Union has also announced a 55% emissions reduction by 2030 on its path to net zero by 2050.[6]

2 Neofytou, H., Nikas, A., & Doukas, H. (2020). Sustainable energy transition readiness: A multicriteria assessment index. *Renewable and Sustainable Energy Reviews, 131*, 109988.

Solomon, B. D. & Krishna, K. (2011). The coming sustainable energy transition: History, strategies, and outlook. *Energy Policy, 39*(11), 7422–7431.

Nalule, V. & Acheampong, T. (2021). Energy Transition Indicators in African Countries: Managing the Possible Decline of Fossil Fuels and Tackling Energy Access Challenges. *The Journal of Sustainable Development, Law and Policy, 12*(1), 1–48.

Bazilian, M., Bradshaw, M., Gabriel, J., Goldthau, A., & Westphal, K. (2020). Four scenarios of the energy transition: Drivers, consequences, and implications for geopolitics. *Wiley Interdisciplinary Reviews: Climate Change, 11*(2), e625.

S&P Global (2021). What is Energy Transition? Available at: www.spglobal.com/en/research-insights/articles/what-is-energy-transition (Accessed: 20 September 2021).

Jackson, F. (2021) Five Market Trends Driving Energy Transition, Forbes. Available at: www.forbes.com/sites/feliciajackson/2021/01/25/five-market-trends-driving-energy-transition/?sh=4bd2e8d15077 (Accessed: 20 September 2021).

3 BBC News (2021). Shell: Netherlands court orders oil giant to cut emissions. Available at: www.bbc.co.uk/news/world-europe-57257982 (Accessed: 20 September 2021).

4 Bacchus, J. (2021). Oil firms face more legal fights on climate change – here's why. Available at: www.weforum.org/agenda/2021/06/oil-shell-exxon-chevron-court-shareholders-climate (Accessed: 20 September 2021).

5 The White House (2021). FACT SHEET: President Biden Sets 2030 Greenhouse Gas Pollution Reduction Target Aimed at Creating Good-Paying Union Jobs and Securing U.S. Leadership on Clean Energy Technologies. Available at: www.whitehouse.gov/briefing-room/statements-releases/2021/04/22/fact-sheet-president-biden-sets-2030-greenhouse-gas-pollution-reduction-target-aimed-at-creating-good-paying-union-jobs-and-securing-u-s-leadership-on-clean-energy-technologies (Accessed: 13 September 2021).

6 European Commission (2021). European Green Deal: Commission proposes transformation of EU economy and society to meet climate ambitions. Available at: https://ec.europa.eu/commission/presscorner/detail/en/ip_21_3541 (Accessed: 13 September 2021).

BBC News (2021). Climate change: EU to cut CO2 emissions by 55% by 2030. Available at: www.bbc.co.uk/news/world-europe-56828383 (Accessed: 13 September 2021).

Likewise, China has set in motion plans to peak its emissions before 2030 and to achieve carbon neutrality by 2060.[7]

The goal is to limit global warming to well below 2 degrees Celsius (°C), preferably to 1.5°C, compared to pre-industrial levels, as documented in the 2015 Paris Agreement at the Conference of the Parties (COP 21).[8] To maintain global temperature to 1.5°C, the world must therefore drastically reduce GHG emissions emanating from its massive dependence on fossil fuels. At present, the world emits about 50 billion tonnes[9] of GHGs every year, an increase of 50% from 1990 to 2018.[10] Overwhelmingly, about 73% of GHGs come from energy use encompassing electricity, heat, and transport.[11] The remainder includes agriculture, forestry, and land use (18.4%), industrial processes (5.2%), and waste (3.2%). Within energy use, the three largest GHG emissions come from mainly burning fossil fuels for industrial energy (24.2%) such as iron and steel; transportation (16.2%) including road transport, aviation, shipping; and, lastly, buildings (17.5%), including both residential and commercial.[12]

Sub-Saharan Africa (SSA) is faced with unique energy challenges, with about 570 million people or 50% of the population unpowered (without electricity access) and another 910 million or 80% of the population without access to clean fuels and technologies.[13] The continent's energy access challenges over the years have meant that industrialisation has been stifled even in the presence of other favourable factors such as trade and capital openness, institutional stability, or financial sector development.[14] Typically, such industrialisation would be powered by carbon-intensive energy sources – such as coal, oil, and other fossil fuel

7 BBC News (2021). Climate change: China aims for 'carbon neutrality by 2060'. Available at: www.bbc.co.uk/news/science-environment-54256826 (Accessed: 13 September 2021).

8 UNFCCC (n.d.). The Paris Agreement. Available at: https://unfccc.int/process-and-meetings/the-paris-agreement/the-paris-agreement (Accessed: 13 September 2021).

9 Ritchie, H. & Roser, M. (2020). CO_2 and greenhouse gas emissions. *Our World in Data*. Available at: https://ourworldindata.org/co2-and-other-greenhouse-gas-emission (Accessed: 13 January 2023).

10 Ge, M., Friedrich, J. & Vigna, L. (2020). 4 Charts Explain Greenhouse Gas Emissions by Countries and Sectors, World Resources Institute. Available at: www.wri.org/insights/4-charts-explain-greenhouse-gas-emissions-countries-and-sectors (Accessed: 13 September 2021).

11 Ritchie, H. & Roser, M. (2020). CO_2 and greenhouse gas emissions. *Our World in Data*. Available at: https://ourworldindata.org/co2-and-other-greenhouse-gas-emission (Accessed: 13 January 2023).

12 Ibid.

13 IEA, IRENA, UNSD, World Bank, WHO (2021). *Tracking SDG 7: The Energy Progress Report*. Washington, DC: World Bank.
SE4ALL (2021). Five takeaways from Tracking SDG7: The Energy Progress Report 2021. Available at: www.seforall.org/news/five-takeaways-from-tracking-sdg7-the-energy-progress-report-2021 (Accessed: 2 December 2021).

14 Haraguchi, N., Martorano, B., & Sanfilippo, M. (2019). What factors drive successful industrialisation? Evidence and implications for developing countries. *Structural Change and Economic Dynamics, 49*, 266–276.

options – as has been observed with the growth of Western nations and, more recently, emerging giants like China and India. It is not surprising that China and India, for example, find these fuels (coal, in this instance) necessary for their development and lobbied for the change in wording at the recent COP 26 in Glasgow, United Kingdom from 'phase out' to 'phase down' to reflect their own needs and challenges.[15] For some groups, this represents a considerable setback to the 1.5°C goal, with some recent analysis suggesting that '60% of oil and fossil methane gas, and 90% of coal must remain unextracted to keep within a 1.5° C carbon budget'.[16] This suggestion is corroborated by the International Energy Agency (IEA) in its 'Net Zero by 2050' report, which determines that achieving the 1.5°C target necessitates 'beyond projects already committed as of 2021, [that] there are no new oil and gas fields approved for development … and no new coal mines or mine extensions are required'.[17] If this happens, Welsby et al. estimate that 51% of Africa's oil reserves, 49% of its fossil methane gas, and 86% of the coal reserves will be unextractable if the 2050 1.5°C scenario is to be met.[18]

In essence, many regions, including SSA, may potentially have a 20- to 30-year window to maximise the value of their hydrocarbon resources. This could have dire implications for governments reliant on these export revenues to deliver public goods such as roads and hospitals but unable to create new industries around the transition. This is highly likely to increase social pressures and challenge ruling governments, democratic or not. On the other hand, the energy transition presents opportunities for countries with mineral resources, such as copper, cobalt, and lithium, which are critical to global value chains, to capture a significant part of the value chain, such as by producing electric vehicles (EVs) battery components, among others. Furthermore, despite the huge renewable energy potential of the African region, many obstacles remain, such as financing, local capacity and technical expertise, among others.[19] For example, less discussed is the present state of

15 Masood, E. & Tollefson, J. (2021). COP26 climate pledges: What scientists think so far. Nature.com. Available at: www.nature.com/articles/d41586-021-03034-z (Accessed: 2 December 2021).

Hook, L., Hodgson, C., & Pickard, J. (2021). India and China weaken pledge to phase out coal as COP26 ends. Available at: www.ft.com/content/471c7db9-925f-479e-ad57-09162310a21a (Accessed: 2 December 2021).

Russell, C. (2021) Coal trajectory is set whether it's 'phase out' or 'phase down'. Reuters. Available at: www.reuters.com/business/cop/coal-trajectory-is-set-whether-its-phase-out-or-phase-down-russell-2021-11-14 (Accessed: 2 December 2021).

16 Welsby, D., Price, J., Pye, S., & Ekins, P. (2021). Unextractable fossil fuels in a 1.5° C world. *Nature, 597*(7875), 230–234.

17 IEA (2021). Net Zero by 2050. Available at: www.iea.org/reports/net-zero-by-2050 (Accessed: 25 October 2021), at p.24.

18 Welsby, D., Price, J., Pye, S., & Ekins, P. (2021). Unextractable fossil fuels in a 1.5° C world. *Nature, 597*(7875), 230–234.

19 Menyeh, B. O. (2021). Financing electricity access in Africa: A choice experiment study of household investor preferences for renewable energy investments in Ghana. *Renewable and Sustainable Energy Reviews, 146*, 111132.

renewable energy (RE) supply chains, which are largely based outside Africa, and the need for some continental production to reduce overdependence and create jobs. Although there are opportunities to establish such industries, many countries are beset with power challenges (price competitiveness and disruptions),[20] making it a classic chicken-and-egg situation.

Amidst these myriad challenges and opportunities, this review adds to the energy transition literature on Africa by exploring two critical but under-explored questions:

1 What is likely to be the longer-lasting implication of the energy transition away from fossil fuels for SSA?
2 Can the region leverage clean energy to drive industrialisation?

Addressing the above questions is crucial given that the continent's economies must now navigate the pandemic-induced changes in demand as well as the energy transition. The rest of this chapter is dedicated to answering these questions. The following section reviews the state of industrialisation in Africa, based on a deep-dive into 50-plus years of industrialisation and development literature in Africa. This is followed by Section 3, where we highlight and propose three policy anchors by which SSA can leverage the clean energy drive to promote industrialisation. Finally, we conclude in Section 4.

2 Review: The State of Industrialisation in Africa

Before discussing the state of industrialisation in Sub-Saharan Africa, it is essential to highlight the continent's pre- and post-COVID-19 economic context. This is more so because the macro-economic environment is a key catalyst for micro-level industrial development. The most recent commodities price boom of the past 10–15 years came with significant opportunities for SSA countries to use the proceeds to diversify their economies. However, that did not happen despite the opportunities therein. Then there was the 2014–2016 commodities price crash, which badly exposed the economic management

2 Alemzero, D., Acheampong, T., & Huaping, S. (2021). Prospects of wind energy deployment in Africa: Technical and economic analysis. *Renewable Energy, 179*, 652–666.

20 Acheampong, T., Menyeh, B. O., & Agbevivi, D. E. (2021). Ghana's Changing Electricity Supply Mix and Tariff Pricing Regime: Implications for the Energy Trilemma. *Oil, Gas & Energy Law, 19*(3).

Yetano Roche, M., Verolme, H., Agbaegbu, C., Binnington, T., Fischedick, M., & Oladipo, E. O. (2020). Achieving Sustainable Development Goals in Nigeria's power sector: Assessment of transition pathways. *Climate Policy, 20*(7), 846–865.

Foster, V., Eberhard, A., & Dyson, G. (2021). The evolution of electricity sectors in Africa: Ongoing obstacles and emerging opportunities to reach universal targets. In *Handbook on Electricity Markets*. Edward Elgar Publishing.

Eberhard, A., Gratwick, K., Morella, E., & Antmann, P. (2017). Accelerating investments in power in sub-Saharan Africa. *Nature Energy, 2*(2), 1–5.

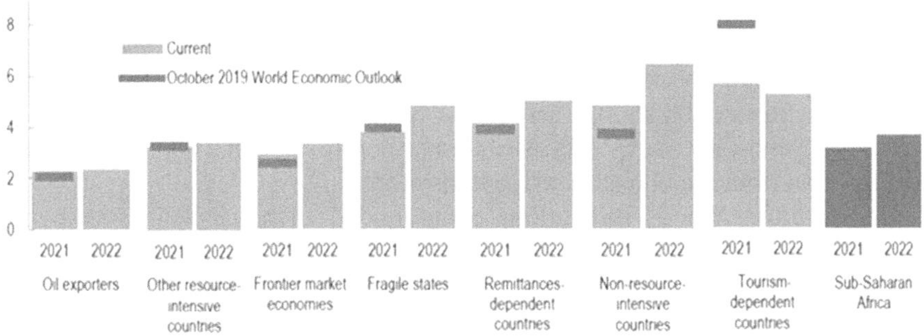

Figure 2.1 Sub-Saharan GDP growth projections, 2021–2022
Source: IMF[21]

approaches used in these countries. Several SSA countries struggled to cope with the impact of the commodities price crash, coupled with the fact that their economies are even more integrated into the global economy. Thus, any global systemic shock impacts local economies and often deepens inequality significantly. Moreover, we have another global downturn with the COVID-19 pandemic, which has exposed SSA economies in their ability to cope with external shocks. This time around, the impact of COVID-19 is likely to even linger for even longer. As a result, per capita output is not expected to return to 2019 levels until after 2022, as Figure 2.1 shows.

Several SSA countries, such as Tanzania, Zambia, Ghana and Nigeria in the early post-independence days of the 1960s, sought to industrialise their economies with varied success. The early post-independence industrialisation policies aimed to position these new nation-states away from the exploitative colonial enterprise. This was predominantly by seeking to add value to their primary commodities before exporting them and import substitution. For example, Ghana, under President Kwame Nkrumah (1959–1966), sought to develop the Volta River Project with the building of the Akosombo dam to provide cheap and reliable electricity to the Volta Aluminium Company's (VALCO) aluminium smelter, exploiting Ghana's bauxite resources.[22] However, many of these initiatives failed for several reasons, including political instability and conflict – successive military coups – which constrained long-

21 IMF (2021). Regional Economic Outlook for Sub-Saharan Africa: Navigating a Long Pandemic. Available at: www.imf.org/en/Publications/REO/SSA/Issues/2021/04/15/regional-economic-outlook-for-sub-saharan-africa-april-2021 (Accessed: 13 September 2021).

22 Acheampong, T. & Mensah, K. A (2018). Towards an Integrated Bauxite and Aluminium Industry in Ghana: Some Policy Considerations. Available at: http://dx.doi.org/10.13140/RG.2.2.17985.40804 (Accessed: 13 January 2023).

New York Times (1964). Ghana Advancing Aluminum Plant; Work Is Scheduled to Begin Soon on Big Smelter. Available at: www.nytimes.com/1964/11/22/archives/ghana-advancing-aluminum-plant-work-is-scheduled-to-begin-soon-on.html (Accessed: 18 September 2021).

term economic policymaking, and exogenous factors such as terms-of-trade shocks like the 1970s oil crises and commodity price decreases.[23] Nevertheless, ineffective or bad policies such as the state-led import substitution with its attendant high protectionism, low managerial capabilities, inefficiencies, and heavy import dependency made African industry poorly prepared for international competition from the Global North and rising Asia.[24]

The Bretton Woods institutions compounded these challenges with the structural adjustment policies of the mid-1980s, which largely disapproved of the import substitution form of industrialisation, among others.[25] These one-size-fits-all neoliberal policy prescriptions of the IMF and World Bank included prioritising fiscal adjustment, spending cuts, and privatisation. Much of the aversion to the IMF of some SSA countries stems from the fact that these policies of the 1980s and 1990s constrained many African countries. Their conditionalities played a role in economic stagnation, with dire political and social consequences. These include decreases in social spending and increased poverty, periods of de-industrialisation, privatising of state-owned enterprises (SOEs), reducing pensions in the public service, and currency devaluations.[26]

The outcome of these historical paradigms is that SSA remains the least diversified continent globally and is highly vulnerable to external shocks – for example, the 2008 financial crisis and the ongoing COVID-19 pandemic. SSA has one of the least complex economic systems globally, as Figure 2.2 shows. Accumulating productive knowledge and using it in more complex industries is a prerequisite for economic development – that is, structurally changing an economy from low- to high-productivity activities.[27] However, in SSA, several economies on the continent are still dominated by primary commodities extraction with little to show in terms of export diversification.[28] UNCTADStats data shows that African countries have a diversification index of more than 0.5, which is higher than Europe (0.18), East Asia (0.30), and selected exporters of manufactured goods and primary commodities (0.28) – see Table 2.1. While the continent's export

23 Page, J. (2021) Africa's Failure to Industrialise: Bad Luck or Bad Policy? Brookings. Available at: www.brookings.edu/blog/africa-in-focus/2014/11/20/africas-failure-to-industrialize-bad-luck-or-bad-policy (Accessed: 19 September 2021).

24 Ibid.

25 Mendes, A. P. F., Bertella, M. A., & Teixeira, R. F. (2014). Industrialisation in Sub-Saharan Africa and import substitution policy. *Brazilian Journal of Political Economy, 34*, 120–138.

26 Signé, L. (2018). The potential of manufacturing and industrialisation in Africa: trends, opportunities, and strategies. Brookings. Available at: www.brookings.edu/wp-content/uploads/2018/09/Manufacturing-and-Industrialization-in-Africa-Signe-20180921.pdf (Accessed: 13 January 2023).

27 Atlas of Economic Complexity (n.d.). Country & Product Complexity Rankings. Available at: https://atlas.cid.harvard.edu/rankings (Accessed: 19 September 2021).

28 Morris, M. & Fessehaie, J. (2014). The industrialisation challenge for Africa: Towards a commodities based industrialisation path. *Journal of African Trade, 1* (1), 25–36.

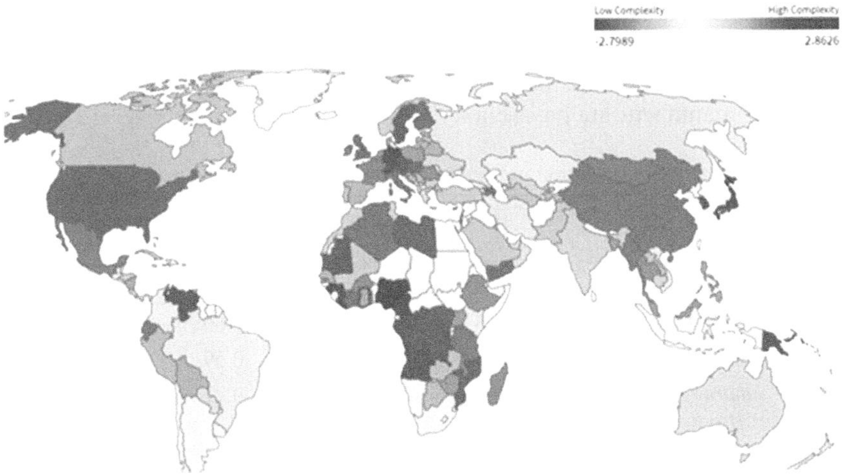

Figure 2.2 Country complexity map 2019
Source: Harvard Economic Complexity Index (ECI) [37]

concentration – which measures the degree to which exports are concentrated on a few products instead of being more homogeneously distributed among several products[29] – has improved compared to 2000 baseline levels (from 0.35 to 0.19), it still lags behind other regions.

Also, overall export diversification has not improved as much (from 0.60 to 0.55). The diversification index measures the extent to which the export structure by product differs from world patterns.[30] This indicates an increasing tendency towards export concentration in Africa at both the sectoral and product level.[31] In other words, the portfolio of exports from Africa is less diversified relative to other regions. The reliance on mostly primary commodities means that the oft-touted higher GDP growth rates on the continent have not translated proportionately into job creation and/or reducing poverty, given that commodity extractions tend to be enclave activities with low employment linkages compared to the broader manufacturing sector.[32] This has stunted the continent's manufacturing and industrial sector and dampened its potential as a critical engine for inclusive growth. Again, this is in

29 UNCTAD Stat (n.d.). Merchandise: Product concentration and diversification indices of exports and imports, annual. Available at: https://unctadstat.unctad.org/wds/TableViewer/tableView.aspx?ReportId=120 (Accessed: 19 September 2021).

30 Ibid.

31 UNDP (2009). Export Dependence and Export Concentration. Available at: www1.undp.org/content/dam/undp/library/Poverty%20Reduction/Inclusive%20development/Towards%20Human%20Resilience/Towards_SustainingMDGProgress_Chapter1.pdf (Accessed: 18 September 2021).

32 UNCTAD (2013). The state of industrial development in Africa: Unexploited opportunities amidst growing challenges. Available at: https://unctad.org/system/files/official-document/presspb2013d1_en.pdf (Accessed: 18 September 2021).

contrast to the findings in the development literature, which show that 'virtually all cases of high, rapid and sustained economic growth in modern economic development have been associated with industrialisation, particularly growth in manufacturing production'.[33]

Table 2.1 Export concentration index and export diversification index of various regions

	Export Concentration Index			*Export Diversification Index*		
Region	**2000**	**2010**	**2020**	**2000**	**2010**	**2020**
Africa	0.35	0.41	0.19	0.60	0.56	0.55
Sub-Saharan Africa	0.34	0.41	0.21	0.59	0.58	0.60
Sub-Saharan Africa excl. South Africa	0.50	0.56	0.31	0.72	0.68	0.69
Northern Africa	0.39	0.42	0.16	0.72	0.65	0.55
Eastern Africa	0.13	0.15	0.15	0.71	0.68	0.65
Middle Africa	0.73	0.81	0.58	0.84	0.81	0.84
Southern Africa	0.15	0.12	0.15	0.55	0.54	0.57
Western Africa	0.64	0.59	0.38	0.78	0.74	0.76
Eastern Asia	0.10	0.10	0.12	0.29	0.33	0.30
South-Eastern Asia	0.18	0.13	0.13	0.38	0.33	0.31
Europe	0.07	0.07	0.06	0.15	0.18	0.18
Emerging markets	0.11	0.11	0.09	0.22	0.18	0.16
Selected exporters of petroleum	0.66	0.65	0.49	0.77	0.71	0.71
Selected exporters of minerals and mining products	0.19	0.27	0.27	0.72	0.72	0.75
Selected exporters of manufactured goods	0.11	0.10	0.12	0.32	0.30	0.27
Selected exporters of manufactured goods and primary commodities	0.09	0.08	0.09	0.30	0.32	0.28

Source: Author's construct based on UNCTADStat Merchandise Trade Matrix[34]

Manufacturing sector contribution to the continent's GDP since 1990 has declined by an average of 0.24% of GDP per annum from a base of 16% of GDP to the current 11% – this is about the same as the 1970s (Figure 2.3).[35] This is despite the marginal improvement by the continent between 2011 and

33 Ibid, at p.1.

34 UNCTAD Stat (n.d.). Merchandise: Product concentration and diversification indices of exports and imports, annual. Available at: https://unctadstat.unctad.org/wds/TableViewer/tableView.aspx?ReportId=120 (Accessed: 19 September 2021).

35 Newman, C., Page, J., & Tarp, F. (2021). Made in Africa – the future of production on the continent. Available at: www.weforum.org/agenda/2017/01/made-in-africa-the-future-of-production-on-the-continent (Accessed: 19 September 2021).

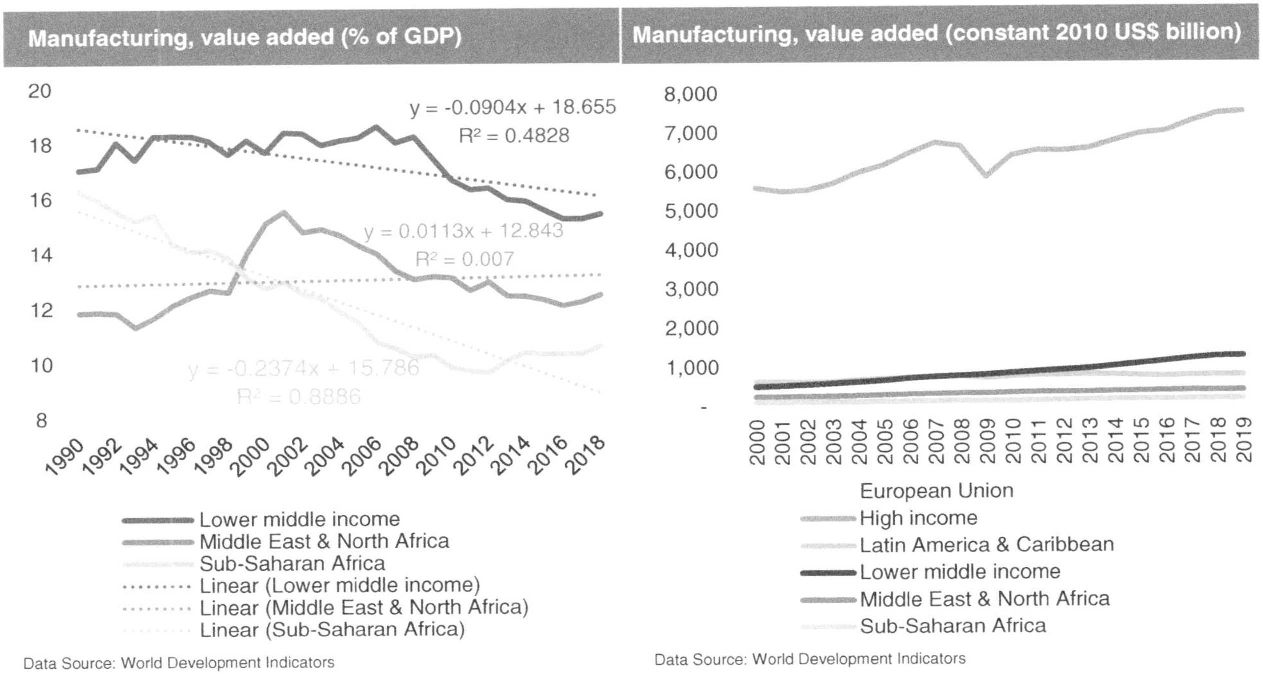

Figure 2.3 Manufacturing as a share of GDP and value-added in selected regions

2018. In contrast, the manufacturing sector contribution for Latin America and Caribbean countries is 16% of GDP; lower middle-income countries (17% of GDP); the European Union (16% of GDP); and East Asia and the Pacific (24% of GDP). This decline in the contribution of manufacturing to GDP has been observed in all of the continent's sub-regions.[36] Likewise, the continent's share of global manufacturing value has declined from about 3% in 1970 to less than 2% currently (Figure 2.3).[37] This means the African continent has been unable to fully take advantage of the emergence of global value chains, which has driven and diversified global manufacturing production with attendant productivity and employment effects. While there is evidence indicating that some East African countries such as Ethiopia and Kenya have captured some low-cost, labour-intensive activities of global value chains, many countries remain primary and intermediate inputs suppliers.[38]

However, since the early 2000s, several countries on the continent have sought to renew or deepen their commitments with a second attempt at industrialising. Part of the reason for this has been to address the various interrelated external shocks, namely: (1) hikes in food prices and energy prices; (2) the 2007–2008 global financial and economic crisis; (3) the 2014–2018 commodities price crash; and, more recently, (4) the COVID-19 pandemic and associated economic crises. This renewed commitment has been implemented through initiatives such as the Africa Union's Action Plan for the Accelerated Industrial Development of

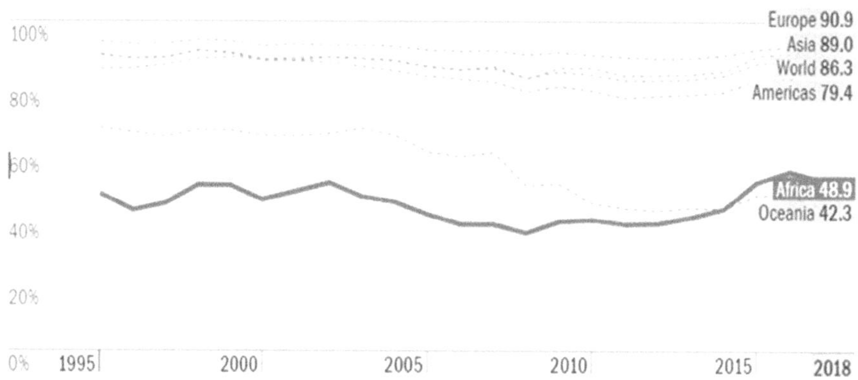

Figure 2.4 Manufactured goods in total merchandise exports (%)
Source: Edwards, 2020[39]

36 Ibid.
37 Edwards, L. (2020). African manufacturing firms and their participation in global trade. Available at: https://iap.unido.org/articles/african-manufacturing-firms-and-their-participation-global-trade (Accessed: 25 September 2021).
38 Ibid.
39 Ibid.

Africa (ADIA)[40] and most recently, the African Continental Free Trade Agreement (AfCFTA), which seeks to remove the barriers to and boost intra-regional trade in manufactured goods as well as catalyse the development of regional value chains.

3 Energy Transition and Industrialisation Options for Sub-Saharan Africa

Despite the preceding, within every crisis lies an opportunity, which makes one wonder if perhaps the COVID-19 pandemic and associated 'green recovery programmes' announced by various governments could also spur the continent to embark on its own green growth industrial transformation agenda. Could SSA have an opportunity to walk the proverbial industrialisation talk after several false dawns of failing to diversify their economies, especially through the resource-based industrialisation of the past several decades? With extractives-based industrialisation, the evidence in several SSA countries shows that extractives continue to remain enclave activities with little spillover into the broader economy.[41] This is despite recent attempts to correct this imbalance through local content and local participation schemes.[42] Likewise, as widely documented in the extant literature, extractives have historically fuelled strife and corruption in several countries on the continent. At the same time, local political elites engage in rent capture with less appetite for serious socio-economic reforms – for example, tax reform – due to easy petro- or mining dollars.

Several SSA countries face a choice between delivering sustained economic growth or growing sustainably.[43] The former encompasses using all available energy resources at its disposal, be it fossil fuels or renewable energy, to address the continent's challenges: large-scale industrialisation, creating jobs

40 African Union (n.d.). Action Plan for the Accelerated Industrial Development of Africa. Available at: https://au.int/web/sites/default/files/documents/30985-doc-plan_of_action_of_aida.pdf (Accessed: 25 September 2021).

41 Hansen, M. W., Buur, L., Mette Kjær, A., & Therkildsen, O. (2016, May). The economics and politics of local content in African extractives: Lessons from Tanzania, Uganda and Mozambique. *Forum for Development Studies, 43*(2), 201–228.

Hansen, M. W. (2014). *From enclave to linkage economies? A review of the literature on linkages between extractive multinational corporations and local industry in Africa* (No. 2014: 02). DIIS Working Paper.

42 Acheampong, T., Ashong, M., & Svanikier, V. C. (2016). An assessment of local-content policies in oil and gas producing countries. *The Journal of World Energy Law & Business, 9*(4), 282–302.

Ovadia, J. S. (2016). Local content policies and petro-development in Sub-Saharan Africa: A comparative analysis. *Resources Policy, 49*, 20–30.

43 Shpitsberg, A., Berkove, D., Tesfay, N., & Bostan, R. (2021). Sub-Saharan Africa's energy transition: A choice between growing sustainably and sustained growth? Available at: https://ihsmarkit.com/research-analysis/subsaharan-africas-energy-transition.html (Accessed: 26 September 2021).

for the millions of its unemployed and underemployed youth, and delivering inclusive growth and sustainable development. The latter is more about using cleaner and renewable forms of energy to power a new industrialisation drive: green industrialisation.[44] The evidence from the development literature shows, for example, that there are no 'existing examples of countries that have pursued a green industrialisation pathway through a deliberate greening policy from the outset'.[45]

Instead, most high-income countries of today developed by polluting their way, using all forms of available energy sources, predominantly coal and hydrocarbons. As a result, environmental considerations were often an afterthought. This raises several questions: first, unlike the previous commodity booms, which have failed to translate into inclusive growth, can Africa take advantage of this energy transition and shift production factors – labour, capital, and entrepreneurship – into the energy and manufacturing industrial sector? In other words, can and must SSA pursue green industrialisation? Second, what factors or conditions are likely to make green industrialisation work on the continent? Third, can green industrialisation co-exist with the traditional resource-based industrialisation, which many countries on the continent have attempted to follow with mixed results? Lastly, must the continent adapt all forms of energy, provided it is cost-competitive, without worrying much about the environmental footprint?

Below, we propose three approaches by which SSA can leverage the clean energy drive to promote industrialisation and critically discuss them.

3.1 Proposal 1: Shifting Production Factors – Labour, Capital, and Entrepreneurship – into Green Industrialisation

While a radical global energy system transformation is necessary to achieve 2050 net-zero targets, this also comes with opportunities. The IEA estimates that annual energy investment needs to rise to between US$4 trillion and US $5 trillion by 2030 from the current US$2 trillion.[46] SSA can position itself to attract a portion of these investments; 10% of the US$5 trillion investment coming into SSA will make much difference on the continent. To put it in comparative terms, this is three times more than the region's estimated US $2.6 trillion GDP; it could spur a major economic boom and sustainable long-term industrialisation.

44 Luken, R. A. & Clarence-Smith, E. (2019). Green Industrialization in Sub-Saharan Africa: A Guide for Policy Makers. Available at: https://media.africaportal.org/documents/Green-Industrialisation-Report_Web_compressed.pdf (Accessed: 25 September 2021).

45 Okereke, C., Coke, A., Geebreyesus, M., Ginbo, T., Wakeford, J. J., & Mulugetta, Y. (2019). Governing green industrialisation in Africa: Assessing key parameters for a sustainable socio-technical transition in the context of Ethiopia. *World Development, 115*, 279–290.

46 IEA (2021). Net Zero by 2050. Available at: www.iea.org/reports/net-zero-by-2050 (Accessed: 25 October 2021), at p.24.

The necessary conditions for green industrialisation work on the continent are the same factors of production – namely, labour, capital, and entrepreneurship – but they are anchored on a clearly defined industrial policy strategy. Industrial policy is essential, but other preconditions are also necessary for diversification. A recent important IMF paper challenges some conventional orthodoxies about industrial policy. Salinas finds that four economy-wide policies (horizontal policies) – namely, governance (institutions), education, infrastructure, and trade policy openness – help to foster diverse exports more than narrowly targeted industrial policies.[47] Within the context of countries with natural resources, Salinas again finds that Chile, one of the world's largest copper exporters, has a similar economic profile to Malaysia's.[48] Strong education and institutions are a significant moderating factor in both countries despite the latter being much closer to the major East Asian global supply-chain hubs of China, Korea, and Japan.[49] Thus, the pursuit of industrial policies to take advantage of the clean energy drive must parallel significant structural horizontal policy reforms, lest it risks becoming another enclave activity with negligible spillover.

With regard to attracting both domestic and foreign capital, Africa can offer targeted tax incentives for the setup of companies dedicated to assembling green energy components such as solar photovoltaic (PV) and wind turbine modules for use on the continent and potential exports. Furthermore, these can take place in special industrial zones to take advantage of learning by doing, innovation and agglomeration effects which are important to increasing firm-level productivity. East Asia's industrial success benefited from deliberate policies[50] to develop the capacity and capabilities of domestic firms, including fostering industrial clusters (agglomeration) via special economic zones such as industrial parks, high-tech zones, and export-processing zones. However, these have either been largely absent or not fully implemented in several SSA countries, and there is a need to fix these if green industrialisation is to work systematically.

A corollary to the requirement for more non-enclave type activities – such as the manufacturing of solar PV components and wind turbines – to have a wider catalytic impact on the economy is the need to deepen research, innovation, and development. Several SSA countries spend less than 1% of their

47 Salinas, G. & Muñoz, S. (2021). Proximity and Horizontal Policies: The Backbone of Export Diversification and Complexity. *IMF Working Papers, 2021*(064).

48 Salinas, G (2021). How Countries Can Diversify Their Exports. Available at: https://blogs.imf.org/2021/09/22/how-countries-can-diversify-their-exports (Accessed: 30 September 2021).

49 Okereke, C., Coke, A., Geebreyesus, M., Ginbo, T., Wakeford, J. J., & Mulugetta, Y. (2019). Governing green industrialisation in Africa: Assessing key parameters for a sustainable socio-technical transition in the context of Ethiopia. *World Development, 115*, 279–290.

50 Edwards, L. (2020). African manufacturing firms and their participation in global trade. Available at: https://iap.unido.org/articles/african-manufacturing-firms-and-their-participation-global-trade (Accessed: 25 September 2021).

GDP on research and development (R&D); this must change.[51] The golden triangle of industry–academic–research institutes offers some pathways on how this could work. First, entrepreneurs, governments, and other donor partners could fund targeted R&D in universities and higher learning institutions localise these RE production systems. China took this approach with solar PVs, and the result is solar PV costs falling by a third in the past decade. Australia-based company SunDrive Solar[52] was formed by a PhD candidate at the University of New South Wales with a simple idea to make solar panels much cheaper by replacing the expensive silver component of the cells with relatively cheaper copper. The research ecosystem benefited from university seed funding and private funding to scale up the idea. Such an ecosystem is not beyond SSA.

SSA countries can build and connect R&D centres of excellence on the continent to contextualise and find new scalable engineering and technical solutions to Africa's energy access challenges. These research, development, and innovation (RDI) centres could, for example, lead the scaling up and mass manufacturing of RE-powered agri-equipment for a new grassroots-led agricultural revolution on the continent. Solar-powered irrigation systems, combine harvesters, tills, shredders, dryers, and warehouses can drastically improve agricultural yields and reduce post-harvest losses while shifting from the 'cutlass and hoe' model of farming into mid-scale commercial or mechanised agriculture in various communities. This will significantly improve livelihoods and reduce poverty.

3.2 Proposal 2: Meeting the Continent's Energy Needs Using All Available Energy Resources

The rapid economic development of any economy is directly linked to modern and reliable energy services provision, especially electricity. Nevertheless, in SSA, access to reliable power is consistently identified as one of the most significant constraints on the continent's development. About 46% or 600 million of the continent's population lack access to electricity, while another 730 million lack clean fuels and cooking facilities.[53] Also, while about 78% of Africa's urban population and 31% of the rural population had electricity in 2019, it is often irregular, unreliable, and expensive in many

51 World Bank (n.d.). Research and development expenditure (% of GDP) – Sub-Saharan Africa. Available at: https://data.worldbank.org/indicator/GB.XPD.RSDV.GD.ZS?locations=ZG (Accessed: 30 September 2021).

52 Vance, A. (2021). Solar Startup Born in a Garage Is Beating China to Cheaper Panel. Bloomberg. Available at: www.bloomberg.com/news/articles/2021-09-09/sundrive-solar-startup-beats-china-s-giant-manufacturers-in-efficiency-test (Accessed: 29 September 2021).

53 Africa Energy Chamber (2021). *Africa Energy Outlook 2021*. Available at: www.whyafrica.co.za/wp-content/uploads/2020/11/AEC-Outlook-2021.pdf (Accessed: 17 November 2021).

countries in the sub-region.[54] Most rural populations continue to rely on traditional biomass and firewood to meet their domestic needs, worsening energy poverty and health disparities.[55]

Within the context of the energy trilemma, enabling access and affordability for bridging the access deficit and for accelerated industrialisation is foremost for the continent. This is because Africa's GHG emissions are among the lowest globally on a volume and per capita basis, excluding South Africa. The continent emitted 1,308 million tonnes of CO_2 in 2019, representing only 2% of global energy-stimulated CO_2 emissions and only a 2% growth in the decade between 2008 and 2018.[56] However, the continent has enormous energy resources – both conventional and non-conventional – which can and must be urgently used to address the access deficit. In a report published in June 2021 to the Norwegian Parliament, the government said that 'Norway's position as an energy nation will be further developed through new initiatives encompassing hydrogen, offshore wind, strengthening the power grid and a low emissions oil and gas sector'.[57] The government is keen on adopting all energy resources available at its disposal to create continued economic growth and new jobs.

Thus, similar to Norway's common-sense approach, we think that SSA's response to the energy transition ought to be premised on using all energy forms at the continent's disposal to catalyse long-term and inclusive economic growth within the framework of the respective country's climate policy and commitments or nationally determined contributions (NDCs) under the Paris Agreement. The Norwegians do not say they will stop oil and gas extraction. Instead, they will focus on making the barrels have more 'green street cred' – that is, lower emissions – so it competes favourably. Furthermore, and interestingly, the ongoing variability of RE supply, especially wind energy[58] in Europe, means that

54 Eberhard, A., Gratwick, K., Morella, E., & Antmann, P. (2017). Accelerating investments in power in sub-Saharan Africa. *Nature Energy, 2*(2), 1–5.

55 Nussbaumer, P., Nerini, F. F., Onyeji, I., & Howells, M. (2013). Global insights based on the multidimensional energy poverty index (MEPI). *Sustainability, 5*(5), 2060–2076.

Ssennono, V. F., Ntayi, J. M., Buyinza, F., Wasswa, F., Aarakit, S. M., & Mukiza, C. N. (2021). Energy poverty in Uganda: Evidence from a multidimensional approach. *Energy Economics, 101*, 105445.

Crentsil, A. O., Asuman, D., & Fenny, A. P. (2019). Assessing the determinants and drivers of multidimensional energy poverty in Ghana. *Energy Policy, 133*, 110884.

56 Ritchie, H. & Roser, M. (2020). CO_2 and greenhouse gas emissions. *Our World in Data*. Available at: https://ourworldindata.org/co2-and-other-greenhouse-gas-emission (Accessed: 13 January 2023).

57 Norwegian Government (2021). Government publishes White Paper on long term value creation from Norway's energy resources. Available at: www.regjeringen.no/en/aktuelt/regjeringen-legger-frem-stortingsmelding-om-verdiskaping-fra-norske-energiressurser/id2860271 (Accessed: 13 September 2021).

58 The Economist (2021). Why has the price of electricity in Europe reached record highs? Available at: www.economist.com/the-economist-explains/2021/09/15/why-has-the-price-of-electricity-in-europe-reached-record-highs (Accessed: 25 September 2021).

several countries are muting plans to restart conventional coal[59] and nuclear power plants.[60] These had hitherto been classified for decommissioning as these did not align with their climate or net-zero commitments. What is evident in these situations is that energy security is the paramount consideration. This cannot be any different for SSA, or must it?

3.2.1 Natural Gas

Natural gas and the gas industry value chain are an integral part of Africa's energy future as they are cost-effective for delivering baseload power and are environmentally friendly. As of 2019, the continent had between 527 and 558 trillion cubic feet (Tcf), making it the fourth-largest gas reserves holder after North America.[61] Countries with gas supplies include established players like Nigeria, Equatorial Guinea, and Ghana, and nascent players like Tanzania, Mozambique, Mauritania, and Senegal. Natural gas is already the primary fuel of choice in power generation for several countries and regions on the continent. This trend is likely to continue into the foreseeable future.[62] Nevertheless, the exploitation of the fuller gas value chain opportunities on the continent for power generation and other industrial uses is stalled by the lack of infrastructure to evacuate gas from the fields to the relevant gas markets. Other constraints hindering the development of gas markets on the continent include regulatory bottlenecks such as ineffective gas pricing and non-payment issues from the inability to recoup costs of gas supplied by upstream companies.

SSA's gas value chain requires significant investments in building trans-regional pipelines to evacuate gas to domestic demand centres, and to an extent, also for liquefied natural gas (LNG) export markets. With an expanded pipeline network in West, East, and Southern Africa, regional gas movements could double by 2030, further displacing coal in some markets. In West Africa, for example, the West African Gas Pipeline (WAGP), which runs from Nigeria's Ikoti export terminal to Takoradi in Ghana, has the potential to export gas further to Côte d'Ivoire. A possible expansion of the WAGP to

59 BBC News (2021). UK fires up coal power plant as gas prices soar. Available at: www.bbc.co.uk/news/business-58469238 (Accessed: 25 September 2021).
Ambrose, J. (2021). Britain's last coal power stations to be paid huge sums to keep lights on. *The Guardian*. Available at: www.theguardian.com/business/2021/sep/13/britain-last-coal-power-stations-to-be-paid-huge-sums-to-keep-lights-on-record-energy-prices (Accessed: 25 September 2021).

60 IRENA (2021). The Renewable Energy Transition in Africa – Powering Access, Resilience and Prosperity. Available at: www.irena.org/-/media/Files/IRENA/Agency/Publication/2021/March/Renewable_Energy_Transition_Africa_2021.pdf (Accessed: 29 September 2021).

61 Africa Energy Chamber (2021). *Africa Energy Outlook 2021*. Available at: www.whyafrica.co.za/wp-content/uploads/2020/11/AEC-Outlook-2021.pdf (Accessed: 17 November 2021).

62 Ibid.

Côte d'Ivoire would bring delivered gas prices far below liquid fuels for power generation and other non-power industrial uses. However, the ability to attract foreign and domestic capital is becoming increasingly difficult in the face of ESG pressures facing traditional financing sources – including concessional and non-concessional loans and guarantees by development finance institutions (DFIs). Nevertheless, some of the more resourced domestic sovereign wealth funds (SWFs) or sovereign-backed infrastructure funds can also enter the void, subject to long-term value-additive and economically profitable projects.

3.2.2 Deploying Renewable Energy Technologies

RE technologies, especially wind and solar, are important in helping bridge the electricity access deficit on the continent. Renewable costs have declined substantially over the past ten years and RE is even out-competing some conventional fuels as the least-cost alternative for new generation capacity. Studies show that utility-scale solar PV and onshore wind costs decreased by 82% and 40%, respectively, between 2010 and 2019.[63] This global trend is also happening in SSA, where there has been a 30% reduction in wind LCOE (levelised cost of energy) costs between 2010 and 2019.[64] The reduction in solar PV tariffs has also been supported by a significant 67% drop in utility-scale battery storage.[65] This means that battery storage solutions can possibly be deployed to address the intermittency challenge of renewables, thereby allowing the provision of dispatchable power. Likewise, the International Renewal Energy Agency (IRENA) reports that 71% of the 2019 generation capacity added were renewable technologies, primarily solar and wind.[66] Research indicates that despite Africa installing only about 1% of its estimated wind capacity as of 2018, wind energy deployment has grown by 740%, from 0.74 Gigawatts (GW) to 5.5 GW between 2009 and 2018.[67] Most of this deployment has been in Southern (South Africa) and Northern Africa (Morocco, Egypt, Tunisia).

63 IRENA (2021). The Renewable Energy Transition in Africa – Powering Access, Resilience and Prosperity. Available at: www.irena.org/-/media/Files/IRENA/Agency/Publication/2021/March/Renewable_Energy_Transition_Africa_2021.pdf (Accessed: 29 September 2021).

64 Alemzero, D., Acheampong, T., & Huaping, S. (2021). Prospects of wind energy deployment in Africa: Technical and economic analysis. *Renewable Energy, 179*, 652–666.

65 IRENA (2021). The Renewable Energy Transition in Africa – Powering Access, Resilience and Prosperity. Available at: www.irena.org/-/media/Files/IRENA/Agency/Publication/2021/March/Renewable_Energy_Transition_Africa_2021.pdf (Accessed: 29 September 2021).

66 Ibid.

67 Alemzero, D., Acheampong, T., & Huaping, S. (2021). Prospects of wind energy deployment in Africa: Technical and economic analysis. *Renewable Energy, 179*, 652–666.

Both policy- and market-based developments will be vital in extending the reach of RE more widely on the continent. Policies to drive RE deployments – renewable obligations standards (ROS), feed-in tariffs (FiTs), auctions, wind investment and tax credit, and net metering —are currently scattered in many countries. Thus, policy coherency is needed, especially in light of the AfCFTA. Furthermore, as we propose in our renewable energy linkages ecosystem shown in Figure 2.5, SSA countries need to be more strategic in harnessing the linkages and value chain opportunities associated with deploying RE.

3.3 Proposal 3: Deepening Linkages in a New Resource-Based Industrialisation Paradigm 2.0

Africa has an abundance of the critical minerals needed for the energy transition. Metals such as copper, lithium, nickel, cobalt, manganese, and graphite are vital to the transition for sustaining battery longevity, performance, and energy density of all-EV motors, solar panels, and wind turbines.[68] The demand for these minerals and other rare earth elements is forecast to increase exponentially in the coming years due to the transition. For example, soaring demand for batteries will require massive amounts of lithium, a key component. The IEA forecasts that the energy sector will become one of the leading consumers of minerals as the energy transition accelerates (Figure 2.6). IMF analysis also

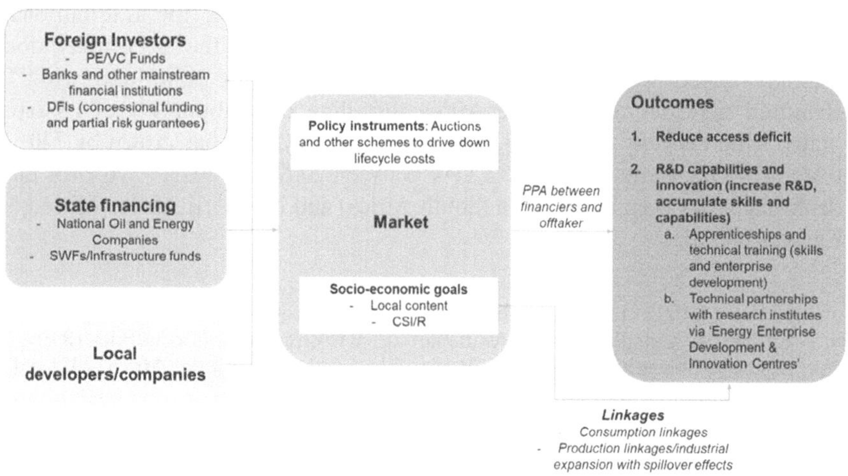

Figure 2.5 Renewable energy linkages ecosystem
Source: Authors' construct

68 IEA (2021). The Role of Critical Minerals in Clean Energy Transitions. Available at: www.iea.org/reports/the-role-of-critical-minerals-in-clean-energy-transitions (Accessed: 13 January 2023).

shows that the combination of soaring demand for metals and slower supply changes could push metals prices to reach historical peaks for an unprecedented length of time.[69] These mineral resources can be found in abundance in Zambia, the Democratic Republic of Congo (DRC), and South Africa, among others (Figure 2.7). This means that there is potential for SSA to play a significant value-adding role in the global supply chains for these critical minerals. However, this must go beyond mere extraction and intermediate processing of the minerals into battery cell manufacturing and assembly to capture a substantial share of the value chain opportunities.

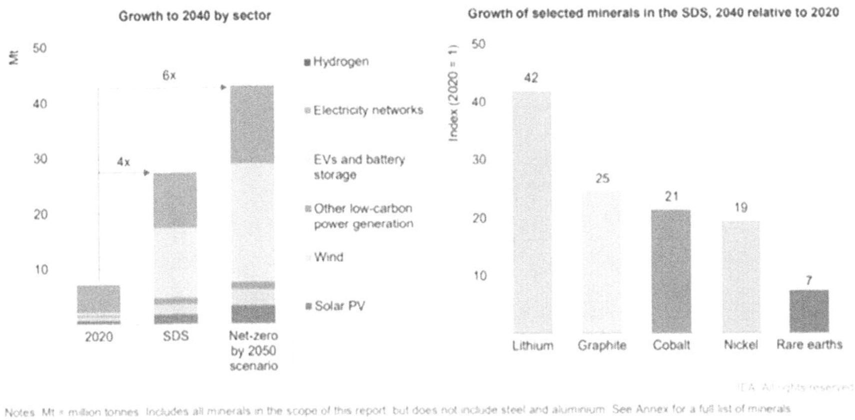

Figure 2.6 Mineral demand for clean energy technologies by scenario
Source: IEA[70]

For SSA to benefit from the transition, the development of value chains and beneficiation is key to avoid a repeat of the 'paradox of plenty' syndrome. Developing value chains also aligns with key continental strategic imperatives such as the African Union (AU) Agenda 2063,[71] the African Continental Free Trade Area (AfCFTA),[72] African Development Bank's (AfDB's) High 5s,[73] and

69 Boer, L., Pescatori, A., & Stuermer, M. (2021). Energy Transition Metals. IMF Working Paper No. 2021/243.

70 IEA (2021). The Role of Critical Minerals in Clean Energy Transitions. Available at: www.iea.org/reports/the-role-of-critical-minerals-in-clean-energy-transitions (Accessed: 13 January 2023).

71 African Union (n.d.). Agenda 2063: The Africa We Want. Available at: https://au.int/en/agenda2063/overview (Accessed: 24 November 2021).

72 African Union (n.d.). African Continental Free Trade Area Questions & Answers. Available at: https://au.int/sites/default/files/documents/36085-doc-qa_cfta_en_rev15march.pdf (Accessed: 24 November 2021).

73 AfDB (2019). The High 5s, African Development Bank – Building today, a better Africa tomorrow. Available at: www.afdb.org/en/high5s (Accessed: 24 November 2021).

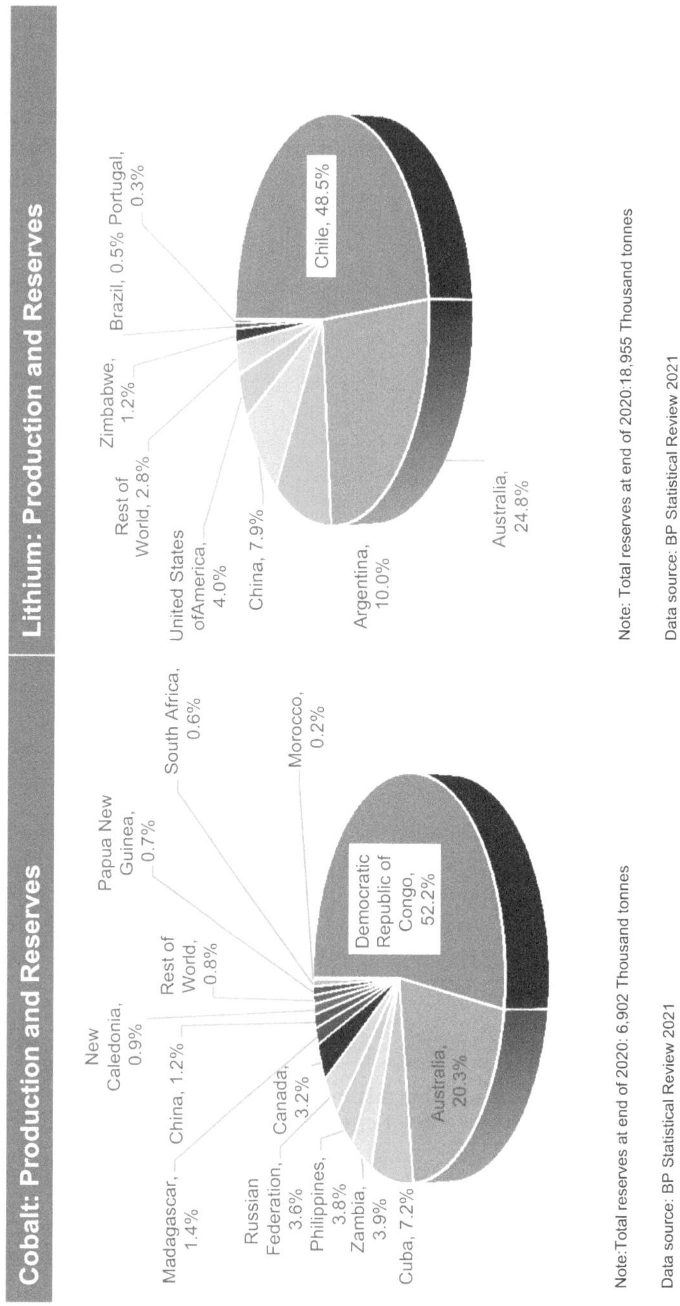

Figure 2.7 Cobalt and lithium production and reserves

the Africa Mining Vision (AMV).[74] These are being championed and funded by regional multilaterals and DFIs such as the AU, AfDB, Afreximbank, the Africa Finance Corporation (AFC), and the UN Economic Commission for Africa (UNECA), among others. In planning to take advantage of future opportunities, past lessons should guide the continent: it is no longer justifiable to export raw minerals. However, to do this, structural bottlenecks need to be significantly addressed. These include the need for cheap and reliable dispatchable power to drive industrialisation. Likewise, good infrastructure such as roads and rail links to connect mines and spread spatial development more evenly is fundamental, especially in stemming the continent's massive urbanisation.

Finally, a stable investment regime and conducive business environment are necessary. Unfortunately, several countries where these mineral resources are abundant on the continent lag on business reforms. While some countries are thinking of special fiscal regimes to guide the extraction of these minerals, which ensures a fair share to the state, these must be balanced with the need for stability and predictability of the fiscal regime, also on a bigger industrial development agenda – that is, a combination of fiscal and industrial policy. The situation whereby Zambia, for example, desperately increased its fiscal take six times between 2012 and 2019 must be avoided, as such moves often serve as a disincentive for investment.[75] Chile's attempt to develop and capture a larger share of the EV value chain can serve as a guide for SSA: In March 2018, the Chilean government signed a deal with South Korea's POSCO and electronics giant Samsung SDI to build factories in the country, which would produce EV battery parts – cathode materials such as nickel cobalt aluminium and nickel cobalt manganese.[76] In return, through CORFO, a government organisation that oversees the country's lithium resources, Chile would provide a guaranteed supply of lithium at an agreed fixed price for 30 years. While reports[77] indicate that the Chilean government has not been able to deliver on the promised volumes, the policy intent or signal is, in our view, the right one.

4 Conclusions

In the SSA context, the energy transition should not just be about reducing emissions or decarbonisation for its own sake. Instead, it crucially needs to be

74 African Union (2021). AMV – Africa Mining Vision. Available at: https://au.int/en/ti/amv/about (Accessed: 24 November 2021).

75 Acheampong. T. (2019). Zambian mineral royalties increase. Available at: https://ihsmarkit.com/research-analysis/zambian-mineral-royalties-increase.html (Accessed: 29 September 2021).

76 Sherwood, D. (2021). How lithium-rich Chile botched a plan to attract battery makers, U.S. Reuters. Available at: www.reuters.com/article/us-chile-lithium-focus-idUSKCN1UC0C8 (Accessed: 29 September 2021).

Electrive.com (2018). Chile's lithium resource redistributed through Corfo. Available at: www.electrive.com/2018/03/12/consortium-to-build-54m-cathode-materials-plant-in-chile (Accessed: 29 September 2021).

77 Ibid.

linked to bridging the energy access gap and encouraging productive uses of energy to create new competitive industrial clusters, which create jobs and sustain economies. The data shows that SSA has the lowest emissions and the lowest energy intensity globally. If our economies are to develop and create better jobs, then the latter number needs to improve significantly using all forms of available energy. The framing of the energy transition matters because millions of lives and jobs are at stake, especially in developing countries such as in SSA. As Nigeria's Vice President argues, 'a global transition away from carbon-based fuels must account for the economic differences between countries and allow for multiple pathways to net-zero emissions'.[78]

While there are clear opportunities for green industrialisation and other 'industries without smokestacks' (IWOSS)[79], such as high-value agriculture, tourism, and business services, these can co-exist with traditional resource-based industrialisation to drive structural transformation. They must not be seen or be promoted as being mutually exclusive. The latter will become even more pertinent as Africa has an abundance of the critical minerals needed for the energy transition, such as copper, lithium, and cobalt, among others. However, there is a need to craft an African strategy – especially in the context of the AfCFTA – and coordinated regional and country-level strategies as industrial orientation will indicate the potential for leveraging the energy transition. SSA needs to increase its manufacturing share and have a genuinely high-tech or advanced industrial sector.

Thus, linkages and value chain development will be critical – beneficiation will be key to avoiding the 'paradox of plenty' syndrome as it is no longer justifiable to export raw minerals. Structural bottlenecks, such as the need for reliable and cheap power, is fundamental. Also, good infrastructure is needed to connect, for example, mines and spread spatial development more evenly. Likewise, a stable investment and business environment is fundamental in attracting domestic and foreign capital. At a firm level, increasing productivity will also be vital. East Asia's industrial success benefited from deliberate policies to build the capacity and capabilities of domestic firms, including fostering industrial clusters (agglomeration) via special economic zones such as industrial parks, high-tech zones and export-processing zones.[80] Unfortunately, these have been largely absent or not fully implemented in several SSA countries.

78 Osibanjo, Y. (2021). The Divestment Delusion – Why Banning Fossil Fuel Investments Would Crush Africa. Available at: www.foreignaffairs.com/articles/africa/2021-08-31/divestment-delusion (Accessed: 1 October 2021).

79 Newfarmer, R., Page, J., & Tarp, F. (2019). *Industries Without Smokestacks: Industrialisation in Africa Reconsidered*. Oxford University Press.

80 Apiko, P., Woolfrey, S., & Byiers, B. (2020). The promise of the African Continental Free Trade Area (AfCFTA) (No. 287). ECDPM Discussion paper.

Acknowledgements

An earlier version of this paper was produced as a policy brief for the Africa Policy Research Institute (APRI) and African Climate Foundation (ACF) ahead of COP 26. We thank Ellen Davies of ACF and two anonymous peer-reviewers for helpful review comments.

Conflicts of Interest

The authors declare no conflict of interest.

3 Africa's Mining Codes

Balancing the Interests

Peter Nuhu, Ishmael Ackah, and Dramani Bukari

1 Introduction

The proportion of gross domestic product (GDP) and gross government revenues that are attributable to mineral receipts is significant in Africa as it is for most developing countries. In 2019, mining contributed approximately 8% to GDP in South Africa.[1] In Ghana, the extractive industry contributed 14% to GDP, 18% to government revenue and 67% to exports in 2018.[2] Similarly, the contribution of mining to both GDP and government revenue has averaged double digits since 1985 in Botswana.[3]

Figure 3.1 shows the historical values of mineral rents in Botswana, Ghana, South Africa, Tanzania, and Zambia.

Sustainable exploitation and appropriate taxation of mineral resources are fundamental to the revenue prospects, growth, and development of many developing countries.[4] This is particularly so when the unique characteristics of the resource are taken into consideration. Minerals are a stock resource; they are depletable, susceptible to price volatilities, immovable, and vary in quality across locations. Their quality and quantity are also unknown initially. As such, sovereign states endowed with such minerals fashion regimes to maximise gains from their exploitation. Yet these regimes must consider the risks presented by the unique characteristics noted above, in addition to the huge initial capital outlay and the long lead time required for exploration and development. The level and structure of fiscal regimes are therefore central to this balancing act because different imposts have different effects on the financial viability and operating parameters of a mine. While the effects of some imposts are attenuated by increased production, others are directly

1 PwC. (2019). SA Mine 2019: In Transition. Johannesburg: PwC Inc.
2 GHEITI. (2018). Extractive Industries Contribution to the Economy. Retrieved from Ghana Extractive Industries Transparency Initiative: https://eiti.org/es/implementing_country/4
3 ANRC. (2016). *Botswana's Mineral Revenues, Expenditure and Savings Policy: A Case Study*. Abidjan: AfDB.
4 ICMM. (2009). *Minerals Taxation Regimes: A Review of Issues and Challenges in Their Design and Application*. London: International Council on Mining and Metals.

DOI: 10.4324/9781003284437-4

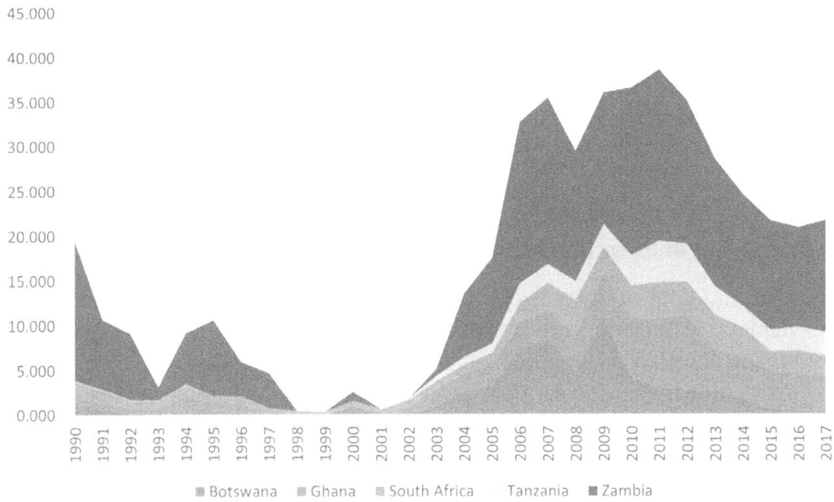

Figure 3.1 Mineral rent as a percentage of GDP
Source: World Bank Development Indicators, 2020

linked to revenues, profits, and profitability, and therefore are essential determinants of the cut-off grade of a project.[5] Accordingly, host nations tend to blend multiple tools, from imposts and incentives to local content regulations, in order to mitigate the financial impact on concessionaires while guaranteeing a considerable proportion of the resource rent for its own development. The essence of this is also to provide a more favourable and secured environment to attract private investment.

This chapter seeks to assess the different economic and fiscal regimes in Africa. Specifically, the chapter reviews the regimes relative to how they allocate risk, balance the timing of revenue receipts, balance efficiency with simplicity and flexibility, provide stability or predictability, and protect project profitability. In addition, the chapter reviews the local content regulations embedded in fiscal regimes and how they are designed to improve the share of mineral rents received by the state. This analysis is done with a view to assessing a regime on its ability to attract investment and generate revenue for the host government.

The chapter is in four sections. In this introduction, the importance of the mining industry to the overall GDP and resource mobilisation drive of developing countries, especially in Africa, is presented. It also provides the spectrum of the study by discussing the pillars upon which each fiscal regime is assessed. Section 1.2 discusses fiscal regimes in the mining sector; zoning in

5 Aryee, B. N. (2018). *A Historical Overview of Ghana's Mining Sector Fiscal Regimes*. Wandsbeck, SA: Reach Publishers.

on the fiscal tools and instruments adopted for the sector. Section 1.3 assesses fiscal regimes in Africa with particular attention to Botswana, Ghana, South Africa, Tanzania, and Zambia. Section 1.4 discusses the local content in Africa's mining sector with particular attention to how African countries have utilised local content regulations for the maximisation of the state's share of mining rents. Section 1.5 concludes the chapter.

2 Fiscal Regimes in the Mining Sector

The cardinal goal of every fiscal regime or policy is to maximise benefits from the depletable natural resource for the state, while attracting the required investment needed to realise such benefits.[6] As noted earlier, the capital requirement for the exploration and development of mineral resources is enormous. At the same time, mineral resources are depletable and their extraction has an impact on the environment. As such, the choice of a fiscal regime is a delicate balancing act or dilemma that resource-rich countries must undertake.[7] To resolve this dilemma, policy must:

i allocate risk between the state and the resource company. As noted, commodity prices are volatile and consequently, resource revenues are also volatile. The critical need for guaranteeing the state's interest in stable revenues must be pursued by crafting regimes that compensate resource companies for shouldering a bigger exposure to revenue volatility. The regime must provide investors with a commensurate return on investment in order to attract the competitive private capital needed.[8]

ii balance the timing for revenue receipts between the state and the resource company. As a stock resource, the more revenue is taken now, the less will be available in the future. Both the state and the resource company prefer frontloaded revenue receipts, taking into consideration the volatility of resource prices. Thus, the state must compensate the company for assuming the risk of postponing its receipts.[9]

iii balance efficiency with administrative simplicity. In reality, realigning the objectives of the fiscal system impacts the ease with which the system is administered. In other words, the instruments that require minimal information or are easiest to administer correctly often have features that disagree with the state's objectives of reallocating risk, controlling timing of revenue receipts, etc.[10]

6 Natural Resource Center, Precept 4. Fiscal regimes and contract terms.

7 Committee of Experts on International Cooperation in Tax Matters. (2016). Proposed Guidance on Fiscal Take in the Extractive Industries. *Thirteenth Session, E/C.18/2016/CRP.21.* New York.

8 Ibid.

9 Ibid.

10 PwC. (2019). SA Mine 2019: In Transition. Johannesburg: PwC Inc.

iv minimise the incentive for successive governments to alter fiscal terms. Mineral resource revenues form a significant portion of the finances of low-income countries. Accordingly, the tendency for successive governments to tamper with existing regimes in response to political pressures is huge. There is a need therefore to be mindful of predictability and the long-term perspective of the regime during the design process.

From the foregoing, it is clear that the principal stakeholders who must be considered in the design of a fiscal regime are the resource holder and the investor. Each of these stakeholders aims at maximising specific interest in the framework to be designed. For the resource holder, mostly the state, the main considerations include the overall fiscal take, timing of receipts, funding concerns, environmental impact of resource extraction, competitiveness of the regime, how the regime interacts with existing laws, and allocation of receipts. On its part, the investor or resource company considers the risk/return balance, stability, competitiveness, and predictability of the regime.[11] Generally, there are two fiscal systems in the extractive industry.[12] These are concessionary and contractual systems. No matter which of these is chosen, the specific tools or instrument adopted are fashioned with the above considerations and trade-offs firmly in mind.

Contractual systems involve the award of a contract covering a designated area to a contractor to operate or produce the resources for the resource holder or state who pays the contractor in a form agreed in the contract. In this case, the resource holder retains ownership of the resource. Under the contractual systems, there are two broad categories: production sharing contracts (PSC) and service contracts.[13] A PSC is a contract between parties detailing the share of production that these participating parties are entitled to after recovery of defined costs and expenses. These types of contracts are predominantly found in the oil and gas industry. In this case, the parties mostly involved are the state, represented mostly by the national oil company, and an international oil company or companies. The contract chooses an operator based on expertise, connections, presence in the country, or number of shares. In PSC systems, government take includes the portion of production allocated to compensate the state for permitting production. It can also include the state's participatory interest, as well as tax.

Under service contracts, the resource holder contracts a service provider to operate or produce the resources at a fee. The service provider holds zero rights to the resources.[14] Accordingly, these types of contracts tend to allocate

11 Committee of Experts on International Cooperation in Tax Matters. (2016). Proposed Guidance on Fiscal Take in the Extractive Industries. *Thirteenth Session, E/C.18/2016/CRP.21.* New York.

12 ICMM. (2009). *Minerals Taxation Regimes, A Review of Issues and Challenges in Their Design and Application: Analysis and Observations.* London: ICMM.

13 Ibid.

14 Ibid.

minimal risk to the service provider except to the extent, in some cases, of its exposure to cost overruns. In the mining industry, in addition to contract mining, support service providers provide services such as drilling, blasting, and equipment rentals. Where the resource holder is the state, in this case, government take encompasses the entire resource produced excluding fees, costs, and expenses. In addition, the government also receives revenues from taxing the service provider. Concessionary systems, on the other hand, license the rights or ownership of the underlying resources to the company; a concession is an agreement in which a government leases exclusive rights of an area, for a specified period of time, to a company for exploration, development, and production of resources at its own expense and risk.[15] These agreements do not themselves detail fiscal features as in the case of PSC but are mostly governed by a host of fiscal instruments. The fiscal take usually covers compensation or royalty for resource extraction, corporate tax, and sometimes profit taxes. Concessions unlike PSC are dominant in the mining industry because of the industry's specific features: higher and front-loaded costs, capital investment throughout mine life, and marketability of mineral products.[16]

2.1 Fiscal Instruments in the Mining Industry

A fiscal regime is a combination of fiscal instruments or tools. These instruments come as imposts and incentives among others.

2.1.1 Special vs Uniform Tax Regimes

There are a number of arguments in favour of and against special taxes for the mining industry. The arguments for this special treatment are both economic and political. Proponents[17] of this position argue that:

i during exploration, there is no income.
ii mine development is capital intensive and requires specialist imports like skills and equipment.
iii the capital invested becomes captive.
iv investors are foreign and transnational.
v mining projects are long-term and therefore open to various instabilities.

15 GHEITI. (2018). Extractive Industries Contribution to the Economy. Retrieved from Ghana Extractive Industries Transparency Initiative: https://eiti.org/es/implementing_country/4

16 ICMM. (2009). *Minerals Taxation Regimes, A review of issues and challenges in their design and application: Analysis and observations.* London, UK: ICMM.

17 Aryee, B. N. (2018). *A historical Overview of Ghana's Mining Sector Fiscal Regimes.* Wandsbeck, SA: Reach Publishers.
Otto, M. J., 2001. *Fiscal Decentralization and Mining Taxation,* s.l.: The World Bank Group Mining Department.

vi there is scale variability in that there are usually very small and very large operations.
vii there are winding-up costs associated with mining projects.
viii mines become costlier to operate as they mature.

Those against special taxes for the mining sector argue that the introduction of additional procedures that come with special taxes complicates tax administration, especially in less developed systems, and creates opportunities for further fiscal leakage.[18] They further argue that standardisation or uniformity precludes lobbying and hijacking of the system to further the interests of privileged groups.[19] In practice, fiscal regimes appear to find a balance of both schools of thought. A review of the fiscal systems as presented in Section 1.2 shows that countries have used uniform instruments but applied different rates and provisions so as to accommodate the unique characteristics of the mining sector.

2.1.2 Tax Instruments

In the mining sector, there are principally two tax classifications: profit-based and production-based taxes. Profit-based taxes are imposts on net revenues – that is, revenues less qualifying expenses. Production-based taxes, on the other hand, are impost on deposits or production inputs and services,[20] Both types of taxes present different advantages and challenges relative to the trade-offs discussed above. Generally, however, while governments tend to favour production-based taxes because they offer early revenues that are politically earmarkable and can be used for public financing, such taxes are regressive and increase the financial risks of companies as they require them to continue tax payments even during loss-making periods. As such, companies prefer profit-based taxes. In practice, the trend now favours progressive tax. Governments are moving from regressive indirect taxes.[21] That said, most states still continue to adopt various royalty taxes.[22]

2.1.3 Fiscal Incentives

As noted earlier, resource extraction is capital intensive, and for resource holders to attract and retain international investor capital, the fiscal regime

18 Otto, J. M., 2017. The taxation of extractive industries. *WIDER Working Paper 2017/75*, March, pp. 1–24.
19 ICMM. (2009). *Minerals Taxation Regimes, A Review of Issues and Challenges in Their Design and Application: Analysis and Observations.* London: ICMM.
20 Ibid.
21 ICMM. (2009). *Minerals Taxation Regimes, A Review of Issues and Challenges in Their Design and Application: Analysis and Observations.* London: ICMM.
22 Otto, J., Craig, A., Cawood, F., Doggett, M., Guj, P., Stermole, F., Stermole, J., & Tilton, J. (2006). Mining Royalties. A Global Study of Their Impact on Investors, Government, and Civil Society. Washington, DC: The World Bank.

Table 3.1 Fiscal imposts/instruments

Classification	*Examples*	*Description*
Profit-based taxes	Corporate income tax Royalty on profit or income Profit taxes on dividends Withholding tax on dividends remitted Resource rent tax	Otto[23] espouses that in designing a profit-based tax system, the things to consider are 'the tax rate, and setting the rules for determining the taxable income base that will be subject to the rate'. Reduces profits and may deter investments in marginal projects. Progressive profit taxes are relatively less deterrent, though. More equitable. Proportional profit taxes fail to recognise capital investments utilised in attaining the profit, aside depreciation allowance. Progressive profit taxes may tax different companies with similar NPVs differently. Does not influence the choice of a cut-off grade. Not easy to administer.[24]
Production-based taxes	Royalty based on unit/ value Sales and excise tax Payroll tax Import and export duties VAT Fees and stamp duty Land rent Property tax Withholding tax on loan interests and services	Imposed on gross revenue realised at mine gate, FOB value, net back value, or net smelter return.[25] Either specific or ad valorem. Mostly adopted as supplementary tax in mining sector. Easy and cost-efficient to administer.[26] Guaranteed minimum revenue as long as production is ongoing. Reduces per unit mined revenue and can deter investment in marginal projects. Influences the choice of a cut-off grade

needs to be competitive. Accordingly, governments generally adopt fiscal incentives that counter or negate the impact of taxes.[27] Aryee describes these as 'negative taxes'.[28] Table 3.2 below lists some of these negative taxes.

23 Otto, M. J. (2001). *Fiscal Decentralization and Mining Taxation*. The World Bank Group Mining Department.
24 ICMM. (2009). *Minerals Taxation Regimes, A Review of Issues and Challenges in Their Design and Application: Analysis and Observations*. London: ICMM.
25 Ibid.
26 ICMM. (2009). *Minerals Taxation Regimes, A Review of Issues and Challenges in Their Design and Application: Analysis and Observations*. London: ICMM.
27 Ibid.
28 Aryee, B. N. (2018). *A Historical Overview of Ghana's Mining Sector Fiscal Regimes*. Wandsbeck, SA: Reach Publishers.

Table 3.2 Fiscal incentives

Incentive	*Description*
Capital allowances (depreciation and amortisation)	Expenses on capital and intangibles are deducted over a period Accelerated depreciation/amortisation quickens payback Improves real discounted profits after tax Reallocates some risks to the State by shifting tax receipts further into the future
Depletion allowances	Reverse royalty
Tax holiday	Period when a mine is exempted from some taxes Allows investors space to utilise needed capital that would have been paid as tax for mine development
Loss carry-forward	Losses carried forward to profit-making years Good cushion against price fluctuations
Allowable Expenses	Expenses incurred during exploration and development are expensed during production years Improves payback period and internal rate of return (IRR)
Interest deductions	Loan interests deducted from taxable income Improves credit cost
Tax credits	Qualifying activities attract tax credits May incentivise reinvestment
Tax stabilization	Assurance not to vary fiscal regime for a period of time to the disadvantage of investor

2.1.4 The Ideal Fiscal Regime

As noted in Section 1.1.2 above, there are two extreme ends on the taxation spectrum – that is, the production- or revenue-based taxes end and the profit-based taxes end. At the revenue-based or lump sum taxes end, the state assumes almost zero risk. Movement along the spectrum thus has impact on the objectives of the fiscal regime. Generally, moving from production-based or lump sum taxes towards profit-based or profitability taxes increases the neutrality of the tax system. The tax system become more responsive to mine profitability. At the same time, such movements also reduce the certainty of government receipts and reallocating more and more risk away from the investor to the state albeit improving the potential for higher government receipts in the future. Furthermore, such movements complicate tax administration.

Accordingly, governments strive to balance the interest of the state with the investors' interest – that is, the government recognises the investors' need for stability, predictability, competitiveness, and risk/return balance while also keeping an eye on the state's own risk aversion. They do this by implementing what Otto[29] calls a 'fair and equitable' system. Although a fair and equitable system is yet to be determined by anyone, the competitiveness of a regime determines this. Fairness and equitability should be pursued by combining different types of imposts and incentives – that is, a fiscal regime should combine the various fiscal tools noted previously. In assessing a regime, therefore, the composite of instruments used must be looked at together and not individually.[30] Indeed, a proper assessment should evaluate the system against its set objectives of revenue raising and apportionment verses investor attraction and retention.

As shown in Table 3.3, the income tax rate averages 30.6% in the five selected countries. Ghana's income tax rate is highest at 35%, with South Africa posting the lowest at 28%. In Botswana, the income tax rate discriminates between resident and non-resident corporations. Whiles resident companies are levied 22%, their non-resident counterparts are levied 30%. In respect of royalty, Ghana is the only country with a uniform 5% on the entire mining sector regardless of mineral type. The rest levy royalties at different rates for different minerals. In Botswana, the royal rate ranges from 3% on copper and coal to 10% on diamond and others. In Tanzania, the rate ranges from 4% on copper to 6% on gold, diamond and others exclusive of a mandatory 1% inspection fee. Zambia charges between 5% and 10% depending on mineral. In South Africa, however, royalties are calculated not at a fixed rate but based on value of extracted mineral. Thus, the South Africa royalty rate ranges from 0.5% to 5% for refined minerals and 0.5% to 7% for unrefined minerals. The royalty base is gross sales revenue, turnover, adjusted revenue, gross value, and nominal/gross value in Botswana, Ghana, South Africa, Tanzania, and Zambia, respectively.

Relative to supplementary imposts, all countries except Zambia charges licence fees.[31] Ghana further charges grounds rent of GHC36.5 per acre or GHC9,109 per square kilometre per annum. Botswana's mining, prospecting, and retention licences are structured just as Ghana's grounds rent in the form of annual payment per area awarded. Zambia is the only country with additional profit or windfall profit tax of 0–15% on copper; Ghana's proposed 10% windfall tax was never operationalised. All five countries levy withholding tax on interest, dividends, royalties, and professional and other services provision. In this respect, too, there is discrimination between residents and non-residents. Again, all countries levy VAT on minerals and others,

29 Otto, J. M. (2000). *Mining Taxation in Developing Countries.* UNCTAD.

30 Otto, J., Craig, A., Cawood, F., Doggett, M., Guj, P., Stermole, F., Stermole, J., & Tilton, J. (2006). Mining Royalties. A Global Study of Their Impact on Investors, Government, and Civil Society. Washington, DC: The World Bank.

31 Maroun, W., Ram, A. J., & Kok, M. (2019). A review of mining taxes in Africa: Tax burden, the strength of democratic systems and levels of corruption. *South African Journal of Business Management* (2078–5976), 1–15.

ranging from Botswana's 12% as the lowest to Tanzania's 18% being the highest.

Each country also offers fiscal incentives of some form. Ghana and Zambia offer capital allowance of 20% and 25% straight line with accelerated depreciation allowed in Zambia. Botswana and South Africa allow 100% expensing of capital expenditure in the year of expenditure. Tanzania offers 100% but capped at a maximum 50% in year 1. In Botswana and South Africa, there is also unlimited loss carry-forward. Ghana permits five years while Zambia allows 10% maximum loss carry-forward for at most ten years. Tanzania allows indefinite forwarding of losses but imposes a 0.3% of turnover charge after three consecutive years of loss making. Other expenses are also deductible in all countries.

2.2 Assessing the Effectiveness of Fiscal Regimes in Africa

Hogan and Goldsworthy[32] set a criterion for assessing fiscal tools which broadly considers economic efficiency, rent allocation, and government risk and cost of administration and compliance. Under economic efficiency, three factors are essential: how does the instrument affect project profitability – that is neutrality, project risk, and stability? Three factors are also considered under rent allocation and government risk – namely, flexibility of revenue collection, fiscal loss, and revenue delay. Figure 3.2 shows a distribution according to the Hogan and Goldsworthy criterion relative to the broad tax classification.

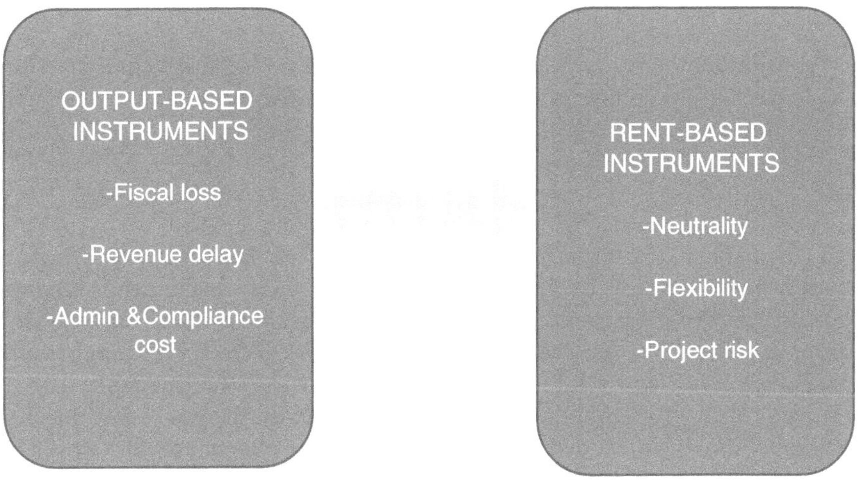

Figure 3.2 Tax Classification and Hogan and Goldsworthy's Criterion

32 Hogan, L. & Goldsworthy, B. (2010). Minerals Taxation: Experience and Issues. In P. Daniel, M. Keen, & C. McPherson (Eds), The Taxation of Petroleum and Minerals: Principles, Problems and Practice. Abingdon: Routledge.

2.2.1 Botswana

Corporate income tax (CIT) is charged discriminately: 22% for resident corporation and 30% for non-resident corporations. Similarly, the royalty regime discriminates on the basis of mineral type, charging 3% for copper and coal, 5% for gold, silver, and platinum, and 10% for diamond and other minerals. The royalty base is gross sales value at mine gate excluding discounts, commissions, and deductions. In addition to these are withholding taxes on dividend, interest, royalty, and management, professional, and consultancy payments, ranging from 7.5% to 15%, as well as a 12% VAT. Yet while Botswana has relatively high royalties, particularly on diamond and others,[33] there is a raft of incentives built into the regime to absorb the impact of these taxes to some extent.

The Botswanan fiscal regime:

- i is simple to administer and monitor because the structure is not too elaborate;
- ii guarantees a reasonable share of rent for the state and therefore satisfies the rent collection and timing criteria (mindful of the fact that 'reasonable share' is a subjective issue);
- iii cannot be said to be neutral given that it does not take into consideration project profitability in determining royalty payment;
- iv is flexible to some extent as it recognises different project and cost structure of different mineral projects;
- v shows the government's willingness to assume some risks as evident in the allowances offered to cover investments, costs, and losses;
- vi can cause high grading and increased selectivity.

In conclusion, the regime balances the interests of both the state and investors by ensuring minimum government receipts while protecting investment. However, it offers little protection for marginal mines.

2.2.2 Ghana

CIT is charged at 35% and a fixed rate royalty of 5%. The royalty base is gross turnover. In addition to these are withholding taxes on dividend, interest, royalty, and management, professional and consultancy payments ranging from 7.5% to 15%, as well as a cumulative VAT, NHIL (National health insurance levy), and GETFL (Ghana education trust fund levy) of 17.5%. There is also a 10% additional/windfall profit tax on the books that has not operationalised. Imbedded in the Ghanaian fiscal regime are allowances on capital, loss carry forward, pre-production expenditure, and forex retention in off-shore accounts. These incentives recognise the risks associated with pre-production investment.

33 ICMM. (2009). *Minerals Taxation Regimes, A Review of Issues and Challenges in Their Design and Application: Analysis and Observations.* London: ICMM..

The Ghanaian fiscal regime:

i is simple to administer and monitor because the structure is not too elaborate;
ii guarantees a reasonable share of rent for the state and therefore satisfies the rent collection and timing criteria (mindful of the fact that 'reasonable share' is a subjective issue);
iii cannot be said to be neutral given that it does not take into consideration project profitability in determining royalty payment;
iv is, to some extent, not flexible as it fails to recognise differences in project and cost structure of different mineral projects;
v shows the government's willingness to assume some risks as evident in the allowances offered to cover investments, costs, and losses;
vi can deter investments in marginal fields and thus increase selectivity.

In conclusion, the regime balances the interests of both the state and investors by ensuring minimum government receipts while recognising risks associated with the huge sunk costs needed to find minerals. But it offers little protection marginal mines.

2.2.3 South Africa

Unlike other mining companies who are taxed 28% CIT, gold mining companies are taxed at a rate of 34 (170/X) – that is, X is the ratio of taxable income to turnover from gold mining, expressed as a percentage. This formula offers protection for the operation of marginal mines.

Similarly, the royalty regime in South Africa is designed to be mindful of the differences in mine characteristics. First, the royalty is charged on adjusted gross sales value of transferred mineral resources and at a rate that considers the profitability of the mine – that is, 0.5 + [EBIT/(gross sales of refined output x 12.5)] x100 or 0.5 + [EBIT/(gross sales of unrefined output x 9)] x 100. The regime also discriminates between refined and unrefined mineral resources. In addition to the built-in cushion, specific relief is provided for small corporations whose production value for the year is below or equal to SAR10 million and estimated royalty of less than SAR100 thousand. This notwithstanding, the formula also guarantees a minimum rate of 0.5% royalty payment for the state. This floor and capping rates ensure that the government shares in not only the rents during favourable periods but also in the risks during busts.[34]

The South African fiscal regime:

i can be said to be neutral given that it takes into consideration project profitability in determining royalty payment;

34 Parliamentary Monitoring Group. (2013, August). Mining Taxation – the South African Context: Economic Tax Analysis. Retrieved from PMG: https://static.pmg.org.za/131106mining.pdf

ii is flexible and recognises variability in cost structure of the mining sector over time;
iii shows the overnment's willingness to share in the risks and rents of mining as shown in the minimum and maximum royalty rates;
iv guarantees a minimum royalty and therefore satisfies the rent collection criterion (mindful of the fact 'reasonable share' is a subjective issue);
v avoids delay in rent payment as collection does not have to wait until projects attain a particular rate of return;
vi is a rent-based regime but whose computation as shown above is pretty straightforward.

In conclusion, the regime balances the interests both the state and investors by ensuring that risk is shared between stakeholders, protecting marginal mines and guaranteeing increased rents for the state in times of boom.

2.2.4 Tanzania

While CIT is charged uniformly at 30%, the royalty rate differs from mineral to mineral. The royalty rate of 3% applies to minerals other than gold, silver, copper, platinum, diamond, and coal. Of these, coal is taxed at 5%, copper on a sliding scale of 4–6%; the rest are charged at 6%. The royalty, which is based on the gross value of minerals at refining or sale point, is also escalated by a 1% inspection cost. In addition, there are withholding taxes on dividend, interest, royalty and management, and professional and consultancy payments ranging from 5% to 15%, as well as a VAT of 18%. Imbedded in the fiscal regime are allowances on capital and loss carry forward. Capital expenditures are deductible, fully, and up to 50% in the first year. Losses can also be carried forward indefinitely.

The Tanzanian fiscal regime:

i is simple to administer and monitor because the structure is not too elaborate;
ii guarantees a reasonable share of rent for the state and therefore satisfies the rent collection and timing criteria (mindful of the fact that 'reasonable share' is a subjective issue);
iii cannot be said to be neutral given that it does not take into consideration project profitability in determining royalty payment;
iv is, to some extent, flexible as it does recognise differences in project and cost structure of different mineral projects;
v shows the government's willingness to assume some risks as evident in the allowances offered to cover investments, costs, and losses;
vi can deter investments in marginal fields and thus increase selectivity.

In conclusion, the regime balances the interests both the state and investors by ensuring minimum government receipts while recognising risks associated

with the huge sunk costs needed to find minerals. However, it offers little protection for marginal mines.

2.2.5 Zambia

While CIT is charged uniformly at 30%, the royalty rate differs from mineral to mineral, just as Tanzania. The royalty rate of 6% applies to gold and diamond, 5% to silver, and 8% to cobalt. Copper is on sliding scale of 5.5–10%. The royalty is based on gross/nominal value of minerals. In addition, there are withholding taxes on dividend, interest, royalty and management, and professional and consultancy payments ranging from 15% to 20%. On dividends, interests, and royalty, there is discrimination on the basis of residency. VAT of 16% is also charged. The fiscal policy attempts a negation of the high imposts with allowances on capital and loss carry forward. Capital expenditures are deductible on a straight line at 25% whiles losses are allowed to be carried forward for up to ten years at a maximum of 10% per annum. Also, there exists no carried interest.

The Zambian fiscal regime:

- i is simple to administer and monitor because the structure is not too elaborate;
- ii guarantees a reasonable share of rent for the state and therefore satisfies the rent collection and timing criteria;
- iii cannot be said to be neutral given that it does not take into consideration project profitability in determining royalty payment;
- iv is, to some extent, flexible as it recognises differences in project and cost structure of different mineral projects;
- v the government's willingness to assume some risks is very minimal as evident from the allowances offered to cover investments, costs, and losses. Nonetheless, it makes up for this with no state participation in the form of free carried interest;
- vi can deter investments in marginal fields and thus increase selectivity.

In conclusion, the regime offers some of the highest imposts but the non-existence of free carried interest by the state negates the effective tax rate, bringing the fiscal regime in line with other countries. It offers little protection for marginal mines, though.

2.3 Comparative Analysis of Fiscal Regimes

From Table 3.3, a crude comparison of fiscal impositions can be made as shown in Table 3.4. In this regard, a value between 1 and 5 is assigned to each country for each major impost; the country offering the lowest rate for an impost is assigned 1 while the country offering the highest is assigned 5. The eight imposts captured in Table 3.4 will therefore yield between 8 and 40. The higher a

Table 3.3 Fiscal Regimes in Africa

Instrument	*Botswana*	*Ghana*	*South Africa*	*Tanzania*	*Zambia*
Average Income Tax (%)	Resident-22, Non-resident-30	35	28	30	30
Royalty (%)	Gold 5, silver 5, copper 3, platinum 5, diamond 10, coal 3, others 10	5	Refined 0.5–5 Unrefined 0.5–7	Gold 6, silver 6, copper 4–6, platinum 6, diamond 6, coal 5, others 3 Inspection fee 1	Gold 6, silver 5, copper 5.5–10, diamond 6, cobalt 8
Royalty base	Gross sales revenue	Revenue (turnover)	Adjusted gross value	Gross value	Nominal value/ Gross value
Fees	Mining licence BWP100/km2 p.a. Prospecting licence industrial minerals BWP5/km2 p.a. (min BWP500 and BWP1000 for others) Retention licence BWP5000/km2 (increases by BWP5000/km2 p.a.)	(cadastral unit/year) Reconnaissance licence $16 Restrict reconnaissance license $10 Prospecting licence $32 Restricted prospecting license $20 Mining license $700 Grounds rent Ghc36.5/acre p.a. (Ghc9,109/km^2 p.a.)		**Prospecting fees** Application $300 Preparation $200 **License fees** Application $5000 (special), $2000 (ordinary) Preparation $2000 (special), $1000 (ordinary)	Fees
Capital allowance	100% allowed in year of expenditure.	20% straight line (5 years)	100% immediate expensing of capital expense	100% allowance; up to 50% in year 1	25% straight line (accelerated depreciation available)

Instrument	*Botswana*	*Ghana*	*South Africa*	*Tanzania*	*Zambia*
Loss carry-forward	Unlimited	5 years limit	Unlimited	Indefinitely but three consecutive tax losses will attract 0.3% of turnover in third year	Max 10% p.a. and not exceeding 10 years
Additional/windfall profit tax (%)		10 proposed but not operational			0–15 (copper)
State participation (%)	15 free carried interest	10 free carried interest	26 of local ownership (HDSA requirement)	Government free carried interest negotiable	No free carried interest
Withholding tax on dividends (%)	7.5	8	20	Listed 5 Unlisted 10	Resident 15 Non-resident 20
Withholding tax on interest (%)	Resident 10 Non-resident 15	8	Resident 0 Non-resident 15	10	Resident 15 Non-resident 20
Withholding tax on non-resident service providers (%)	15	20	0	15	20
Withholding tax for local professionals and consultancy (%)	0	7.5	0	5	15
Withholding tax on royalty (%)	Resident 0 Non-resident 15	15	Resident 0 Non-resident 15	15	Resident 15 Non-resident 20
Value added tax (VAT)/Sales tax (%)	12	VAT 12.5 National Health Insurance Levy (NHIL) 2.5 GetFund Levy (GETFL) 2.5 Flat rate 3	15	18	16

Instrument	*Botswana*	*Ghana*	*South Africa*	*Tanzania*	*Zambia*
Capital gains tax (%)	Resident 22 Non-resident 30	15	0–28		
Duties and levies (%)		Import duty on plants and machinery 0 Fiscal Stabilisation levies 1.5 Stamp duty 0.25–1		**Stamp duty** Transfer of immovable property 1.5 Share transfer 1	Import duty on copper and cobalt concentrates 5 Export duty on precious metals, gemstones, manganese ore, and concentrates 15
Treatment of expenses	Expenses on production of assessable income is deductible	Pre-production expenses can be capitalised subject to ring-fencing rules	Immediate expensing	Expenses on production of assessable income are deductible	Immediate expensing
Ring-fencing		Yes	Yes		No
Others	Capital transfer tax 12.5% Transfer duties on immovable property 30% (5% for residents with first BWP1million exempt)				Skill development levy 0.5% of emoluments Property transfer tax 10%

Table 3.4 Comparison of fiscal imposts in five African countries

Instrument	*Botswana*	*Ghana*	*South Africa*	*Tanzania*	*Zambia*
Income tax	3	5	1	4	4
Royalty	4	2	1	3	5
Withholding tax on dividends	1	2	5	3	4
Withholding tax on interest	4	1	3	2	5
Withholding tax on non-resident service providers	4	5	1	4	5
Withholding tax for local professionals and consultancy	1	4	1	3	5
Withholding tax on royalty	2	3	2	3	4
VAT/sales tax	1	4	2	5	3
Total	**20**	**26**	**16**	**27**	**35**

country's score, the higher the cumulative rate in the country for the eight imposts. The results show that South Africa offers the lowest cumulative rate as against Zambia that offers the highest. In between, Botswana, Ghana, and Tanzania follow in that order.

The above notwithstanding, it is essential to remember that a fiscal regime is an envelope of both imposts and incentives. It would also be recalled, as stated earlier, that fiscal incentives negate fiscal imposts. As such, the average effective tax rate (AETR), which takes all fiscal charges into consideration, provides a better picture of the total tax burden of a regime. AETR is the ratio of tax expense to profit before tax. Using the Fiscal Analysis of Resource Industries model (FARI) developed by the International Monetary Fund (IMF) and assuming the base parameters offered in the model for all the countries, AETR, government of share of total benefits, and post-tax internal rate of return (IRR) on total funds are calculated. The following parameters were utilised for the calculation.

From Figure 3.1, Tanzania and Ghana are the countries with the relatively highest tax regimes with AETR of 60% and 57%, respectively. South Africa offers the lowest taxing regime with 53% AETR. Botswana followed by Zambia are in the middle.

Table 3.4 and Table 3.3 disagree to the extent that the positions of Tanzania, Ghana, and Zambia have changed. This is explainable by the fact that the non-existence of free carried interest in Zambia is a bigger incentive relative to the incentives offered in Tanzania and Ghana. On the other hand, the negotiable free carried regime in Tanzania, based on which the current range is 16%–50%, increases both the AETR and government share of total rents.

Table 3.5 Calculation parameters

Mineral		*Gold*
Financing	Share of development cost borrowed	70%
	Repayment period	5 years
	Real interest rate	5%
Economic	Discount rate (governmentt)	10%
	Discount rate (investor)	12.5%
	Inflation rate	2%
	Gold price	US$1,300/ounce
	Transport post-fiscal point	US$25/ounce
	Smelting and refining charge post-fiscal point	US$50/ounce
Mining project (total)	Production gold (000)	1,900 ounces
	Transport post-fiscal point	US$48m
	Refining and processing post-fiscal point	US$95m
	Exploration costs (expensed)	US$10m
	Development costs – intangibles (expensed)	US$50m
	Development costs – tangibles (depreciable)	US$200m
	Replacement capital costs (depreciable)	US$50m
	Operating costs – mining	US$380m
	Operating costs – milling	US$380m
	Operating costs – other	US$114m
	Decommissioning costs	US$20m

Furthermore, the AETR of the five countries falls between 53% and 60%; offering an average AETR in the five countries of 56%. This range can therefore be classified as the competitive and equitable tax rate. Thus, tax rates that exceed 60% may be uncompetitive, just as rates below 53% may be inequitable.

From the investor's perspective, the post-tax IRR on total funds exceeds the market rate of return in all five countries. As such, the high AETR notwithstanding, investments in mining remain profitable in all the countries. That said, the AETR drives the post-tax IRR on total funds in each country, as to be expected. Consequently, the country with the lowest AETR, South Africa, has the highest IRR, followed by Botswana. Tanzania has the lowest IRR to match its highest AETR.

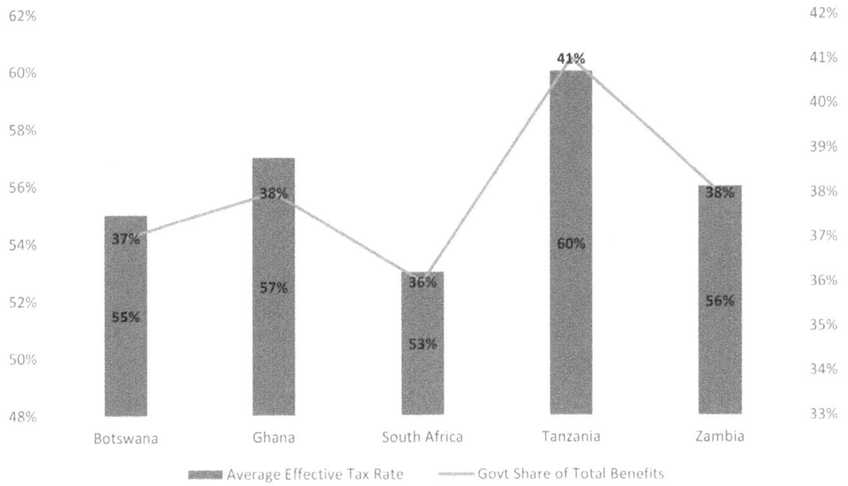

Figure 3.3 AETR and government share of five African countries

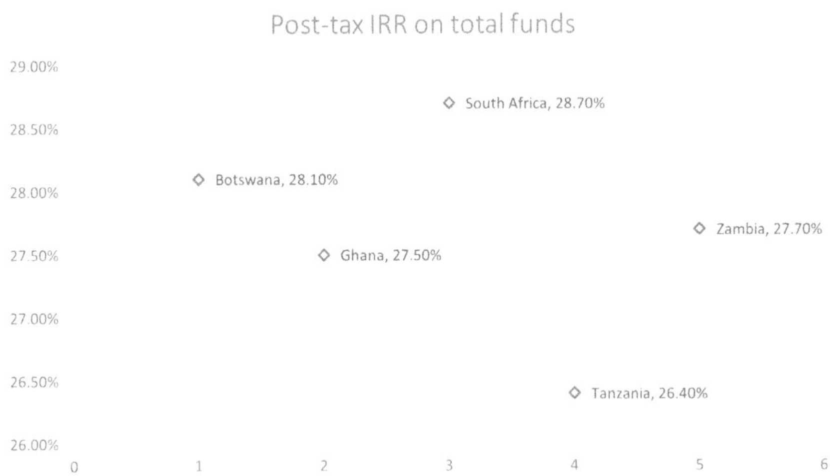

Figure 3.4 Investor's IRR in five African countries

3 Local Content in Africa's Mining Sector

In the past, the fiscal regime, as discussed in the previous sections, was the means through which African countries benefited from their mineral resources. Often countries depend on mineral rents, which is often at the mercy of mineral prices. Yet mineral prices are volatile and exogenously determined, leading to procyclical expenditure.

Lately, in addition to the existing fiscal regimes, African countries now rely on local content regulation to improve their overall fiscal take from mineral rents. However, the design and approach of these regulations have differed from one country to the other. For instance, while some countries aggregate all related regulations and laws into one comprehensive legislation to guide the sector, others rely on disjointed existing pieces of laws. In the five countries discussed, this difference is palpable.

3.1 Botswana

Unlike other mineral-producing African countries, Botswana does not have a specific regulation on local content.[35] However, two important regulations, the Mines and Minerals Act of 1999 and the Competition Law of 1999, have provisions on local content. The Mines and Minerals Act includes provisions on local content requirements and how exploitation permits are granted. The Act specifies the granting of permits for industrial minerals only to citizens of Botswana, unless exception is given by the minister. In addition, the Competition Law of 2009 provides competition incentives for enterprises to train and employ Botswana citizens and use Botswana goods. Under the Competition Law of 2009, proposed mergers are to be evaluated, in part, by whether the merger is likely to promote citizen empowerment initiatives and citizen-owned enterprises. In addition to these regulations, the economic diversification drive seeks to promote local procurement in the mining sector.[36] The Chamber of Mines also launched the Business Development Initiative in 2014. The initiative seeks to encourage the procurement of goods and services by the private sector. For Botswana, 'local' procurement in the case of goods focuses on goods and products 'made in Botswana' – a value-added criterion.

A review by the Intergovernmental Forum on Mining, Minerals, Metals and Sustainable Development[37] found that Botswana local content efforts is supply-side driven with a number of limitations. First, the definition of local content does not cover local participation, while technology transfer is not adequately addressed.

3.2 Ghana

The objectives of Ghana's local content regulations in the mining sector are to create employment and grow domestic manufacturing industries. In order

35 Columbia Center on Sustainable Investment. (2016, September). *Local Content: Botswana-Mining*. Retrieved from: http://ccsi.columbia.edu/files/2017/07/Local-Content_Botswana_Sept2016.pdf

36 Nickerson, E. & Geipel, J. (2018). Botswana: Upstream linkages—The challenge of aligning supply-side policy options with the needs of local companies (Case Study). IGF Guidance for Governments: Leveraging Local Content Decisions for Sustainable Development. Winnipeg: IISD: Intergovernmental Forum on Mining, Minerals, Metals and Sustainable Development (IGF).

37 Ibid.

to achieve these objectives, local content provisions have been made in the Minerals and Mining Act, 2006 (Act 703) and L.I. 2173, 2012. Until recently, the number of goods that companies were obliged to procure from Ghanaian manufacturers was eight.[38] Indeed, instead of manufacturing, some Ghanaian companies resorted to 'import and sell', depriving the country of the benefits of value addition. This was expanded to 19 and then to 29. This has led to increased expenditure on Ghanaian goods and services. Figures available at the Minerals Commission show that value of procurement of goods and services locally for use in the mining industry increased by 286.64% to US$1,440.52 million in 2019 from US$372.57 million in 2017. The sharp growth was attributed to the expansion of items on the local procurement list from 19 to 29 effective 2019 and enforcement of provisions in the Minerals and Mining (General) Regulations, 2012 (L.I. 2173). To deepen local participation through job creation and promote growth of domestic manufacturing, a revised Minerals and Mining (Local Content and Local Participation) Regulations has been drafted.

3.3 South Africa

The Department of Trade is mandated to designate specific industries where tenders prescribe that only locally manufactured products with a minimum threshold are considered. This is in line with the local content provisions in the amended Preferential Procurement Policy Framework Act. The minimum threshold is 30% and can rise to 100% in some cases. Since 2012, South Africa has designated 27 products for local production. The four most recent additions include plastic pipes, bulk material handling, lead acid batteries, and MV switchgear. While a number of successes have been chalked, there are challenges. These include the implementation of manipulation of the bid price to meet local content thresholds, non-compliance with the local content requirements once the tender has been awarded (e.g. placing of orders), and inadequate funding of local content verification.

3.4 Tanzania

The Mining (Local Content) Regulations, 2018 mandates companies to comply with local content requirements which were introduced for the first time by the Written Laws (Miscellaneous Amendments) Act No. 7 of 2017 which amended the Mining Act, 2010.[39] According to the Regulations as amended, an indigenous company is defined as a company incorporated

38 Joy Online. (2015, June 26). Mining: Ghana to increase local content items to 18. Retrieved from: www.myjoyonline.com/mining-ghana-to-increase-local-content-items-to-18

39 Breakthrough Attorneys. (2019, February 28). Mining Law Update: Unpacking the Mining (Local Content) Regulations, 2018 And the Mining (Local Content) (Amendments) Regulations, 2019. Retrieved from: http://breakthroughattorneys.co.tz/mining-law-update-unpacking-the-mining-local-content-regulations-2018-and-the-mining-local-content-amendments-regulations-2019

under the Companies Act No. 12 RE. 2002 which has at least 20% of shares owned by Tanzanians and has at least 81% of Tanzanians in executive- and senior-level positions and 100% of Tanzanians in non-managerial and other positions. Despite these provisions and much progress, Ovadia posits that implementation of local content regulations in Tanzania is characterized by inconsistency, confusion, and uncoordinated donor interventions.[40]

3.5 Zambia

Zambia has a number of laws that promote local content. For instance, the Citizen Economic Empowerment (CEE) Act (2006) seeks to empower Zambians to take part in economic development. Unfortunately, most of these regulations are partially implemented while preferential procurement only applies to public entities.[41] For instance, the International Council on Mining and Metals (2014) revealed weak backward linkages from mining to manufacturing in Zambia. In order to close these gaps, the Zambia Mining Local Content Initiative (ZMLCI) has been developed. The aim of the initiative is to make mining work for Zambia. It aims to enhance local content and the use of locally manufactured inputs in the mining industry.

3.6 Have African Countries Utilized Local Content Regulations to Maximize the State's Share of Mining Rents?

Implementation of local content regulations helps countries to build and expand the manufacturing base, enhance recruitment of indigenous staff, and take control of the mining value chain. These benefits complement revenue sources from royalties, corporate tax, and other payments from mining companies to the government. Ablo posits that local content laws are essential for promoting positive synergies between extractive industries and broader economies.[42] Often the regulations cover local participation, which represents equity shareholding by the citizens of the country where the mineral has been discovered and produced. The local content represents a percentage of goods and services reserved to be provided by local firms or individuals that manufacture these goods or provide these services in the country.

The evolution of regulations is not a recent phenomenon; and this trend can be categorised into four phases. The first phase of mining reforms in Africa can be traced to the liberalisation that took place in Ghana beginning in 1986 through the implementation of the International Monetary Fund and

40 Ovadia, J. S. (2017). Local Content in Tanzania's Gas and Minerals Sectors: Who regulates? CMI Brief.

41 Copenhagen Business School. (2014, October 29). Resource-Led Development in Zambia: The Role of the Suppliers to the Mines. SAFIC Policy Brief.

42 Ablo, A. D. (2016). From Local Content to Local Participation: Exploring Entrepreneurship in Ghana's Oil and Gas Industry. PhD thesis, University of Bergen.

Table 3.6 Benefits From Local Content

	Botswana	*Ghana*[1]	*South Africa*	*Tanzania*	*Zambia*
Some benefits of implementing local content regulations	Mining contributed to 3.2% of total employment in 2015.[2] Due to local procurement prioritisation, 80% of the country's annual procurement budget is spent locally.[3] In addition, beneficiation programs have created 3,000 new jobs in Botswana.	The procurement of goods and services locally for use in the mining industry increased by 286.64% to US$1,440.52 million in 2019 from US $372.57 million in 2017. Local procurement of mining products manufactured in-country grew from US$143.36 million in 2014 to US$ 252.93 million in 2018.[4]	The value of black economic empowerment (BEE) transactions put together by mining companies between 2000 and 2014 amounted to R205 billion. Indeed, Chamber of Mines estimates that R47 billion in dividends has been paid out to 7.1 million 'historically disadvantaged South Africans'.[5] The Chamber further estimates that while mining contributed R291 billion to GDP directly, mining companies spent R245 billion on purchases of goods and services from the other sectors of the economy.[6]	As at 2018, there were 390 mining firms in the country; out of which 378 (97%) were domestically owned. There are only ten foreign-owned mining firms in the census, and two jointly owned firms. Indeed, foreign-owned firms employ 8,785 individuals in total, and jointly owned firms employ 1,187. Meanwhile, domestically owned firms employ 8,706 workers. Foreign-owned firms spend US$916,282 on training on average.[7]	About US$1.8 billion annual expenditure is made on manufacturing-related inputs going into the mining supply chain. However, only about US$100 million (about 6%) is directed to the local manufacturing industry.[8] About 50 local value-adding mine suppliers provide Zambia's mines with pumps and valves, vehicle wear or spare parts, information and communication technology, storage equipment, transport services, health and safety equipment, and civil engineering which were imported in the past.[9]

1 Ghana Chamber of Mines. (2017, August). Local Content in Ghana's Mining Industry. Retrieved from: http://ghanachamberofmines.org/wp-content/uploads/2017/08/GCoM_LOCAL-CONTENT-IN-THE-MINING-INDUSTRY.pdf

2 World Bank Group. (2016, August). *Botswana Mining Investment and Governance Review*. Retrieved from: https://openknowledge.worldbank.org/bitstream/handle/10986/25225/109316.pdf?sequence=8&isAllowed=y

3 Suzman, J. (January/February, 2010). Diamonds, De Beers and Defying the Resource Curse. *Mining Environment Management*.

4 GhanaWeb. (2020, August 24). Local procurements under mining local content increases exponentially. Retrieved from: www.ghanaweb.com/GhanaHomePage/business/Local-procurements-under-mining-local-content-increases-exponential-1041040

5 FSE. (2018, February 10). *The Impact of Mining on the South African Economy and Living Standards*. Retrieved from the Federation of a Sustainable Environment: http://fse.org.za/index.php/item/593-the-impact-of-mining-on-the-south-african-economy-and-living-standards

6 FSE. (2018, February 10). *The Impact of Mining on the South African Economy and Living Standards*. Retrieved from the Federation of a Sustainable Environment: http://fse.org.za/index.php/item/593-the-impact-of-mining-on-the-south-african-economy-and-living-standards

7 Ellis, M., & McMillan, M. S. (2018). Optimal local content for extractive industries: How can policies best create benefits for Tanzania? *WIDER Working Paper*, No. 2018/133. The United Nations University World Institute for Development Economics Research (UNU-WIDER), Helsinki, http://dx.doi.org/10.35188/UNU-WIDER/2018/575-6.

8 Kabeta, J. (2020). Extractive Hub Policy Brief for Zambia: Strengthening Local Content in Zambia's Mining Sector. *Extractive Hub*, 1–14.

9 Kragelund, P. (2017). The making of local content policies in Zambia's copper sector: institutional impediments to resource-led development. *Resource Policy, 57*, 57–66. https://doi.org/10.1016/j.resourpol.2016.11.008.

World Bank's structural adjustment program that recommended the adoption of liberalisation strategies to attract foreign capital into the mining sector. The second phase emerged in the middle of the 1990s in countries such as Guinea, which supplies half of global bauxite, and continued with the liberalisation and privatisation agenda of the Bretton Woods institution. In the third phase, the role of the state in regulating and enforcement was enhanced. Indeed, countries such as Mali, Madagascar, and Tanzania are representative of this process of increased state authority in the mining sector. The final phase involves the promulgation of regulations that facilitate effective governance and enhanced benefits to citizens of the mineral-producing countries. These emerging reforms include the development of local content regulations.

Local content regulations can lead to backward, forward, and fiscal linkages (Ovadia, 2016, 2017; Bloch and Owusu, 2015; Ablo, 2016). Fiscal linkages come in the form of tax receipts, royalties, carried and participating interests, which are important to the central government. Forward linkages are associated with benefits derived from the processing of mineral commodities for domestic consumption or export. Backward linkages, on the other hand, arise from activities established in the supply chain to provide inputs into the production of a commodity.

Indeed, the development and implementation of local content regulations is in line with the African Mining Vision which calls for linkages by helping to establish additional local suppliers and manufacturing industries in mineral-producing countries in Africa. Local content regulations help to reverse the resource curse phenomenon and overcome the concept of enclavity associated with mining. Hirschman (1958) indicates that extractive resource production can have three main linkages to the economy. These are fiscal (taxes and royalties for industrial development), consumption (output from other sectors via income from commodities sector), and production (backward and forward linkages) that are crucial to an economy's growth. The last two are often guaranteed by the implementation of an effective local content strategy. Table 3.6 shows some of the benefits that Botswana, Ghana, South Africa, Tanzania, and Zambia have obtained through the implementation of local content regulations.

4 Conclusion

Africa is blessed with many natural resources, hosting about 30% of the global mineral reserves and accounting for the global production of 80% platinum, 77% cobalt, 51% manganese, 46% of diamonds, 39% chromium, and 22% of gold (estimates from African Development Bank and the United States Geological Survey). These should have essentially translated into economic development. However, limitations such as value addition, poor contracting, rent seeking, and price volatility have affected the continent's quest to benefit more from mineral resources. This mismatch between development and the stock of mineral deposits has resulted in socio-political agitations by citizens. The usual cause revolves around a juxtaposition of local circumstance with other places endowed with similar resources. For instance, it is not rare to hear the residents of Obuasi, Ghana, compare the development of their area with Cape Town, South Africa. The conclusion, usually, from this comparison is that government has failed, either through collusion or ineptitude, to protect the interest of the state. And for that matter, the share of mineral rents coming to the state is unfairly tiny relative to what goes to the mining companies. This perception has mostly impacted political discussion and, to a large extent, is responsible for the need for mineral agreement reviews and, in some extreme cases, nationalisation.[43] Nonetheless, while this perception was somewhat true previously, it is to a large extent not true presently. As shown in Section 1.3, the share of government take of total mineral rents is significant in many African countries. Through a combination of fiscal tools, nation states are now able to elicit up to about 41% of total mineral rents. Indeed, when one factors in the benefits driven from local content regulations, as currently being adopted in most African countries, the government's share is going to be even bigger than noted here.

The focus therefore should be on how governments distribute mineral revenues and development at the sub-national level. In unitary states like Ghana, where mineral rents obtained from one side of the nation are consolidated with other funds for the development of the entire nation, it is mostly difficult for mining areas to properly ascertain what proportion of mineral rents are dedicated for their development. This is particularly so when there are significant disparities in development across the country. In addition, local content regulations must focus on building capacities and creating employment opportunities in mining areas. It is through this that the political pressure from these communities can be abated.

Furthermore, mineral exploitation in Africa has been going on for at least a century. Over time, the continent has gained a geological advantage that gives host governments the muscle on the negotiation table. This is likely why the adoption of freezing stabilisation clauses appears to be giving way to adaptation clauses whose focus is on balancing the economic equilibrium of agreements. Aside from this, the gradual increase in expenditure on and procurement of locally produced goods and services is not only a testament to the state's growing share in the benefits from mineral exploitation, but also marks the improved capacity of African governments on the negotiation table.

Also, the subject of mineral revenue monetisation is one that needs to be explored in detail. In this regard, the Ghanaian government recently mooted this idea through the Agyapa Royalties Monetisation deal. That arrangement seeks to monetise Ghana's gold royalties for the next 25 years. The essence is to leverage future gold royalties in the form of equity for today's development. Although the deal fell through because of its alleged opaqueness and poor

43 Otto, J., Craig, A., Cawood, F., Doggett, M., Guj, P., Stermole, F., Stermole, J., & Tilton, J. (2006). Mining Royalties. A Global Study of Their Impact on Investors, Government, and Civil Society. Washington, DC: The World Bank.

valuation of the country's royalties, the concept in itself is worth looking at. Indeed, the advantages of such an arrangement go beyond just frontloading accruals from the gold industry but also present a huge potential for the economy in the future. Such vehicles, like the Norwegian Sovereign Fund, can build a portfolio that can turn the fortunes of an economy.

In sum, this chapter concludes that African countries attract a sizable share of mineral rents using a combination of fiscal instruments. In fact, even with an increased adoption of local content regulations, the fiscal regimes so established remain competitive. The focus, now, should be how to maximise and transparently distribute these benefits and development, uniformly.

Bibliography

Ablo, A. D. (2016). *From Local Content to Local Participation: Exploring Entrepreneurship in Ghana's Oil and Gas Industry*, PhD thesis, University of Bergen.

Ambe-Uva, T. (2017). Whither the state? Mining codes and mineral resource governance in Africa. *Canadian Journal of African Studies/Revue canadienne des études africaines*, 51(1), 81–101.

ANRC. (2016). *Botswana's Mineral Revenues, Expenditure and Savings Policy: A Case Study*. Abidjan: AfDB.

Aryee, B. N. (2018). *A Historical Overview of Ghana's Mining Sector Fiscal Regimes*. Wandsbeck, SA: Reach Publishers.

Bloch, R. & Owusu, G. (2012). Linkages in Ghana's gold mining industry: Challenging the enclave thesis. *Resource Policy*, 37(4), 434–442. https://doi.org/10.1016/j.resourpol.2012.06.004

Breakthrough Attorneys. (2019, February 28). Mining Law Update: Unpacking the Mining (Local Content) Regulations, 2018 and the Mining (Local Content) (Amendments) Regulations, 2019. Retrieved from Breakthrough Attorneys: http://breakthroughattorneys.co.tz/mining-law-update-unpacking-the-mining-local-content-regulations-2018-and-the-mining-local-content-amendments-regulations-2019

Chandler, J. (2013). International Mining for Development Centre Action Research Report. IM4DC.

Columbia Center on Sustainable Investment. (2016, September). Local Content: Botswana-Mining. Retrieved from CCSI: http://ccsi.columbia.edu/files/2017/07/Local-Content_Botswana_Sept2016.pdf

Committee of Experts on International Cooperation in Tax Matters. (2016). Proposed Guidance on Fiscal Take in the Extractive Industries. Thirteenth Session, E/C.18/2016/CRP.21. New York.

Copenhagen Business School. (2014, October 29). Resource-Led Development in Zambia: The Role of the Suppliers to the Mines. SAFIC Policy Brief.

Deloitte. (2020). Guide to fiscal information: Key economies in Africa 2020. Deloitte.

Deloitte. (2020). International Tax South Africa Highlights 2020. Deloitte.

Ellis, M. & McMillan, M. S. (2018). Optimal local content for extractive industries: How can policies best create benefits for Tanzania? *WIDER Working Paper*, No. 2018/133. The United Nations University World Institute for Development Economics Research (UNU-WIDER), Helsinki, http://dx.doi.org/10.35188/UNU-WIDER/2018/575-6

FSE. (2018, February 10). The Impact of Mining on the South African Economy and Living Standards. Retrieved from the Federation of a Sustainable Environment: http://fse.org.za/index.php/item/593-the-impact-of-mining-on-the-south-african-economy-and-living-standards

Ghana Chamber of Mines. (2017, August). Local Content in Ghana's Mining Industry. Retrieved from: http://ghanachamberofmines.org/wp-content/uploads/2017/08/GCoM_LOCAL-CONTENT-IN-THE-MINING-INDUSTRY.pdf

GhanaWeb. (2020, August 24). Local procurements under mining local content increases exponentially. Retrieved from: www.ghanaweb.com/GhanaHomePage/business/Local-procurements-under-mining-local-content-increases-exponential-1041040

GHEITI. (2018). Extractive Industries Contribution to the Economy. Retrieved from Ghana Extractive Industries Transparency Initiative: https://eiti.org/es/implementing_country/4

Government of Botswana. (2021). Mining Licence Application (Department of Mines, Botswana). Retrieved from: www.gov.bw/mining/mining-licence-application

Hogan, L. & Goldsworthy, B. (2010). Minerals Taxation: Experience and Issues. In P. Daniel, M. Keen, & C. McPherson (Eds), *The Taxation of Petroleum and Minerals: Principles, Problems and Practice.* Abingdon: Routledge.

ICMM. (2009). *Minerals Taxation Regimes: A Review of Issues and Challenges in Their Design and Application.* London: International Council on Mining and Metals.

IMF: Fiscal Affairs Department. (2015, September). South Africa: Technical Assistance Report-Fiscal Regimes for Mining and Petroleum: Opportunities and Challenges. Retrieved from www.elibrary.imf.org/view/IMF002/22798-9781513514802/22798-9781513514802/22798-9781513514802_A002.xml?redirect=true.

Joy Online. (2015, June 26). Mining: Ghana to increase local content items to 18. Retrieved from: www.myjoyonline.com/mining-ghana-to-increase-local-content-items-to-18

Kabeta, J. (2020). Extractive Hub Policy Brief for Zambia: Strengthening Local Content in Zambia's Mining Sector. Extractive Hub, 1–14.

Koola, P. (2019, May 14). Mining fiscal regime. PwC. Retrieved from: www.pwc.co.tz/press-room/mining-fiscal-regime.html

Kragelund, P. (2017). The making of local content policies in Zambia's copper sector: Institutional impediments to resource-led development. *Resource Policy*, 57, 57–66. https://doi.org/10.1016/j.resourpol.2016.11.008

Maroun, W., Ram, A. J., & Kok, M. (2019). A review of mining taxes in Africa: Tax burden, the strength of democratic systems and levels of corruption. *South African Journal of Business Management* (2078–5976),1–15.

Mkokweza, C., Zulu, M. A., & Shandavu, N. (2019, July 4). Mining duties, royalties and taxes in Zambia. Lexology. Retrieved from www.lexology.com/library/detail.aspx?g=d5f6c22c-35fc-4bf9-8ad2-8b2ee7e12a3f

Natural Resource Center. (n.d.). Precept 4. Fiscal regimes and contract terms.

Nickerson, E. & Geipel, J. (2018). Botswana: Upstream linkages – The challenge of aligning supply-side policy options with the needs of local companies (Case Study). IGF Guidance for Governments: Leveraging Local Content Decisions for Sustainable Development. Winnipeg: IISD: Intergovernmental Forum on Mining, Minerals, Metals and Sustainable Development (IGF).

Otto, J. M. (2000). *Mining Taxation in Developing Countries.* UNCTAD.

Otto, J. M. (2017, March). The taxation of extractive industries. WIDER Working Paper 2017/75, pp. 1–24.

Otto, J., Craig, A., Cawood, F., Doggett, M., Guj, P., Stermole, F., Stermole, J., & Tilton, J. (2006). *Mining Royalties. A Global Study of Their Impact on Investors, Government, and Civil Society.* Washington, DC: The World Bank.

Otto, M. J. (2001). *Fiscal Decentralization and Mining Taxation.* Washington, DC: The World Bank Group Mining Department.

Ovadia, J. S. (2016). Local content policies and petro-development in Sub-Saharan Africa: A comparative analysis. *Resource. Policy*, 49, 20–30. https://doi.org/10.1016/j.resourpol.2016.04.003

Ovadia, J. S. (2017). Local Content in Tanzania's Gas and Minerals Sectors: Who regulates? CMI Brief.

Parliamentary Monitoring Group. (2013, August). Mining Taxation-the South African Context: Economic Tax Analysis. Retrieved from: https://static.pmg.org.za/131106mining.pdf

Patricia, A. (2020). Implementing local content under the Africa Mining Vision: An achievable outcome?. *Canadian Journal of Development Studies/Revue canadienne d'études du développement*, 41(3). https://doi.org/10.1080/02255189.2020.1809360

PwC. (2019). *SA Mine 2019: In Transition.* Johannesburg: PwC Inc.

PwC. (2020, July 31). Botswana: Corporate - Deductions. PwC. Retrieved from https://taxsummaries.pwc.com/botswana/corporate/deductions

Sipemba, T. M. & Matiko, J. (2020, August 12). Mondaq connecting knowledge & people. Mondaq. Retrieved from www.mondaq.com/energy-and-natural-resources/975742/mining-comparative-guide

Sipemba, T. M. & Matiko, J. (2020, August 12). Mondaq Connecting Knowledge & People. Retrieved from: www.mondaq.com/energy-and-natural-resources/975742/mining-comparative-guide

Suzman, J. (2010). Diamonds, De Beers and Defying the Resource Curse. *Mining Environment Management.*

World Bank Group. (2016, August). *Botswana Mining Investment and Governance Review.* Retrieved from: https://openknowledge.worldbank.org/bitstream/handle/10986/25225/109316.pdf?sequence=8&isAllowed=y.

Zambia EITI. (2019). *Report for the year ended 31 December 2017.* Tunis: BDO Tunisia Consulting.

Part II

Case Studies Explored in African Mining

4 Addressing the Persistent Issues in Artisanal and Small-Scale Mining

The Impact of Kenya's Mining Laws on ASM Formalization

Cynthia Andiabutswa Opakas[1]

1 Introduction

Although there is no standardized global definition of Artisanal and Small-Scale Mining (ASM), there is no doubt as to what characterizes ASM and, therefore, what it is. The Extractive Industries Transparency Initiative (EITI), in its due diligence guidelines for responsible mining in high-risk areas defines ASM as formal or informal mining operations with simplified forms of exploration, extraction, processing, and transportation.[2] ASM is defined by labour-intensive and rudimentary techniques of mineral extraction that are often surrounded by disorganization and various illegalities. For a long time, Kenya's economic sector was dominated by agriculture, and little prominence had been given to the country's mining industry. Resultantly, the mining industry is underexplored and remains a relatively small industry in Kenya. Kenya has various minerals spread out across the Coast, Nyanza, Western, Rift Valley, Northeastern, and Central Provinces. The resources are both in the form of non-metallic commodities such as soda ash, gypsum, and metallic minerals such as titanium, gold, iron ore, and some rare earth metals.[3] Notably, Kenya is so far Africa's largest producer of soda ash in Africa.

In many countries in Sub-Saharan Africa (SSA), ASM has become an integral part of the rural economies in areas that are endowed with these resources. Kenya is no exception. For many years, ASM has and continues to be an informal sector that is marred by negative perceptions and gross human

1 Cynthia Opakas is an Advocate of the High Court of Kenya, currently practising at TripleOKLaw LLP in Nairobi, Kenya. She is also a Fellow of the Chartered Institute of Arbitrators and a mediator with the Kenyan judiciary. She holds a Master of Laws (LLM) in International Energy Law and Policy as well as training in mining law at CEPMLP, University of Dundee.

2 OECD, *OECD Due Diligence Guidance for Responsible Supply Chains of Minerals from Conflict-Affected and High-Risk Areas: Third Edition* (OECD 2016). www.oecd.org/daf/inv/mne/OECD-Due-Diligence-Guidance-Minerals-Edition3.pdf, accessed 11 February 2021. Page 65.

3 Kenya Mining Investment Handbook 2016. www.tralac.org/documents/resources/by-country/kenya/1928-kenya-mining-investment-handbook-2016/file.html, accessed 11 February 2021. Page 19.

DOI: 10.4324/9781003284437-6

rights violations. Informality means that the sector is not regulated, and its participants operate without legal titles.[4] Despite these challenges, ASM has been a source of livelihood for many households and cannot therefore only be chastised for its shortcomings, but also recognized for the changes it has made for communities that engage in it. ASM silently contributes to the attainment of the SDGs through the fact that income from ASM goes towards provision of food, education, healthcare, and. in general, the alleviation of poverty. The contribution of mining towards achieving the SDGs is gaining great recognition and was acknowledged by the United Nations Environment Assembly in 2019.[5]

Recognizing that ASM has had and can have an even greater contribution to the livelihood of the communities and the economy at large, the Government of Kenya is taking steps towards formalization of the sector. After about seven decades, Kenya enacted a new Mining Act in 2016. Formalization, in the broadest possible sense, refers to the incorporation of ASM into the country's mainstream legal framework. Arguably, formalization cannot be said to have occurred merely by the enactment of a law, but the practical activities in the sector must speak to formalization. ASM is an example of an industry that may be described by economist Hernando De Soto as an 'extralegal economy'.[6] De Soto argues that efforts to eradicate such economies will certainly be futile, and therefore in order to alleviate poverty and avoid conflict between these two economies, it is best to adopt them.[7] Accordingly, many governments in whose countries ASM is practised are now deeming it wise to incorporate the sector rather than outlaw it. The Government of Kenya has now also ensured that development of the mining industry has been captured in the country's long-term vision, Vision 2030.[8] There is therefore renewed focus on developing the mining industry in Kenya. The Act now recognizes ASM, and this chapter seeks to dissect how that law serves the formalization of the sector and the challenges so far.

Kenya's mining industry is ASM dominated. This is despite Kenya's ASM sector still being in its nascent stages compared to other countries such as Ghana where Artisanal and Small-Scale Gold Mining (ASGM) directly employed about one million people and supported around 4.5 million others as at 2016.[9]

4 Victoria R Nalule, *Mining and the Law in Africa: Exploring the Social and Environmental Impacts* (Palgrave Macmillan 2020). Page 55.

5 United Nations Environment Programme, '4/19. Mineral Resource Governance – Resolution Adopted by the United Nations Environment Assembly on 15 March 2019'. https://wedocs.unep.org/xmlui/handle/20.500.11822/32839, accessed 11 February 2021.

6 Shefa Siegel and Marcello M Veiga, 'Artisanal and Small-Scale Mining as an Extra-legal Economy: De Soto and the Redefinition of "Formalization"' (2009) 34 Resources Policy 51.

7 Ibid. Page 53.

8 Kenya Vision 2030. http://vision2030.go.ke/about-vision-2030, accessed 11 February 2021.

9 James McQuilken and Gavin Hilson, 'Artisanal and Small-Scale Gold Mining in Ghana'. www.jstor.com/stable/resrep02683.1, accessed 16 January 2023.

The nascency of Kenya's ASM sector can best be highlighted by the fact that Ghana had its first ASGM legislative frameworks in 1989.[10] Ghana has made significant positive steps in its ASM sector but continues to face various challenges to date and this is a testament to the fact that ASM formalization is indeed a process. Due to the smaller operational scale, the challenges faced in Kenya's ASM are not as aggravated as the challenges being witnessed in Ghana or even the resource-abundant DRC.[11] ASM continues to be a breeding ground for various social, legal, and environmental concerns in Kenya even subsequent to the enactment of the Mining Act.

This chapter seeks to dissect whether this legal framework is adequate in terms of formalizing the sector and how the legal regime in Kenya seeks to address the challenges in ASM. The chapter further seeks to analyze the practical implementation of these laws and any institutional measures that the government has adopted. In looking into the implementation of the laws, the chapter seeks to determine the impact of legalization on Kenya's ASM industry and whether this has made any difference in the practical operations of ASM in the country. The hypothesized position is that despite there being a satisfactory legal framework, there is a lot to be desired in the practical implementation of these regulations. The study is based on desk research of the ASM landscape and the attendant legal regime in Kenya while relying on primary and secondary sources such as peer-reviewed articles of scholars.

The first section of this chapter looks into the legal and institutional framework of ASM in Kenya. Subsequently, the chapter delves into the positive contributions and the challenges within Kenya's ASM industry. These include participation of women and children, health and safety, environmental protection, and the violation of various human rights. As is customary for any activity, ASM is making a great contribution to the communities engaged in it and, at the same time, experiences many challenges. ASM has further been seen to provide a 'natural' environment for the perpetuation of environmental and human rights breaches which are now deemed to be a natural occurrence in the industry. The discussion is then merged to assess the implementation of these resources and how best to move forward towards improving the conditions in ASM and achieving formalization.

2 Legal and Institutional Framework

Although artisanal mining has been ongoing in Kenya for a very long time, the mining legal regime before 2016, being the Mining Act, Cap 306, Laws of Kenya, did not recognize ASM. This is because, despite several amendments

10 Gavin Hilson, 'A Contextual Review of the Ghanaian Small-Scale Mining Industry' (International Institute for Environment and Development 2001). www.iied.org/sites/default/files/pdfs/migrate/G00722.pdf, accessed 16 January 2023.

11 Sara Geenen, 'A Dangerous Bet: The Challenges of Formalizing Artisanal Mining in the Democratic Republic of Congo' (2012) 37 *Resources Policy* 322.

to the Act, Cap 306 was a colonial piece of legislation enacted in 1940, and because of the place of mining in the government agenda, it took over seven decades for a new law to be enacted. Formalization is indeed a process and legalization is the first step of that process. It is therefore critical that the laws are there to streamline the processes and create a roadmap for the sector.

The Constitution of Kenya, 2010 ('the Constitution') states that all minerals and mineral oils as recognized by the law shall constitute public land.[12] The Constitution further states that all the land bearing mineral resources shall vest in the national government and shall be administered by the National Land Commission (NLC) on behalf of the Kenyan people.[13] At the moment, mining matters fall under the Ministry of Petroleum and Mining. This creates an instance of division of attention and sparks the debate as to whether mining should have a sole ministry dedicated to it. Pursuant to Article 66 of the Constitution, parliament was tasked to enact legislation to govern mining investments and operations with a view to ensuring that these investments benefit the local communities and their economies. Article 71 of the Constitution tasks parliament to enact legislation on the grant of natural resources rights. To operationalize these constitutional provisions, the Mining Act, No.12 of 2016 ('the Mining Act') was enacted.[14]

In fresh efforts towards formalization of ASM, the Mining Act now recognizes artisanal mining and small-scale operations. The Mining Act speaks to both artisanal mining and small-scale operations as somewhat separate categories. Artisanal mining is defined as traditional and customary mining operations using traditional or customary ways and means.[15] An activity shall be classified as a small-scale operation where the prospecting area does not exceed 25 contiguous blocks in the case of prospecting and two contiguous blocks in the case of mining.[16] Further, in small-scale operations, the techniques employed in prospecting and mining should not employ specialized prospecting, mechanized mining technologies, or chemicals such as mercury and cyanide.[17] The practical operations of artisanal mining and small-scale operations are intertwined and often overlap.

2.1 The Mining Act on Small-Scale Operations

There are three categories of permits that may be issued with respect to small-scale operations: reconnaissance permits, prospecting permits, and mining permits.[18] To be eligible for any of these permits that relate to small-scale

12 The Constitution of Kenya, 2010. Article 62(1).
13 Ibid. Article 62(3).
14 Mining Act 2016.
15 Ibid. Section 4.
16 Ibid. Second schedule.
17 Ibid.
18 Ibid. Section 123.

operations, one must be a citizen of Kenya or have 60% local shareholding in the case of a corporate body.[19]

Reconnaissance permits are issued for non-intrusive searches for mineral resources through techniques including but not limited to geochemical and geophysical surveys. The applicants for this permit must make applications to the Cabinet Secretary (CS); applications should have the applicant's details, the minerals against which the permit is sought, and the area. The applicant is also expected to pay the prescribed fee. Upon receiving the permit, the permit holder is expected to comply with the terms contained therein and make every effort to protect the environment.

Prospecting permits apply to the search operations for the location and extent of mineral resources both onshore and offshore. Applicants for this permit apply to the CS and pay the prescribed fee. The application must also have details of the area and mineral(s) against which the permit is sought, the details of the applicant and particulars of financial and technical ability. A prospecting permit can be issued for a maximum period of five years and is renewable once. The permit can only be issued with respect to a maximum area of 25 contiguous blocks. Similarly, the permit holder has an obligation to comply with the terms of the permit, ensure environmental protection, and comply with record-keeping obligations.[20]

The final category of permits in small-scale operations is the mining permit which allows the holder to undertake the actual extraction of the resource. An application for this permit is similarly lodged with the CS and must detail all the statutory particulars such as the mining experience, financial resources available, details of the applicant and the directors in the case of a body corporate, and details of the minerals as well as the area. The term of the mining permit shall not exceed five years and is renewable once for a similar term or the remainder of the life of the mine, whichever is shorter. The permit holder shall have, among others, the obligation to pay royalties and mining taxes, submit to the CS quarterly reports on mine development and mineral product, and take all necessary measures to protect and restore the environment.

Applications for mining permits are considered on a first come, first served basis. Where the CS intends to decline a permit application, notice must be given to the applicant of the intention to reject the application. If the defect in the application can be remedied, the applicant shall be given an opportunity to do so.

2.2 The Mining Act on Artisanal Mining

Artisanal mining is provided for in sections 92–100 of the Mining Act. Essentially, the Mining Act places management of artisanal mining under the county governments. Section 93 provides that the CS may establish an office

19 Ibid. Section 124.
20 Ibid. Section 133.

of the ministry at the counties and designate representatives of the director of mines to the respective county offices. The functions of these representatives shall include to grant, renew, and revoke artisanal mining permits, compile registers of artisanal miners, supervise operations of artisanal miners, promote fair trade of artisanal miners, advise, and provide training facilities among other functions.[21]

To assist the director's representative in performing his/her functions, each county is expected to have an artisanal mining committee ('the committee'). The role of the committee is to advise the representative of the director on the grant, renewal, and revocation of permits. For a person to qualify for an artisanal mining permit in Kenya, they must be a citizen of Kenya who has attained majority age and may be a member of an artisanal mining cooperation association.

An applicant for the artisanal mining permit in Kenya must attach the landowner's consent to the application where the area is not designated as an ASM mining area. The duration of the artisanal mining permit is three years, and it is only renewable once. Thereafter, the applicant must graduate to small-scale operations if they still intend to proceed with mining activities. The Mining Act provides that holders of the artisanal mining permit should observe the conditions of the permit, uphold good mining practices, observe health and safety, and observe protection of the environment.

Applications for ASM permits are submitted online through the mining cadastre portal, and once the applications for ASM permits are approved, they are issued digitally.[22] This presents the first challenge to ASM formalization. The people who participate in ASM are mostly resident in rural areas and have little to no formal education with comparatively little income.[23] The industry is therefore attractive to many people as they are able to earn a living without necessarily having any technical or formal skills. Requiring these miners to apply for permits through an online platform makes it somewhat inaccessible for many ASM miners to do so. The online application platform does not encourage many ASM practitioners to apply for the permits as they perceive the process to be complicated and tedious.

In the case of artisanal mining, the representative of the director is advised on the grant of permits by the committee. So far, only four counties in Kenya – Kitui, Migori, Siaya, and Kakamega – have gazetted these committees and therefore allowed the artisanal to apply for and obtain permits. It is evident that the implementation of the Mining Act is happening at quite a slow pace, particularly with respect to ASM.

21 Ibid. Section 93.

22 Mining (License and Permit Regulations), 2017. http://kenyalaw.org:8181/exist/kenyalex/sublegview.xql?subleg=No.%2012%20of%202016#KE/LEG/EN/AR/M/NO.%2012%20OF%202016/SUBLEG/HC_LN872017, accessed 11 February 2021.

23 Gavin Hilson, 'Small-Scale Mining and Its Socio-Economic Impact in Developing Countries' (2002) *26 Natural Resources Forum* 3. Page 6.

3 Positive Consequences of Artisanal and Small-Scale Mining

At the very heart of the Sustainable Development Goals (SDGs) is the alleviation of extreme poverty as enshrined in SDG 1.[24] The African Mining Vision also recognizes ASM as a critical mining sub-sector whose potential needs to be harnessed as it has helped and continues to help many people in the continent to escape poverty.[25] ASM also makes a direct contribution to satisfying SDG 2 which targets ending hunger and achieving food security. ASM also plays an indirect role in the achievement of other SDGs such as promoting healthy lives and equitable education as the resources from the sector promote these goals.

To a degree that so far remains unquantified, ASM continues to play a role in the development of rural economies in Kenya. It has further made it possible for people to diversify and supplement their sources of income in areas where people were originally only doing agriculture and fishing.[26] ASM has been a critical industry in creating employment and this has the general net effect of reducing poverty levels both directly and indirectly by facilitating food security and access to education as well as healthcare. The data on ASM participation in Kenya, the number of people that the industry employs as well as supports, still remains unclear as its informality has hampered the ability to maintain accurate and factual data.

4 The Challenges in Artisanal and Small-Scale Mining

Traditionally, the informality related to ASM has been deemed to be the source of various negative social, environmental, and fiscal impacts.[27] In Kenya, ASM has been recognized as a legal activity; however, total implementation of these formalization frameworks remains a work in progress. As in many developing countries, ASM is a conundrum as it is a source of employment for many people. but, at the same time, grave human and environmental violations have thrived in those environments. The ASM industry in Kenya continues to face various challenges, most of which are regulated against, but they continue to occur in ASM in what appears to be a vicious cycle. These include environmental degradation, health and safety, gender equity violations, and child labour. As shall be highlighted below, despite all

24 The 2030 agenda for Sustainable Development, A/RES/70/1. https://sustainabledevelopment.un.org/content/documents/21252030%20Agenda%20for%20Sustainable%20Development%20web.pdf, accessed 11 February 2021

25 Africa Mining Vision. https://docs.igihe.com/IMG/pdf/africa_mining_vision_english.pdf, accessed 11 February 2021.

26 Gavin Hilson and Roy Maconachie, 'Entrepreneurship and Innovation in Africa's Artisanal and Small-Scale Mining Sector: Developments and Trajectories' (2020) 78 *Journal of Rural Studies* 149. Page 152.

27 Thomas Hentschel, *Artisanal and Small-Scale Mining: Challenges and Opportunities* (IIED 2003). Page 11.

these areas being regulated through both national and international frameworks, these ills persist with a worrying degree of notoriety.

4.1 The Environmental Challenges of ASM

One of the greatest challenges associated with ASM is environmental degradation.[28] Although the Mining Act bestows upon ASM practitioners the obligation to take all measures to protect the environment, it does not create the obligation for these practitioners to submit any environmental protection guarantees or impact assessment reports as is the case for large-scale operations. Although the area in which ASM activities are undertaken is smaller than what is permissible for large-scale operations, the environmental impacts of these activities cannot be overlooked.

Poor site management is a great environmental challenge for many communities where ASM is practised. In Taita Taveta, for instance, where gemstone mining is undertaken, ASM practitioners are highly migratory, and they often move in search of the next resource-filled location.[29] Both the prospecting and mining activities involve the drilling and digging of pits. These miners barely rehabilitate the environment to the state in which they found it once they finish their activities. They move on, and it is difficult to trace who they are so they can be held accountable. Poor site management is also common in salt mining which is undertaken in Kilifi County, where the salt harvesters create ponds before beginning their activities. It often occurs that once the miners conclude the mining or prospecting activities, they move on to the next location without employing the necessary site restoration measures. These open pits destroy the physical landscape of the land and often lead to other physical and health hazards.[30] The interference with the landscape also increases the chances of soil erosion.[31] During the rainy seasons, the soil dug from the trenches is carried away by rainwater, sweeping away valuable soil nutrients. This is further exacerbated by the fact that deforestation will have taken place before the mining activities begin. As the miners leave one site to the next, they leave behind a bad environmental situation with open pits, stagnant waters, and no trees.[32] Unless the miners have been employed by a company or are part of a cooperative or artisanal miners

28 Gavin Hilson, '"Creating" Rural Informality: The Case of Artisanal Gold Mining in Sub-Saharan Africa' (2013) 33 *SAIS Review of International Affairs* 51. Page 53.

29 Kipsang Bernard Rop, 'Economic and Job Creation Potential of Artisanal and Small-Scale Mining In Taita Taveta County' (United Nations Development Programme 2014). Page 52.

30 Ibid. Page 53.

31 Florence Apollo and others, 'Feasibility and Acceptability of Environmental Management Strategies among Artisan Miners in Taita Taveta County, Kenya' (2017) 16 *Journal of Sustainable Mining* 189.

32 Karl M Wantzen and Jan H Mol, 'Soil Erosion from Agriculture and Mining: A Threat to Tropical Stream Ecosystems' (2013) 3 *Agriculture* 660.

association, it becomes difficult to locate the perpetrators of these environmental ills and hold them accountable.

Chemical pollution poses a great environmental concern in ASM. This is particularly notorious in ASGM where the miners use mercury in amalgamation processes.[33] The use of mercury in gold mining continues to cause pollution to the atmosphere, soil, and waterways of the greenstone belt of Western Kenya.[34] This chemical pollution in the waterways causes contamination of the water bodies and affects aquatic ecosystems. One area in which ASGM is practised in Kenya is Migori, which is close to Lake Victoria, the largest freshwater lake in Africa in which fishing is a major activity. The contamination of soil and water bodies causes a danger to human health, as shall be highlighted below. The levels of mercury and the extent of damage caused by the various levels is a scientific question. That notwithstanding, it is critical to employ safe mining practices and keep monitoring the mercury concentration levels in the soil, fish, and water in these areas since it is well known to be an unsafe metal.[35]

Section 140 of the Mining Act expressly imposes an obligation on miners to not use chemicals such as mercury and cyanide. Kenya has signed the Minamata Convention on Mercury ('the Minamata Convention') whose objective is to protect the environment as well as human health from the anthropogenic emissions and releases of mercury and its compounds.[36] Ratification of the Minamata Convention is still pending. Article 7(2) of the Minamata Convention requires state parties to take steps to reduce and, where possible, eliminate the use of mercury and its compounds. Under the Minamata Convention, where a state party determines that the ASGM activities in its territory are more than insignificant, it ought to notify the secretariat and have a progress review every three years. Kenya is one such country that has reported more than insignificant ASGM activities in its territory.

Noise pollution is another environmental concern that arises in ASM. This is mostly associated with the use of explosives in the mining activities. Corporate entities that operate quarries employing the services of artisanal miners have got into legal run-ins with the National Environmental Management

33 Benjamin Okang' Odumo and others, 'Impact of Gold Mining Associated with Mercury Contamination in Soil, Biota Sediments and Tailings in Kenya' (2014) 21 *Environmental Science and Pollution Research* 12426. Page 12433.

34 Joseph Muchiri Githiria and Moshood Onifade, 'The Impact of Mining on Sustainable Practices and the Traditional Culture of Developing Countries' (2020) 10 *Journal of Environmental Studies and Sciences* 394. Page 396.

35 Linda Campbell, D Dixon and R Hecky, 'A Review Of Mercury in Lake Victoria, East Africa: Implications for Human and Ecosystem Health' (2003) 6 *Journal of Toxicology and Environmental Health*. Part B, Critical reviews 325. Page 352.

36 Minamata Convention on Mercury, www.mercuryconvention.org/Portals/11/documents/Booklets/COP3-version/Minamata-Convention-booklet-Sep2019-EN.pdf, accessed 15 February 2021.

Authority (NEMA).[37] These instances have occurred where the terms of the Environmental Impact Licence are not complied with, leading to an order of closure of the site. Similarly, such enforcement is difficult for ASM miners who are operating as individuals and without permits.

All over the world, environmental preservation and protection is at the core of forward planning. The Constitution recognizes this very crucial responsibility to preserve the environment for the benefit of both existing and future generations.[38] In a bid to lower environmental degradation and increase environmental sensitivity among ASM miners, it is necessary for the government to put in place sensitization programmes that educate the miners on the importance of environmental conservation and employing sustainable mining practices. This can be done by the introduction of environmental management training and support centres by county governments that are closer to these miners.[39] Another action item that presents a burden to the government is the provision of alternatives to mercury in ASGM that are available and affordable to these miners. Prohibiting the use of mercury without promoting an alternative does not solve the problem as the miners will simply continue to use it illegally.[40] Fast-tracked action in issuing permits, particularly to artisanal miners, will be instrumental in enforcing environmental guidelines and provisions of the Environmental Management and Coordination Act. This will also be crucial in the practical enforcement of sanctions.

4.2 Health and Safety

Health and safety concerns are not to be unexpected given that the miners operate with primarily lowly mechanized tools. The first health challenge associated with ASM arises from the use of mercury in ASGM. Mercury is considered one of the most highly toxic metals to human health.[41] The use of mercury in ASGM causes it to be released into the atmosphere where it is transformed by anaerobic organisms into methylmercury, which is highly toxic to human health.[42] Once in the ecosystem, this substance contaminates the soil, atmosphere, and water, and eventually finds its way into the food

37 *HCCC 54 of 2011; R Versus National Environmental Management Authority ex parte Hakika Transport Services Limited* Kenya Law (High Court).

38 The Constitution (n 12) Chapter 5.

39 Gavin Hilson, 'Small-Scale Mining in Africa: Tackling Pressing Environmental Problems with Improved Strategy' (2002) 11 *The Journal of Environment & Development* 149. Page 166.

40 Laura Maria Barreto and others, 'Economic Contributions of Artisanal and Small-Scale Mining in Kenya: Gold and Gemstones' (2018). https://assets.publishing.service.gov.uk/media/5a392bb8e5274a79051c9d7c/Kenya_case_study.pdf, accessed 16 January 2023. Page 32.

41 J Githiria and others, 'Environmental and Health Effects in Artisanal and Small-Scale Gold Mining in Kenya' (2020) 1 *African Journal of Mining, Entrepreneurship and Natural Resource Management* 7.Page 3.

42 Ibid.

chain. It is then likely to be consumed not only by the miners and the surrounding communities but also by other communities due to the migratory nature of trade activities and waterways.

The studied effects of mercury on human health include renal failure, visual and auditory impairment, and negative effects on cardiovascular and reproductive systems.[43] Mercury has also been studied for its effect in causing delays in children's growth patterns.[44] Reduction of the use of mercury in mining needs serious attention at both central and county government levels. Although it is factual that mercury is a dangerous metal, it is important to understand the needs and perceptions of the miners and engage them in sensitization and friendly discussions on how to move forward.[45] Although the goal is to eliminate the use of mercury, it is difficult to find 'magical solutions' whose efficiencies and availability have been tested and proven because the ASM miners must obtain their livelihoods. Recognizing this fact and acknowledging that the road to eliminating mercury begins with its reduced use, scholars have recommended that small steps be taken, such as amalgamating the gold concentrate rather than the whole ore.[46]

The second health and safety challenge in ASM is the lack of protective clothing and equipment such as dust masks, boots, gloves, and helmets. ASM workers engage in highly labour-intensive and unmechanized tasks which can tend to be strenuous, unsafe, and prone to accidents.[47] Despite these downsides, there is a real pressure for these miners to earn their livelihoods and hence the tendency to overlook taking precautionary measures while undertaking these activities. The lack of protective equipment may be the result of negligence, poverty, ignorance, lack of proper enforcement, and lack of training on the importance of having protective equipment.[48] Although this has been flagged as a problem, it is important to note that some miners, particularly those working under corporate entities or artisanal miners' cooperative associations do have protective gear, and failure to use them is attributed to negligence or lack of enforcement.

This protective equipment helps to protect the miners from the adverse working conditions, protect against the effects of the toxic chemicals that are used or arise from the operations, and reduce the impact from accidents that may arise during the mining operations. Mining sites often experience

43 Farhana Zahir and others, 'Low Dose Mercury Toxicity and Human Health' (2005) 20 *Environmental Toxicology and Pharmacology* 351.

44 Ibid.

45 Marcello M Veiga, Gustavo Angeloci-Santos and John A Meech, 'Review of Barriers to Reduce Mercury Use in Artisanal Gold Mining' (2014) 1 *The Extractive Industries and Society* 351. Page 359.

46 Ibid.

47 Nicole M Smith and others, 'Human Health and Safety in Artisanal and Small-Scale Mining: An Integrated Approach to Risk Mitigation' (2016) 129 *Journal of Cleaner Production* 43. Page 44.

48 Rop (n 29). Page 76.

potentially fatal accidents such as the collapse of tunnels and shafts as well as rock falls. These accidents are often, but not always, occasioned by the poor underground site management due to the lack of technical skills and mechanization in ASM.[49]

Kenya has enacted the Occupational Safety and Health Act (OSHA) which provides for the health, safety, and welfare of workers and all persons lawfully present at workplaces.[50] Pursuant to Section 178(1) of the Mining Act, provisions of the OSHA on health and safety shall be applicable to the workers in mining operations. The OSHA provides that employees and self-employed persons shall take all necessary precautions to ensure their own health and safety and that of other people in the work environment. Section 6 of the OSHA further imposes obligations on employers to ensure the health and safety of persons working at their workplace by adopting, among other things, preventive and protective measures.[51]

Section 178(2) of the Mining Act provides that further to provisions of the OSHA, the CS shall make regulations on the safety and health of all persons employed in mines so that they can undertake their operations in sanitary, safe, and proper conditions. These regulations are yet to be enacted. Subsidiary to the Mining Act on the Kenya law platform are the outdated regulations which were ancillary to the repealed Mining Act. It is critical that updated regulations that are specific to the mining sector be enacted to supplement the OSHA as anticipated in the Mining Act. Further, sensitization of the miners on the importance of protective gear and equipment is necessary – where possible, providing this equipment at a subsidized cost because affordability is one of the reasons why some miners do not have the equipment. Penalties should also be enforced on the miners who have the equipment but fail to use it. These measures should strictly be provided for under the regulations and enforced at the county level for artisanal mining as part of the state's responsibility to protect its people.

The third health and safety concern in ASM is poor hygiene and sanitation. This is often occasioned by scarcity of water and sanitation facilities such as toilets and waste-dumping sites.[52] Given that some mining operations are migratory and the further fact that ASM miners often do not own the land, they are unlikely to take the initiative to build sanitation facilities. In other cases, the mining sites are such huge areas, and the sanitation facilities are not in close proximity to the location of the miners, leading to situations where the miners dispose waste within their operating environments. This often causes an influx in hygiene-related diseases such as cholera.[53]

49 Hilson, 'Small-Scale Mining and Its Socio-Economic Impact in Developing Countries' (n 23). Page 8.
50 The Occupational Safety and Health Act 2007.
51 Ibid. Section 6
52 Rop (n 29). Page 52.
53 Franklin W Schwartz, Sangsuk Lee and Thomas Darrah, 'A Review of the Scope of Artisanal and Small Scale MiningWorldwide, Poverty, and the Associated Health Impacts' [2020] *Geohealth*. Page 8.

Poor site management resulting in environmental degradation often also leads to health concerns. Stagnant water, which is a common scene in gold mining areas, provides a breeding ground for insect-borne diseases such as malaria.[54] Further, ground breaking and rock breaking during ASM often releases dust particles which place the miners at a higher risk of contracting respiratory complications and diseases such as silicosis.[55] Some of the health consequences in ASM are gender specific. Women are particularly vulnerable to some of the health and safety risks.[56] Some of the toxic chemicals from these activities have been found to have adverse effects on women's reproductive health. These gender-specific health effects occur particularly in the gold and salt mining sectors due to methylmercury and the emission of ammonia gas and hydrogen sulphide respectively.

4.3 Gender Justice in ASM

There is not a single globally accepted definition of what gender justice is, but what it encompasses is well understood. Gender justice demands that no one should carry a higher burden without compensation or enjoy lesser benefits because of their gender.[57] Gender justice is predominantly about ensuring equal power and opportunities between the genders. The push for ensuring equal opportunities between the genders has received global attention over the years, leading to its inclusion in the SDGs.[58]

ASM is one such predominantly informal industry that has a noticeable participation of women in Kenya. The participation of women is, however, heavily influenced by the perceptions of traditional African societies. Despite there being participation of women in ASM, the roles are highly gendered. Women are often relegated to undertaking ancillary and 'lighter' duties in the mining operations such as processing, cooking, and being service providers.[59] Because these kinds of jobs are deemed to be lighter and hence inferior to those undertaken by men, such as digging, their income is also lower because their contribution is deemed to be inferior to that of men. This certainly creates a power imbalance. There are women who undertake these menial jobs, but they also perform other domestic duties in the ASM communities such as childcare and domestic roles. Essentially, these women work longer hours in

54 David Villar and David J Schaeffer, 'Disarmament Is the New War, Gold Is the New Opium, and Ecohealth Is the Historic Victim' (2019) 13 *Environmental Health Insights*. Page 9.

55 Barreto and others (n 40). Page 97.

56 Smith and others (n 47). Page 45.

57 Anea Gheaus, 'Gender Justice' (2011) 6 *Journal of Ethics & Social Philosophy* 1. Page 5.

58 The 2030 agenda for Sustainable Development (n 24). Goal 5.

59 Jennifer Hinton, Christian Beinhoff, and Marcello Veiga, 'Women and Artisanal Mining: Gender Roles and the Road Ahead' in Gavin Hilson (ed.), *The Socio-Economic Impacts of Artisanal and Small-Scale Mining in Developing Countries* (Taylor & Francis 2003). Page 2.

support of the ASM industry than their male counterparts, yet these efforts are hardly ever recognized.[60] In informal mining systems, women are often considered to be cheap sources of abundant labour with little to no control over their work environments.[61]

In Osiri in Kenya, for instance, where ASGM is undertaken, women are greatly disadvantaged by the gender-based allocation of roles in the sector.[62] Women are hardly present in the shafts as they mostly undertake the supporting roles. Once the diggers, who are often male, have the ore, they either process it themselves or sell it to the women to process it.[63] It is from this stage in the value chain that there is an increased presence of women. In buying the ore, women often fall prey to dishonest practices where the diggers, who are mostly men, opt to process the good-quality ore and sell the lesser-quality ore to women.[64] Another powerful role that can be undertaken by women in these mines is owning the shafts, which needs financial muscle and support of the men. The men generally hold the view that women are inferior to them, and therefore the men working under female supervision tend to disobey the orders or directions of their female supervisors.[65] Others simply prefer not to work under the leadership of a woman. This power imbalance is also witnessed in gemstone mining in Taita Taveta County. Here, men wield control and ownership over most assets and tools in the industry.[66] Further, even where men and women perform similar tasks, the women receive less compensation, a classical characteristic of gender injustice.[67]

The gendered roles in mining are often justified by the explanation that women are generally weaker compared to the men and therefore unable to undertake the physically intensive duties in ASM such as shaft digging.[68] While this could be true in some instances, it is a blanket conclusion that denies women who are able and willing to undertake these roles the choice and opportunity to do so. It is not impossible that some women are physically capable and interested in undertaking these roles, but the system is already designed to restrict them to certain roles that have already been defined for them.

The gender imbalance cannot only be attributed to the natural biological abilities of men and women but to the traditional perceptions and practices

60 Rop (n 29). Page 87.
61 Kuntala Lahiri-Dutt, 'The Feminisation of Mining' (2015) 9 *Geography Compass* 523. Page 529.
62 Doris Buss and others, '"Remember the Women of Osiri": Women and Gender in Artisanal and Small-Scale Mining in Migori County, Kenya' (2020) 54 *Canadian Journal of African Studies/Revue canadienne des études africaines* 177.
63 Ibid. Page 184.
64 Ibid. Page 186.
65 Hilson, 'Small-Scale Mining and Its Socio-Economic Impact in Developing Countries' (n 23). Page 7.
66 Rop (n 29). Page 86.
67 Ibid.
68 Buss and others (n 62). Page 184.

on the roles of women. If physical capabilities were anything to go by, women would still have equal power and opportunities. This can be achieved by ensuring an equal playing field in ownership of land and key assets and being supported in having critical positions within the industry. The lack of equality in power and opportunities often puts the women in desperate positions, and some opt to give sexual favours in exchange for this power and opportunities.[69]

Given that gender biases are sometimes rooted in deep personal practices such as culture, sensitization and having laws of gender equality are not enough. Affirmative action is one of the ways to increase the participation of women in ASM and in positions of power such as those enjoyed by their male counterparts. The Constitution, being a transformative charter, makes provisions for gender equality and equity. Article 27(3) provides that men and women shall have the right to equal treatment and equal opportunities in political, cultural, social, and economic spheres of life. It further provides that the state takes legislative and other measures to implement the principle that no more than two-thirds of the members of appointive or elective bodies shall be of the same gender.[70] This is what is commonly referred to as the 'two-third gender rule' which is still awaiting statutory implementation; however, there is nothing to stop institutions from implementing it since it is an express constitutional provision. This rule ought to be implemented from the smallest institutions in the various industries in the economy right up to the national institutions.

The Mining Act recognizes the formation or artisanal miners' associations and cooperative groups. From a leadership perspective, these associations should be keen to implement the two-third gender rule so as to achieve balance between the genders. Training and participation of women also needs to be made a priority. These trainings should be extended to financial literacy, knowledge on minerals, value addition as well as skills on competing for opportunities with the men without necessarily having to give sexual favours. Many women in Kenya work through women's groups and these create a great platform to encourage and teach women in the industry about the relevant skills, innovations, and marketing and value addition strategies, giving them an opportunity to also encourage each other among themselves.[71] Recognizing that this problem cannot be resolved quickly or easily as they are at times perpetuated by cultural beliefs and personal choices which are difficult to turn around, the goal is to make the situation better and give women more inclusivity, power, and choice, even if slowly.

69 Chris Huggins, Doris Buss and Blair Rutherford, 'A "Cartography of Concern": Place-Making Practices and Gender in the Artisanal Mining Sector in Africa' (2017) 83 *Geoforum* 142.

70 The Constitution (n 12) Article 27(8).

71 Maarifa Ali Mwakumanya, Mwikamba Maghenda and Hamida Juma, 'Socio-Economic and Environmental Impact of Mining on Women in Kasigau Mining Zone in Taita Taveta County' (2016) 15 *Journal of Sustainable Mining* 197. Page 203.

4.4 Children in ASM

The participation of children in ASM in Kenya is not as grave as in other regions in SSA, but it is there and it is happening.[72] Research done in Taita Taveta County, for instance, revealed that most children in mining sites were about 15 years old.[73] In other areas such as ASGM sites in Western Kenya, there is a notable presence of children, which is certainly a cause for concern.[74] Because people earn a little more from ASM than they do from agriculture, some families involve their children in ASM when they are not in school. For others, the children just opt to work in the sector part-time either to provide for themselves, supplement the family's income, or just earn some extra income. The involvement and/or participation of children is largely poverty driven.[75]

The Constitution provides that children have the right to be protected from, among other things, hazardous and exploitative labour.[76] Under Kenya's legislative framework, a child is one who has not attained the age of 18 years.[77] Children are protected from economic exploitation and hazardous jobs that are likely to interfere with their education, health (physical and mental), and moral and social development.[78] The Employment Act at Section 58(2) prohibits the employment of children on any opencast or sub-surface workings that are entered by means of a shaft or an adit. It further imposes a penalty on persons who engage children in activities that constitute worst forms of child labour.[79] Worst forms of child labour are defined in the Employment Act to include work which, when carried out, is likely to harm the health, safety, and morals of the child.

Although there are circumstances in which the Employment and Children's Acts permit engagement of children under the age of 18 to undertake light duties such as apprenticeship, the environment in which ASM takes place is not ideal for the involvement of children. ASM environments suffer from health, safety, and moral challenges, and engaging a child in these activities could easily constitute worst forms of child labour. Involvement of children is also likely and has been seen to interfere with children's education whereby many drop out of school. Participation of children in ASM also hinders their

72 Barreto and others (n 40). Page 25.
73 Ibid.
74 'Kenya: Child Labour Prevalent in Artisanal Gold Mining, Says Columnist'. Business & Human Rights Resource Centre. www.business-humanrights.org/en/latest-news/kenya-child-labour-prevalent-in-artisanal-gold-mining-says-columnist, accessed 7 March 2021.
75 Gavin Hilson, '"A Load Too Heavy": Critical Reflections on the Child Labour Problem in Africa's Small-Scale Mining Sector' (2008) 30 *Children and Youth Services Review* 1233. Page 1236.
76 The Constitution (n 12). Article 53 (1)(d).
77 Children Act 2002. Section 2.
78 Ibid. Section 10(1).
79 Employment Act 2007. Section 64(2).

ideal social development as these children are made to grow up before their time. Children engage in ASM whether directly or indirectly, and this directly offends against the legal provisions. Some of the children, mostly infants, are in ASM environments not as workers but because their mothers are in ASM and cannot afford to leave them with anyone else.[80] These children are also exposed to health risks.

The involvement of children in ASM is a delicate balance because poverty is at the centre of it all. This notwithstanding, it should, ideally, be the responsibility of the adult population to make provision for their families. The children could assist their families in other ways and in better environments that do not jeopardize their school attendance, health, safety, and social growth. Depending on the circumstances, legal enforcement against guardians and parents might be one effective way to discourage inclusion of children in ASM activities.

4.5 ASM's Impact on Society's Moral Fabric

The question of morality has been the subject of many scholarly debates. What is certain, however, is that morality is heavily defined or shaped by the society in question.[81] Kenya, as an African society, has its standards on what is considered moral, and the happenings in ASM often offend against these standards. ASM is often associated with excessive leisure and hence depreciation of the moral fabric in the societies within which it is undertaken. It has been found that there is more pronounced use of alcohol and drugs as well as sex trade in ASM areas.[82] The spread of sexually transmitted infections including HIV/AIDS in Western Kenya, for instance, has been observed to be noticeably higher in mining areas such as Migori County.[83] This has similarly been observed in Taita Taveta County where there is heightened sex trade and subsequently, accelerated spread in HIV/AIDS and other sexually transmitted diseases.[84]

The morally degraded environment that many ASM environment presents has also had some adverse effects on the family unit. The miners practising ASM tend to be migratory and hence spend long days away from their families and tend to create a vibrant sex trade industry.[85] In a study done on

80 Nalule (n 4). Page 63.

81 Jonathan Haidt, 'Morality' (2008) 3 *Perspectives on Psychological Science* 65. Page 71.

82 Rop (n 29). Page 48.

83 J S Ogola, Winnie V Mitullah, and M A Omulo, 'Impact of Gold Mining on the Environment and Human Health: A Case Study in the Migori Gold Belt, Kenya'. http://erepository.uonbi.ac.ke/handle/11295/38728, accessed 8 March 2021. Page 156.

84 Irene Wakio Mwakesi, Raphael Githaiga Wahome, and Daniel Weru Ichang'i, 'Mining Impacts on Society: A Case Study of Taita Taveta County, Kenya' (2020) 11 *Journal of Environmental Protection* 986. Page 996.

85 Githiria and others (n 41). Page 407.

mines, migration, and HIV/AIDS in Southern Africa in 2012, it was established that migrant miners between the ages of 30 and 44 are 15% more likely to be HIV-positive and their partners are 8% more likely to be HIV-positive.[86]

These factors also make it undesirable for children to be in such environments. Social ills have a way of dragging behind the development of a society because the resources are diverted to excessive leisure and sometimes, eventually, healthcare. Kenya has a robust legal framework prohibiting discrimination against people living with HIV/AIDS, but it is critical to do everything possible to stop the spread. The government and NGOs have done and continue to do a commendable job in educating the masses on preventive measures and providing healthcare solutions. Drugs and excessive consumption of alcohol are also destructive practices that essentially take up resources that could be channelled to more productive uses. Empowering the miners on financial management, the dangers of drug use, and personal health issues are causes that should be strengthened by government structures such as county governments. These are some of the initiatives that county governments should be empowered to do because they are closer to the artisanal miners.

4.6 Market Challenges

ASM practitioners also suffer market challenges which often subject the miners to exploitation. The Mining Act provides that a person shall not be allowed to deal in minerals except with a mineral dealer's licence or permit. A mineral dealer's licence is an annual licence permitting the licence holder to engage in the trade of minerals both locally and internationally. A mineral dealer's permit on the other hand allows the holder to buy and sell minerals only locally.[87] The holder of a mineral dealer's permit must be a citizen of Kenya or, in the case of a body corporate, have at least 60% local shareholding.

Section 100 of the Mining Act provides that the CS shall enact regulations on the sale of minerals won by artisanal miners. Under the dealings in minerals regulations, artisanal miners should sell their extracted resources to either category of mineral dealers or the National Mining Corporation.[88] Artisanal miners can only sell their resources to any other persons as the CS may gazette. Further, artisanal miners can only export minerals with the approval of the CS.[89] This is of course far-fetched, idealistic, and arduous to the miners. The miners need quick money, so the minerals often end up crossing the border to neighbouring countries without any regulation. The industry is also full of brokers who deal in minerals but are not licensed

86 Lucia Corno and Damien de Walque, 'Mines, Migration and HIV/AIDS in Southern Africa'. https://openknowledge.worldbank.org/handle/10986/3251, accessed 8 March 2021.

87 Mining Act. Section 164.

88 Mining (Dealings in Minerals) Regulations, 2017. Regulation 25.

89 Ibid.

mineral dealers. Further, because of this desperation and lack of the technical know-how to correctly value the minerals, the miners end up being exploited by these brokers.[90] This susceptibility to exploitation is made worse by the fact that most miners lack formal education.

It is important to create favourable market structures and conditions that enable the miners to benefit, as they rightfully should, from their work. The pricing for the various minerals should be regulated so that there are minimum permissible prices for the various categories of minerals in their various states. Price regulation should be done subject to a consultative process by the national government, county government, and the respective miners' associations for the various minerals. It is important also to train and equip miners with negotiation skills, financial management skills, and, most importantly, value addition to the minerals. Most of the precious stones extracted in Kenya, for instance, are not cut and refined in the country.[91] This will increase the value of these resources and create more employment opportunities.

5 Is the Sector Formal or Informal?

Even with an improved regulatory framework, the ASM industry in Kenya largely remains informal. Even with a new Mining Act and laws on health and safety, the environment, and children and gender, daylight violation of these laws continues to be witnessed in ASM sites. Recognition of artisanal mining in the Mining Act is certainly a step in the right direction towards incorporating this industry into the mainstream economy, but this is not without challenges. These challenges are due to shortcomings on the part of the government in terms of stifled implementation.[92] Regrettably, for many years, mining has been in the bottom cadre of the priority list of the government and political class – hence, the enactment of an updated mining law over four decades post-independence. This is, however, changing, especially as mining is now featuring among the critical conversations in Kenya. The subjection of mining to political priorities is now being witnessed in county governments where the majority of the counties in which artisanal mining is practised are yet to constitute the committees. The bureaucracy and costs involved in the application for permits are catalysts for informality rather than formalization.[93] The licensing processes should be made simple and accessible. The government should also put in place measures to assist the miners in obtaining the permits, understanding that a good number of them are illiterate.

90 Ibid. Page 29
91 Barreto and others (n 40). Page 26.
92 Gavin Hilson and others, 'Artisanal and Small-Scale Mining (ASM) in Sub-Saharan Africa: Re-Conceptualizing Formalization and "Illegal" Activity' (2017) 83 *Geoforum* 80. Page 82.
93 Ibid. Page 84.

The delay in implementing formalization structures such as mining permits is also partially attributed to the reluctance by the miners to do so. Miners fear formalization predominantly because of suspicion of government intentions and fear of taxation.[94] For this reason, it is crucial to make the miners understand the objective of formalization and help them understand that being part of the regulated framework is more streamlined and predictable than operating in an illegal environment where they have to pay bribes as well as other hidden costs. Training of the miners and providing technical assistance are a critical part of the journey.[95]

Even with there being a legal framework, the miners need to be aware and have access to funding options to be able to support and grow their industry.[96] For this, it is important for the government to do everything possible to ensure that the miners and their communities benefit and are empowered.[97] Understanding that alleviation of poverty is a key but not sole driver in ASM formalization, the sector ought to be closely monitored, streamlined, and supported. The government should ensure that the communities reap the maximum benefit of the resources in their area. Formalization is indeed a process, a long-term one, and it is for this reason that countries that have had ASM at the top of their agenda, such as Ghana, are still yet to attain peak formalization. The journey towards formalization is not an easy one and can only be achieved with government focus, commitment of time, expertise, and resources.

6 Conclusion

ASM formalization needs support from all relevant stakeholders as it is an integral part of growing rural economies. The push towards poverty alleviation is greatly hinged on empowering these rural communities. The national government should prioritize actual formalization of ASM because it is a process and cannot be said to have been achieved merely by incorporation into the laws. Incorporation of ASM into the legal framework is only the beginning. As has been witnessed, ASM continues to suffer similar challenges and remains a largely informal sector six years after enactment of the Mining Act. Formalization efforts should be done in cooperation with county governments which have now been incorporated into the licensing procedures and are close to the industry players. The industry operators (the miners) should also be involved in this process with the aid of the county government by sensitizing them on the benefits of formalization, financial management, and how to mitigate the social and environmental impacts. To be able to realize

94 Siegel and Veiga (n 6). Pages 52, 54.
95 Gavin Hilson and Roy Maconachie, 'Formalising Artisanal and Small-Scale Mining: Insights, Contestations and Clarifications' (2017) 49 *Area* 443. Page 7.
96 Siegel and Veiga (n 6).
97 Hilson and others (n 92). Page 82.

this, the government needs to take its commitment to mining activities a notch higher and have a ministry dedicated to mining so the sector can be given the necessary attention.

ASM formalization is not an easy target, but gradual steps are the goal. This target is not easy because it requires a delicate balance in bridging the gap between poverty and formalization. Formalization of ASM should be done by employing methods and policies that are friendly to the miners, and this can only be done by involving them through their associations. The process of effecting the regulations such as acquisition of permits should further be made simple and easily accessible. To achieve this, political will is at the centre of it all. Very little can be achieved without government support and goodwill as all these programmes require resources and influence to succeed. The ASM sector in Kenya, although existing, has barely reached its full-blown capacity, and this is a good opportunity for the government to shape the sector and learn from the successes and struggles of other countries.

5 Contemporary Issues in the Governance and Regulation of Artisanal and Small-Scale Mining (ASM) in Africa

Antoine Floribert Mulaba-Bafubiandi, Michael Uche Ukponu, Austin Ukpata Odo, Nirdesh Singh, and Beejo Blessing Chauke

1 Introduction

Africa is blessed with abundant reserves of various mineral resources which are increasingly in global demand.[1] They include diamond, copper minerals, cobalt minerals, uranium minerals, gold minerals, tin minerals, limestone, zinc minerals, iron ore, lead minerals, nickel minerals, bauxite, coal, tantalite, and gypsum.[2] Where properly harnessed, Africa's mineral resources have the potential to contribute immensely to socio-economic empowerment and inclusive and sustainable development for Africans through the creation of direct and indirect employment, and the generation of additional revenues for African governments to fund pressing developmental programmes.[3] However, due to varying factors to be highlighted in the course of this chapter, this has not been the case for most resource-rich African countries, the vast majority of whose citizens, especially in the indigenous host mining communities, continue to wallow in squalor and economic underdevelopment despite their national governments generating substantial revenues from mineral extraction projects.

There are primarily three types of mining, namely large-scale mining (LSM), artisanal mining, and small-scale mining, although artisanal mining and small-scale mining are usually classified as one type of mining – Artisanal and Small-scale Mining (ASM). ASM has its roots in pre-colonial and colonial times, well before the advent of large-scale mining (LSM) as it is known in many African countries today.[4] The definitions of artisanal mining and

1 Kathryn Sturman, Perrine Toledano, Charles Akayuli, and Mzamose Gondwe, 'African Mining and the SDGs: From Vision to Reality' in Maano Ramutsindela and David Mickler (eds), *African and the Sustainable Development Goals* (Springer Link, 2020) 59.

2 Ibid; Oxfam, *From Aspiration to Reality: Unpacking the African Mining Vision* (Briefing Paper) 4.

3 Sturman et al. (n 1) 59.

4 Robyn d'Avignon, 'Primitive Techniques: From "Customary" to "Artisanal" Mining in French West Africa' (2018) 59(2) *The Journal of African History* 179; Showers Mawowa, 'The Political Economy of Artisanal and Small-Scale Gold

DOI: 10.4324/9781003284437-7

small-scale mining and the dichotomy between the two terms has been the subject of major debates for a long time as there is yet to be a consensus on its meaning and context. However, for the purposes of this chapter, ASM could be defined as the informal, semi-formal, or formal extraction of minerals by the labour of unskilled, semi-skilled, or skilled miners applying either rudimentary or low-level mechanized methods and equipment. We de-emphasize the issue of legality or otherwise in the definition of ASM because it is possible to have illegal LSM operations. Also, while large-scale commercial mining is usually carried out through LSM, the ASM sub-sector is beginning to witness efforts to escalate its commercialization as a result of the realization of its potential to influence socio-economic growth.[5] The World Bank indicates that there are about 100 million artisanal miners globally. According to Oxfam, 8 million of these artisanal miners operate in African countries, and ASM supports 30 million indirect jobs, accounting for 20% of gold and all gemstones (except diamond) in Africa.[6] This is why there is an escalating push by many resource-rich developing economies to officially recognize, formalize, and legalize this sub-sector.[7]

Despite the valuable benefits that ASM portends for socio-economic empowerment, increased revenue, and sustainable development in Africa, especially for rural communities and indigenous peoples, there is a lack of consensus among stakeholders in the African mining sector with respect to a generally accepted definition of ASM. This situation is attributable to divisions as to whether the sub-sector is entirely informal or illegal. Also, due to erratic, seasonal, and/or often illegal ASM activities, there is significant information asymmetry among stakeholders (especially regulators and civil societies) with respect to its scale and fiscal benefits, among other indices. Thus, a precise and generally acceptable definition of ASM which mirrors and synthesizes all plausible aspects of the sub-sector is essential as this will inspire the substantial integration and regulation of ASM across the board for optimum socio-economic impact in Africa.

Several other challenges abound that inhibit the optimal leverage of the ASM sub-sector for inclusive and sustainable development in resource-rich African countries in contemporary times. This chapter aims to discuss these challenges, including ASM formalization and legalization; mining technology and digitization; gender inclusion; health, safety, and environmental issues; and mining for sustainable development within the ambit of the Sustainable Development Goals (SDGs). The chapter projects that the harmonization of

Mining in Central Zimbabwe' (2013) 39(4) *Journal of Southern African Studies* 921, 924–925.

5 Ibid 61 and 65; Oxfam (n 2) 10; Gavin Hilson, 'The Africa Mining Vision: A Manifesto for More Inclusive Extractive Industry-led Development?' (2020) 41(3) *Canadian Journal of Development Studies* 417, 424–425.

6 Oxfam (n 2) 10.

7 During colonial and post-colonial periods in many cases, ASM was informal, not legally recognized and largely unregulated. See Mawowa (n 4) 922.

national mining/minerals governance and regulatory frameworks with the African Mining Vision (AMV), as well as their substantial implementation and effective monitoring, will empower African governments to maximize the potentials of ASM in leveraging Africa's minerals for inclusive and sustainable development.

2 Artisanal and Small-Scale Mining in Africa: Identification and Analysis of Contemporary Issues and Challenges

Harnessing ASM for the benefit of Africa and Africans continues to be a major topic of discourse at a number of national and international forums. In the quest to harness the sub-sector, there are certain issues that African policymakers should bear in mind when trying to forge a pathway for ASM development across the continent. These issues, which include the formalization of ASM, improvement of mining technology and digitization, the role and challenges of women in ASM, and mining for sustainable development, are examined in the chapter's ensuing sub-sections.

2.1 The Formalization (and Legalization) of ASM

ASM activities are often outside the legal, fiscal, and institutional mining regulatory frameworks in many African countries, thus causing difficulties in effectively tackling the many environmental, social, and economic challenges associated with it.[8] Between 70% and 80% of ASM operators worldwide are informal (or illegal as the case may be),[9] and it is our understanding that in resource-rich African countries such as Nigeria, Ghana, Ethiopia, South Africa, Tanzania, Zambia, and Zimbabwe, the statistics are equally significant.[10]

8 Oxfam (n 2) 13.

9 Morgane Fritz et al, *Global Trends in Artisanal and Small-Scale Mining (ASM): A Review of Key Numbers and Issues* (Report, Intergovernmental Forum on Mining, Minerals, Metals and Sustainable Development, 2017) v.

10 Allwell Okpi, 'Digging Up The Numbers on Informal Mining in Nigeria' (Africa Check, 3 October 2018), https://africacheck.org/fact-checks/reports/digging-numbers-informal-mining-nigeria (accessed 19 September 2021); Punam Chuham-Pole et al, 'Mining in Africa: Are Local Communities Better Off?' (Study, World Bank, 2017) 70, https://openknowledge.worldbank.org/bitstream/handle/10986/26110/9781464808197.pdf (accessed 19 September 2021); Extractive Industries Transparency Initiative (EITI) 'Getting Down to the Small-Scale Mining Level in Ethiopia' (22 November 2016), https://eiti.org/blog/getting-down-to-smallscale-mining-level-in-ethiopia (accessed 19 September 2021); United Nations Economic Commission for Africa/African Minerals Development Centre, 'South Africa ASM Profile', https://knowledge.uneca.org/ASM/sa (accessed 19 September 2021); Enzanayi Extractives Governance, 'Holding a Mirror for Artisanal and Small-Scale Mining Taxation Challenges in Zimbabwe' (Mining & Sustainable Development, 11 September 2021), https://mukasirisibanda.wordpress.com/2021/09/11/holding-a-mirror-for-artisanal-and-small-scale-mining-taxation-challenges-in-zimbabwe (accessed 19 September 2021).

Having robust frameworks which legalize ASM into the formal mining sector helps to improve its integration and regulation as well as the deployment of strategic interventions by governments and development partners to catalyze its growth.[11] These ultimately lead to the effective management of the many challenges associated with ASM.[12] Legalizing ASM activities is usually aimed at guiding a transition of ASM from a level of informality to that of formality through a process generally termed formalization. Singo and Seguin opine that:

> Formalization of artisanal mining is a process that can include the introduction of legal and regulatory frameworks, providing legal access to minerals, information about geological data, organizing miners into flexible and dynamic organizations, and providing access to capital, equipment, and technical assistance.[13]

Similarly, the Intergovernmental Forum on Mining, Minerals, Metals and Sustainable Development (IGF) defines formalization as a 'process of bringing informal income-earning activities and economies such as ASM into the formal sector through legal, regulatory and policy frameworks'.[14]

Formalization contributes to economic integration and strengthening of local economies, offering ASM operators opportunities to maximize profits through improved access to markets, finance, and modern mining technologies. Verbrugge and Besmanos note the school of thought that formalization is critical to transforming the sector towards realizing its maximum potential.[15] In addition, we posit that formalization gives ASM the certainty of recognition, regularization, legalization, institutionalization, and regulation, a guarantee that investors look for in the mining sector. ASM operators who formalize their operations through mining cooperatives, enterprises, and associations can benefit from different support services including capacity trainings, grants, and low-interest loans which governments and their partners

11 Nina Collins and Lynda Lawson, *Investigating Approaches to Working with Artisanal and Small-scale Miners: A Compendium of Strategies and Reports from the Fields* (Report, International Mining for Development Centre, 2014) 16–24; Johanna Carstens, *The Artisanal and Small-scale Mining (ASM) Sector and Its Importance for EU Cooperation with Resource-rich Developing and Emerging Countries* (Policy Brief, November 2017) 8.

12 Matti Salo et al, 'Local Perspectives on the Formalization of Artisanal and Small-scale Mining in the Madre De Dios Gold Fields, Peru' (2016) *The Extractive Industries and Society* 1058, 1060; Remi Pelon and Gotthard Walser, *Mining Together: Large-scale Mining Meets Artisanal Mining* (Guidebook, Communities and Small-scale Mining – CASM, March 2009) 27–29.

13 Patience Singo and Katy Seguin, *Best Practices: Formalization and Due Diligence in Artisanal and Small-Scale Mining* (White Paper, IMPACT, May 2018) 7.

14 Fritz et al (n 9) 49.

15 Boris Verbrugge and Beverly Besmanos, 'Formalizing Artisanal and Small-scale Mining: Whither the Workforce? (2016) 47(3) *Resources Policy* 134.

offer to improve the ASM operators' operational, technical, and management capacities. Consequently, this leads to the creation and retention of wealth by ASM operators through earnings of competitive prices for mineral products. For example, the Association of Zambian Women in Mining (AZWIM) has been a formalized association offering a solid platform for women engaged in gemstone mining to sell their products at competitive prices internationally. In 2011, members of these group sold approximately 135kg of emerald to a Canadian company.[16] Their individual earning potentials have been significantly improved because of their membership of AZWIM.

However, the success of any formalization process depends on certain factors, including the formulation and commitment to unambiguous ASM policies and laws; the existence of scientific and institutional capacities; access to adequate funding, sufficient data to inform initiatives, plans, and programmes aimed at supporting ASM; and robust strategies that allow effective engagement and collaboration between all relevant stakeholders.

Figure 5.1 shows the increasing trend in ASM in African countries between 1993 and 2017. Although there has been an upward trend in ASM operations, this has not necessarily translated to the formalization/legalization of ASM in most African countries. Insufficient data and reporting may be responsible for this undesirable index. Thus, a lot more needs to be done in data collation, and fully informed policies and regulations need to be developed to ensure that the benefits of ASM are fully recognized.

The African economy is dominated by persons working informally, offering employment to roughly 90% of employees in sectors such as agriculture and

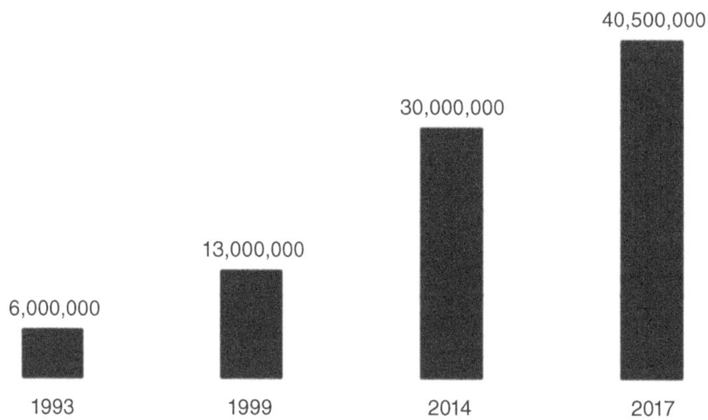

Figure 5.1 Upward trend in number of ASM operators in Africa.[17]

16 Fritz et al (n 9) 49.
17 Ibid 2.

mining.[18] Tackling poverty and attaining sustainable livelihoods cannot be achieved as long as ASM is still being operated in the informal sector.[19] Strategically incorporating ASM activities into the mainstream economic system through formalization can help to address these issues and position the sub-sector to contribute significantly to sustainable development.

Several factors act as barriers to the formalization process. The first are legal barriers. Existing legislative and fiscal policy frameworks can either help efforts succeed or frustrate them altogether. For example, under the Nigerian Minerals and Mining Act (MMA), the ASM Department in the Ministry of Mines and Steel Development (MMSD) was 'discretionarily' created to formalize and support ASM in Nigeria.[20] The challenge, however, is that the roles and responsibilities of the ASM Department are not clearly defined in the Act unlike those of the Mines Inspectorate and Mines Environmental Compliance Departments.[21] The absence of a clear, legally defined role for the ASM Department may be indicative of the political and/or regulatory bias or preference towards LSM operations. Furthermore, even with formalization, the certificate of formalization issued to ASM operators states boldly that the document is not a mining permit. The mining permit caveat disincentivizes the process and discourages other ASM operators from coming forward as they may think that the end result is ultimately not worth the stress of the process.

Another major barrier is the highly centralized administrative power structures for mineral resources governance in many African countries. In Ethiopia, Ghana, and Nigeria, all minerals are vested with the federal/central government with very little or no participation/control of the state and local governments[22] which are better placed to manage ASM due to being the closest governments to the grassroots. Of course, this takes away the opportunity for a wider range of stakeholders to be effectively involved in resources management, and by extension ASM management. With the central government having such disproportionately huge control over mineral resources, it is

18 Weng Xiaoxue, *The Rural Informal Economy: Understanding Drivers and Livelihood Impacts in Agriculture, Timber and Mining* (Working Paper, International Institute for Environment and Development, April 2015) 4.

19 Organization for Economic Cooperation and Development (OECD), 'Removing Barriers to Formalisation' in *Promoting Pro-Poor Growth: Policy Guidance for Donors* (OECD Publishing, 2007) 75, 76.

20 Nigeria Minerals and Mining Act (MMA) 2007, s 16.

21 MMA, ss 16, 17 and 18.

22 In Nigeria, a representative of a State Ministry in charge of minerals/land matters, agriculture/forestry, environment; a representative each from the State agency/department in charge of the environment and the State Surveyor-General; and a representative of a Local Government Council (only when matters affecting the said Local Government Area are being considered by the Committee) may be members of the Mineral Resources and Environmental Management Committee subject to the supervision of the Federal Minister in charge of mines: Nigeria Minerals and Mining Act (NMMA), s 19(2)(b),(d),(e),(f),(g), and (h).

arguable that a basic natural principle of land law is being violated. This principle states that *whoever owns the land, it is theirs up to the heavens and down to hell*[23] – although the state governments own the land, the federal government owns the mineral resources therein. This issue has been a subject of conflict between the federal and state governments in Nigeria. Comparatively, in advanced mining jurisdictions such as Australia, with resource control enabled by law, the states/territories have the powers to manage the resources within their territorial jurisdictions,[24] and this increases the chances of having a wider buy-in of the people, effective governance, and the optimal performance of the mining sector. Also, there are complaints of bogus and complicated bureaucratic processes and inefficient institutions involved in the entire process which frustrates ASM operators' efforts to formalize their operations.[25] This makes the expense and time in completing the entire process cost-ineffective, and therefore a disincentive to the formalization process. To effectively harness ASM, African countries need to give greater room for resource control by the lower tiers of government, with the central government focusing more on creating enabling environments for macroeconomic growth and foreign direct investments (FDIs).

Another major issue is institutional corruption and mismanagement, which results in the illegal diversion of funds meant to incentivize formalization. An example is the alleged mismanagement by government officials of about US $75 million grant under the defunct Sustainable Management of Minerals Resources Programme to help upscale the productive capacity of artisanal miners in Nigeria.[26] With such abuse of office, ASM operators have become wary of government officials and the promises they make and do not consider them genuine and committed; hence, most of them prefer to remain outside the legal frameworks.

23 *Cuius est solum, eius est usque ad coelum et ad inferos.*

24 In Australia, natural resources, including minerals, onshore (land and inland water ways) and up to three nautical miles offshore are owned and regulated by the state/territory governments; beyond three nautical miles offshore and up to the continental shelf, they are owned by the Commonwealth government. This is the extant resource management arrangement by virtue of the Offshore Constitutional Settlement between the state/territory and Commonwealth governments, the contents of which are reflected in relevant federal laws. See generally: Australian Government – Attorney-General's Department, *Offshore Constitutional Settlement: A Milestone in Cooperative Federalism*, www.ag.gov.au/international-relations/publications/offshore-constitutional-settlement-milestone-cooperative-federalism (accessed 20 September 2021); Sea and Submerged Lands Act 1973 (Cth); Coastal Waters (State Powers) Act 1980 (Cth); Coastal Waters (State Title) Act 1980 (Cth); Coastal Waters (Northern Territory Powers) Act 1980 (Cth); Coastal Waters (Northern Territory Title) Act 1980 (Cth).

25 Hilson (n 5) 424.

26 World Bank Group, *Nigeria – Sustainable Management of Mineral Resources Project (English)* (Report No. AC439, 2004), http://documents.worldbank.org/curated/en/929331468292245720/Nigeria-Sustainable-Management-of-Mineral-Resources-Project.

Another nagging issue is inadequate resources and logistics to enable the formalization process. In most African countries, data collection and management are a big challenge[27] – ascertaining the actual number of artisanal miners, the minerals they mine, the number of children working in the mines, and the volume of ore reserves in ASM host communities. This makes planning and appropriate deployment of ASM support services very difficult. With respect to logistics, there are well-documented cases of African countries not having adequate and well-trained personnel to help monitor ASM activities as well as non-complex matters like inadequate vehicles and equipment to enable staff carry out field operations.[28]

In most African countries, traditional landowners have very strong ties to communal ancestral and cultural relics, which in effect make traditional ownership of lands a major issue for African governments to navigate with a view to successfully formalizing artisanal miners laying claim to such lands.[29] These traditional landowners usually operate with traditional, often informal rules and regulations, which are based on cultural norms. In Ghana, the traditional rulers, chiefs, and elders have firm control over rent and tax.[30] This makes local miners and other land users in these areas unwilling to pay another 'legal fee' to the central government which they consider illegal, coupled with the impression that the central government is not even in touch with socio-economic yearnings at the grassroots.

In African countries, the lack of security of lives and properties stemming from terrorism, insurgencies, banditry, and kidnapping seem to have become major shortcomings of their respective governments. This can potentially be worse in mineral-rich areas where militant groups take advantage of economic misgovernance, unfair treatment, and social neglect to illicitly mine and trade in mineral resources.[31] The result is that most artisanal mining operations in such

27 Florent Bedecarrats, Jean-Pierre Cling, and Francois Roubaud, 'The Data Revolution and Statistical Challenges in Africa: Introduction to the Special Report' (2016) 258(2) *Afrique Contemporaire* 9.

28 Salvadore Mondlane, *Report on Artisanal and Small-scale Mining in Africa: Selected Countries Policy Profile Review on ASM* (African Minerals Development Centre, 2017) 66–71.

29 Liz Alden Wily, *Community Property in Africa* (Brief § 2, June 2018) 10.

30 Janine Marisca Ubink, *In the Land of the Chiefs: Customary Law, Land Conflicts, and the Role of the State in Peri-Urban Ghana* (Leiden University Press, 2008) 41–46.

31 Maurice Ogbonnaya, 'Mining and Extractives: Illegal Mining Drives Nigeria's Rural Banditry and Local Conflicts' (ENACT, 28 May 2020), https://enactafrica.org/enact-observer/illegal-mining-drives-nigerias-rural-banditry-and-local-conflicts (accessed 25 September 2021); Muauz Gidey Alemu, 'Artisanal Small-scale Opal Mining (ASOM) Insecurity in the Delanta Wereda, Ethiopia: The Shifting Landscape of Multidimensional Insecurity in the Face of Emergent ASOM Wealth' (2018) 27(1) *Africa Security Review* 61, 70–71; Thierry Vircoulon, 'Behind the Problem of Conflict Minerals in DR Congo: Governance' (International Crisis Group, 19 April 2011), www.crisisgroup.org/africa/central-africa/democratic-republic-congo/behind-problem-conflict-minerals-dr-congo-governance (accessed 25 September 2021).

areas are controlled by the militants and can elude the watch of government officials as they continue operating informally/illegally. Thus, security barriers to formalization can manifest when government officials are unable to access mineral-rich areas where government control has been significantly weakened.

2.2 Gender Issues

The role of women in mining is a hot topic inspired by scholarly and policy debates bordering on questions of discriminatory and exclusionist actions to the detriment of women in the mining sector – hence the need to bring female miners up to par with their male counterparts. We increasingly see measures deliberately designed and implemented to favour women, a sort of 'positive discrimination' referred to as affirmative action.[32] Women now seek increased employment opportunities and better working conditions in the African mining sector which is dominated by men.[33] The International Labour Organization (ILO) has reported in the past that up to 80% of the people working for ASM operations in Africa, Asia, and South America operate outside the formal regulatory framework.[34] These are mostly women and children working in very dangerous conditions and exposed to many of the social, economic, and environmental risks associated with ASM. Women ASM workers tend to make up larger portions of ASM workers in Africa.[35] Women in Ghana and Malawi respectively account for more than 50% of the national ASM mining workforce.[36] They are mostly uneducated or poorly educated in both contexts of literacy and basic rights.

Women's associations such as Women In Mining are being formed across the continent to promote the role of women in mining and are doing very well in this regard.[37] Thus, the promotion of affirmative action and the rights and welfare of women is key to establishing an all-inclusive and just platform for participants to flourish, irrespective of gender. Countries like Namibia, South Africa, and Uganda have enacted laws to promote affirmative action. Some mining companies are entrenching gender equity and inclusion in their

32 CL Bacchi, *The Politics of Affirmative Action: 'Women', Equality and Category Politics* (Sage Publications, 1996) x.

33 MP Dikane, *The Implementation of Employment Equity and Affirmative Action as a Tool of Balancing the Injustices of the Past in the Mining Industry* (Master Degree Thesis, North-West University, October 2006) 33, http://dspace.nwu.ac.za/bitstream/handle/10394/16/dikane_matthewsp.pdf;sequence=1 (accessed 25 September 2021).

34 'Small-scale Mining on the Increase in Developing Countries' (International Labour Organization, 17 May 1999), www.ilo.org/global/about-the-ilo/newsroom/news/WCMS_007929/lang–en/index.htm (accessed 25 September 2021).

35 Oxfam (n 2) 10.

36 African Minerals Development Centre, *African Women in Artisanal and Small-Scale Mining: A Special Report* (Special Report, African Union, March 2015) 1–2.

37 See generally: International Women in Mining, https://internationalwim.org (accessed 29 September 2021).

mission and vision statements and recruitment policies.[38] However, questions abound as to whether such corporate policies are genuinely implemented, which can be deciphered through effective regulatory monitoring and evaluation. Namibia's Employment Equity Commission published the Affirmative Action Act Annual Reports of 2011/2, 2012/3, 2013/4, and 2014/5 which revealed that Namibia's mining industry employment rates are still skewed unfavourably against women despite the implementation of affirmative action policies.[39]

The African Mining Vision envisages a sustainable and well-governed mining sector that is safe, healthy, gender and ethnically inclusive, environmentally friendly, socially responsible, and appreciated by surrounding communities.[40] But many African governments relegate gender inclusion in mining to the background in favour of accruable resource revenue from international mining corporations who appear content with 'imported' personnel from their own countries. Where a central government fails, the host communities and non-governmental organizations can implement the AMV to promote the inclusion of women via alternative pathways such as community development agreements (CDAs), which have proven to fill legislative lacuna or complement government action in this regard.[41] Implementing gender-based principles in mining is a challenge, but as scholarly and mining policy progressives, we posit that the institutionalization of these principles in the mining sector would be easier with the formalization of ASM.

2.3 Health, Safety, and Environmental Concerns

There are several cases pointing to the negative impacts of unregulated ASM cutting across health, safety, and environmental dimensions. In the unfortunate Zamfara lead-poisoning incident of 2010, where a high mortality rate of around 43% occurred,[42] at least 400 children in different communities within Zamfara State, Northern Nigeria, died because of the crude techniques

38 For example, a South African mining firm called PetraDiamonds created a Diversity and Inclusion Policy for the recruitment and training of female staff: PetraDiamonds, 'Equality and Diversity', www.petradiamonds.com/sustainability/diversity-equality (accessed 29 September 2021).

39 P Mwetulundila and N Indongo, 'Trend Analysis on Gender Disparities in Namibia's Mining Sector: Evidence from The Affirmative Action Reports of 2011–2015' (2018) 7 *Journal for Studies in Humanities and Social Sciences* 87.

40 African Union, *African Mining Vision* (February 2009) v, https://au.int/sites/default/files/documents/30995-doc-africa_mining_vision_english_1.pdf (accessed 20 September 2021).

41 David Brereton, John Owen and Julie Kim, *World Bank Extractive Industries Source Book: Good Practice Notes on Community Development Agreements* (Report, World Bank Group, 20 September 2011) 146.

42 'Nigeria: Mass Lead Poisoning from Mining Activities, Zamfara State – Update 1' (World Health Organization, 11 November 2011), www.who.int/emergencies/disease-outbreak-news/item/2011_11_11-en (accessed 25 September 2021).

deployed in the mining and processing of gold.[43] The local miners ground the lead-laden gold with rocks, causing lead dust to come off and settle on the soil and other surfaces. The locals spread their grains on the ground to dry, and this inadvertently exposed them to lead contamination, resulting in very sick children who consumed the lead-contaminated food.[44] Exposure to lead poisoning can cause long-term physical and mental health problems, including aggressive behaviour and cognitive impairment in children.[45]

In Ghana, there are instances of conflicts between ASM operators and government agencies, LSM operators and/or fellow community dwellers who compete intensely for access to land, water, and other natural resources. Conflicts surrounding the access to and use of land between ASM and LSM companies is an important example to draw upon here. Due to the potential huge inflow of FDI and revenues, the government's pecuniary interest is drawn towards favouring LSM operators with allocations of mining titles over extensive pieces of land and the sometimes forcible ejection of ASM operators (called *Galamsey* or illegal miners) from those lands. This has led to conflicts in locations such as Tarkwa and Ashanti.[46] There have been violent clashes between *Galamseys* and security forces protecting LSM company workers at the Obuasi mining site of the Ashanti Goldfields resulting in the loss of over $1 million.[47]

These examples indicate the extent to which unregulated mining can negatively affect societies in different ways. Unregulated ASM leads to health and safety risks for both the miners and the host communities through exposure to the preventable discharge of contaminants during rudimentary mining and processing. Arable land, hitherto for agricultural purposes, would now be rendered inappropriate for such purposes. Some of these contaminants could even compromise a community's only source of water. This biohazard can lead to the death of entire community populations that depend on this water body for drinking and other subsistent uses. Also, the indiscriminate drilling/digging of land over time precipitates environmental degradation, erosion, and biodiversity loss.

43 Medecins Sans Frontieres (MSF), *Lead Poisoning Crisis in Zamfara State, Northern Nigeria* (Briefing Paper, May 2012) 1.

44 Simba Tirima et al, 'Food Contamination as a Pathway for Lead Exposure in Children During the 2010–2013 Lead Poisoning Epidemic in Zamfara, Nigeria' [2018] 67 *Journal of Environmental Services* 260.

45 Center for Disease Control and Prevention (CDC), 'Lead Poisoning Investigation in Northern Nigeria' (25 October 2016), www.cdc.gov/onehealth/in-action/lead-poisoning.html (accessed 20 September 2021).

46 Gavin Hilson, 'An Overview of Land Use Conflicts in Mining Communities' (2002) 19 *Land Use Policy* 65, 67.

47 Gavin Hilson, *A Contextual Review of the Ghanaian Small-scale Mining Industry* (Report, International Institute for Environment and Development, September 2001) 19.

Happily, at least 38 African countries, including Nigeria, have signed the *Minamata Convention*,[48] an international treaty agreed by more than 100 countries to protect human health and the environment from anthropogenic emissions of mercury compounds. In Nigeria, following the Zamfara lead-poisoning incident, there is now greater awareness among miners of the need to engage in safe mining and environmental management practices.[49] This improvement also can be attributed to the Safer Mining Programme of the Ministry of Mines and Steel Development, which formalized miners under the banner of mining cooperatives and small-scale mining associations benefit from.

2.4 Mining Technologies and Equipment

ASM is mostly unregulated and rife with operators who lack modern technical mining skills,[50] especially in geology, exploration, extraction, and minerals processing. ASM operators are generally unaware or negligent about the environmental and health hazards of their activities.[51] It is not uncommon for ASM operators to leave behind thousands of pits and trenches dug when xcavating for minerals,[52] as a consequences of which rich agricultural lands and biodiversity are lost[53] and the sites could become a breeding habitat for vectors or parasites causing prevalent illnesses such as malaria.[54] Environmental and health impacts of ASM result from the use of basic tools and machines such as drills, excavators, and

48 Minamata Convention on Mercury (adopted 10 October 2013, entered into force 16 August 2017, No. 54669), www.mercuryconvention.org/en (accessed 20 September 2021).

49 Lisa Goldman et al, *Artisanal and Small-Scale Gold Mining in Nigeria: Recommendations to Address Mercury and Lead Exposure* (Report, Environmental Law Institute, November 2014) 1, 4, and 11–13.

50 PF Ledwaba, 'The Status of Artisanal and Small-Scale Mining Sector in South Africa: Tracking Progress' (2017) 117 *The Journal of the Southern African Institute of Mining and Metallurgy* 33, 38.

51 YF Yewuhalashet, AG Abraha, and YK Hailemariam, 'Assessing The Main Challenges of Artisanal and Small-scale Mining Operations in Africa: A Simple Overview' (2020) 7(2) *Geominetech: The Indian Mineral Industry Journal* 1, 3; Mining, Minerals and Sustainable Development (MMSD), 'Artisanal and Small-Scale Mining' in *Breaking New Ground: Mining, Minerals and Sustainable Development* (International Institute for Environment and Development, 2002) 314, 320.

52 Jan Macháček, 'Typology of Environmental Impacts of Artisanal and Small-Scale Mining in African Great Lakes Region' (2019) 11 (11) *Sustainability* 3027, 3034 and 3036.

53 LJ Sonter, SH Ali, and JEM Watson, 'Mining and Biodiversity: Key Issues and Research Needs in Conservation Science' (2018) *Proceedings of the Royal Society B: Biological Sciences* 285, 287.

54 Francois Dao et al, 'Burden of Malaria in Children Under Five and Caregivers' Health-Seeking Behaviour for Malaria-Related Symptoms in Artisanal Mining Communities in Ghana' (2021) 14(1) *Parasites and Vectors* 418, 419.

crushers;[55] unsafe use of hazardous chemicals;[56] lack of or disregard for personal protective equipment (PPE);[57] and the absence of plans for reclamation of degraded lands.[58]

A lack of appropriate tools for mineral prospecting results in a lot of time spent on exploring for minerals in an area with little or no success. Even where some mineral prospects are located, the chances that they are commercially viable reserves are usually very low, making these mineral prospecting activities highly unreliable and unsustainable. Generally, ASM relies on manual labour, and common equipment used by labourers includes hammers, chisels, diggers, and shovels for excavation, while it is not uncommon to use harmful chemicals such as mercury and cyanide for gold extraction.[59] The implications of using such rudimentary tools include inefficient recovery of minerals, which leads to lost time and high chances of the miners exploring more lands and causing further environmental degradation. Regarding the use of harmful chemicals such as mercury and cyanide, severe air, land, and water pollution can typically occur and persist in the environment for extended periods, having deleterious, long-term consequences on the health of miners and members of host and adjoining communities.[60]

Africa is widely known to be very rich in precious metals, non-metallic and metallic minerals, and gemstones. Gold and gemstones are known to rapidly generate high revenue and are the most commonly mined on the continent.[61] Artisanal and Small-Scale Gold Mining (ASGM) is particularly important because it has a high revenue contribution profile, yet uses dangerous chemicals such as mercury and cyanide for mineral extraction. Research reveals that East Asia and Sub-Saharan Africa contribute the most to mercury-related emissions globally.[62] Because of this, it is instructive to focus on ASGM in

55 KJ Bansah, AB Yalley, and N Dumakor-Dumpey, 'The Hazardous Nature of Small Scale Underground Mining in Ghana' (2016) 15(1) *Journal of Sustainable Mining* 8, 9–11.

56 Nicole M Smith et al, 'Human Health and Safety in Artisanal and Small-Scale Mining: An Integrated Approach to Risk Mitigation' (2016) 129 *Journal of Cleaner Production* 43, 48.

57 *Improving Occupational Safety and Health in Artisanal Mining*, iTSCi Phase 3: Social Value [9], www.pactworld.org/sites/default/files/Occupational%20Safety%20%26%20Health%20summary.pdf (accessed 26 September 2021).

58 Collins and Lawson (n 11) 41–42.

59 Boris Verbrugge, Cristiano Lanzano, and Matthew Libassi, 'The Cyanide Revolution: Efficiency Gains and Exclusion in Artisanal- and Small-Scale Gold Mining' (2021) 126 *Geoforum* 267–274.

60 Louisa J Esdaile and Justine M Chalker, 'The Mercury Problem in Artisanal and Small-Scale Gold Mining' (2018) 24 *Chemistry: A European Journal* 6905, 6911; Kevin M Rice et al, 'Environmental Mercury and Its Toxic Effects' (2014) 47 *Journal of Preventive Medicine and Public Health* 74.

61 'Africa: Resources' (National Geographic), www.nationalgeographic.org/encyclopedia/africa-resources (accessed 26 September 2021).

62 Niladri Basu et al, 'Integrated Assessment of Artisanal and Small-Scale Gold Mining in Ghana – Part 1: Human Health Review' (2015) 12(5) *International Journal of Environmental Research and Public Health* 5143, 5148.

Africa, with details given on the various technologies commonly used in this sub-sector, and the alternatives that exist.

2.4.1 Artisanal and Small-Scale Gold Mining (ASGM)

Some notable methods used in ASGM in Africa are mercury amalgamation, cyanidation, sluicing, and panning. Mercury amalgamation is the most widely used method because of its low cost and ease and speed of use. The steps involved in the mercury amalgamation process include extraction of gold ore, crushing of gold ore, concentration of gold ore, addition of mercury to gold to form an amalgam, melting of the amalgam to form gold dore, and refining of gold dore to produce gold.[63]

Mercury amalgamation poses dangerous environmental and health risks when mercury is transformed to methylmercury and released into the environment.[64] The *Minamata Convention on Mercury* was adopted in 2013 to reduce, control, and, where possible, eliminate the use of mercury[65] in a range of industries, including ASGM. ASGM is the biggest user of mercury in the world, accounting for approximately 1,400 tonnes of the total volume of mercury used in 2011 alone.[66] ASGM jurisdictions, who are signatories to the *Minamata Convention*, have begun developing national action plans (NAPs), detailing strategies for a phased elimination of mercury use and promoting the use of alternative and safer technologies in ASGM and other mining-related activities.[67]

It is also important to talk briefly about the use of cyanide in gold extraction. It is based on the process referred to as cyanide leaching, whereby cyanide is directly added to the ore or to tailings where mercury had already been added.[68] This makes cyanide extremely dangerous.[69] Cyanide leaching is most commonly

63 Kevin Telmer and Daniel Stapper, *Reducing Mercury Use in Artisanal and Small-Scale Mining: A Practical Guide* (Research Report, UN Environment, 2012) 11; UN Environment, *Developing a National Action Plan to Reduce and, Where Feasible, Eliminate Mercury Use in Artisanal and Small-Scale Gold Mining* (Guidance Document, September 2017) 12–13. Note: Regarding Step 3, mercury can either be applied to the whole ore (whole ore amalgamation) or to the concentrate (concentrate ore amalgamation) which is better as it limits the quantity of mercury required for processing. See Marcello M Veiga and Aaron J Gunson, 'Gravity Concentration in Artisanal Gold Mining' (2020) 10 *Minerals* 1026, 1058–1059.

64 Nadine Steckling et al, 'The Burden of Chronic Mercury Intoxication in Artisanal Small-Scale Gold Mining in Zimbabwe: Data Availability and Preliminary Estimates' (2014) 13 *Environmental Health* 111, 112.

65 Minamata Convention on Mercury, art 1.

66 Alex Kirby et al, *Mercury – Time to Act* (Report, UN Environment, 2013) 17.

67 'National Action Plans' (UN Environment), https://web.unep.org/globalmercurypartnership/our-work/reducing-mercury-artisanal-and-small-scale-gold-mining-asgm/national-action-plans (accessed 26 September 2021).

68 Minamata Convention on Mercury, annex c (1)(b)(iv).

69 Samuel J Spiegel and Marcello M Veiga, 'International Guidelines on Mercury Management in Small-Scale Gold Mining' (2010) 18(4) *Journal of Cleaner Production* 375, 379.

used in large-scale gold mining (LSGM), but is now increasingly used for ASGM in African countries as a result of its ability to efficiently recover gold from ore.[70] As a highly toxic and dangerous substance, it is of the utmost importance that it is handled with extreme care and stored correctly to avoid serious health and environmental hazards.[71] Cyanide contamination is particularly severe on the human heart and brain.[72] The *International Cyanide Management Code* developed in 2002 is in force and is intended to assist with controlling the risks associated with cyanide.[73]

2.4.2 Alternative Approaches to Mineral Processing in ASGM

Governments and international development partners are promoting the adoption of 'zero chemicals' in ASGM. While safer alternatives to mercury and cyanide exist, there are problems with their adoption because of a range of factors such as cost of the products and a lack of awareness by ASM operators. Alternative and safer approaches to safer ASGM practices include the use of retorts, fume hoods, and mercury activation. In addition, ASGM operators can adopt panning, sluicing, centrifuges, spiral concentrators, vortexes, direct smelting using fluxes such as borax, and iGoli using dilute hydrochloric acid and sodium metabisulphate.

2.5 Digitalization in Mining

African governments and other organizations still rely heavily on obsolete or ineffective record-keeping and information management systems which makes the tracking of key indices largely unreliable. Thus, inefficiencies regarding licensing, planning, and overall management of the sector are still very widespread, worsened by the dismal infrastructural state of most African states, especially in the areas of power supply.[74] The promotion of good governance

70 Verbrugge, Lanzano and Libassi (n 59) 268–271; Astrid M Knoblauch et al, 'Potential Health Effects of Cyanide Use in Artisanal and Small-Scale Gold Mining in Burkina Faso' (2020) 252 *Journal of Cleaner Production* 1, 1–2.

71 Government of Western Australia Department of Commerce, *Cyanide – Information on Handling, Storage and Hazards* (Fact Sheet A18039643) 1–4, www.commerce.wa.gov.au/sites/default/files/atoms/files/cyanide.pdf (accessed 27 September 2021).

72 Artisanal Gold Council, *Chemical Hazards in the Artisanal Gold Sector: Impacts of Mercury, Cyanide and Silica Dust on Human Health and Environment* (Fact Sheet, 2020) 8, www.artisanalgold.org/wordpress/wp-content/uploads/2020/07/AGC_Health-brochure_Chemical-hazards-in-ASGM.pdf (accessed 27 September 2021).

73 International Cyanide Management Code For the Manufacture, Transport, and Use of Cyanide In the Production of Gold ('Cyanide Code'), https://cyanidecode.org/the-cyanide-code (accessed 24 September 2021).

74 Victoria R Nalule, *Energy Poverty and Access Challenges in Sub-Saharan Africa: The Role of Regionalism* (Springer Nature Switzerland AG, 2019) 21–39.

principles – namely, rule of law, transparency, and accountability – is significantly assisted by authentic and available records.[75] With effective digitization, challenges with information management in the mining sector can be easily addressed.[76] Digital systems such as portals, websites, and 'apps' can support the timely and hassle-free acquisition of relevant mining-related permits and are important for correct reporting of activities, sharing of information and generally for strategic planning and decision making. With the advent of blockchain technologies, traceability of mineral purchases made online has been improved,[77] putting mineral smuggling and mineral-induced conflicts in check and effectively improving revenue generation and the prevention of violence in mining communities. Most of these are readily applicable to LSM, but duly formalized ASM can better be monitored with such digital systems and social vices in the sector easily tackled.

2.6 Formal and Informal Mining for Sustainable Development

Sustainable development 'meets the needs of the present without compromising the ability of the future generations to meet their own needs'.[78] It is geared towards ensuring socio-economic development and environmental sustainability[79] through good governance measures.[80] Thus, economic development, social inclusion, environmental sustainability, and good governance are the pillars on which sustainable development stands.[81] Economic development entails the use of resources and assets to generate wealth and enhance the capacity of businesses, especially small and medium-scale

75 Olefhile Mosweu and Donald Rakemane, 'The Role of Records Management in Ensuring Good Governance in Africa: Impediments and Solutions' (2020) 53 *Journal of the South African Society of Archivists* 103.

76 Iqra Atif, Frederick T Cawood, and Muhammad A Mahboob, 'The Role of Digital Technologies That Could Be Applied for Prescreening in the Mining Industry During the COVID-19 Pandemic' (2020) 5 *Transactions of the Indian National Academy of Engineering* 663, 671; see also TS Golosinki, *Use of the Internet and Information Technology in Mining* (Conference Paper, 17th International Mining Congress and Exhibition of Turkey, 2001) 13, www.maden.org.tr/resimler/ekler/109af321d970c64_ek.pdf (accessed 27 September 2021).

77 RCS Global, *Blockchain For Traceability in Minerals and Metals Supply Chains: Opportunities and Challenges* (Study, 20 December 2017) 4.

78 World Commission on Environment and Development (WCED), *Our Common Future* ('Brundtland Report') (Report, WCED, 1987) 41, www.un-documents.net/our-common future.pdf.

79 United Nations Environmental Programme (UN Environment) and United Nations Development Programme (UNDP), 'Managing Mining for Sustainable Development' (Sourcebook, UN Environment and UNDP, April 2018) 26, www.undp.org/content/undp/en/home/librarypage/poverty-reduction/Managing-Mining-for-SD.html.

80 Society for Mining, Metallurgy and Exploration (SME), 'The Mining Industry and Sustainable Development' (Report, SME, April 2016) 2, www.eenews.net/assets/2016/06/29/document_pm_01.pdf.

81 Ibid 2–3.

enterprises.[82] Achieving social development requires strategic capital expenditure on critical infrastructure, equitable distribution of benefits, preservation of human rights, and promotion of intergenerational equity.[83] To ensure a sustainable environment, mineral and natural resources will have to be developed in a responsible and eco-friendly manner by applying precautionary, preventive, and reparative measures.[84] Equally important are strong governance structures that enforce the preceding three pillars through viable policy and regulatory frameworks, institutional transparency and accountability; citizens' participation in resource governance, as well as access to information and justice.[85]

In 2015, 193 United Nations member states gathered to adopt and commit to the Sustainable Development Goals (SDGs),[86] the central aim of which is to eradicate hunger by 2030 as it calls on all stakeholders – national and subnational governments, local communities, business enterprises, development partners, and civil society organizations – to undertake actions to ensure prosperity and a sustainable environment for the present and future.[87] The SDGs are targeted at achieving socio-economic growth and sustainable development by ending poverty, providing and investing in education, health, and security, and creating employment opportunities while addressing climate change and environmental degradation.[88]

Often, mining is associated with several global challenges such as conflict, illegal minerals extraction and trade (especially with regard to ASM), poverty, disease, human rights abuses, gender discrimination, child labour, and environmental degradation.[89] However, at the same time, mining can be leveraged

82 Ibid; R Eggert, 'Mining, Sustainability and Sustainable Development' in P Maxwell (ed), *Mineral Economics* (Australian Institute of Mining and Metallurgy, Carlton, 3rd ed, 2013) 215, 216.

83 SME (n 80) 2–3; R Eggert (n 82) 216.

84 SME (n 80) 2–3; R Eggert (n 82).

85 SME (n 80) 2–3.

86 'Transforming Our World: 2030 Agenda for Sustainable Development' ('SDGs') (United Nations), www.un.org/sustainabledevelopment/sustainable-development-goals (accessed 26 September 2021).

87 Ibid; SME (n 80) 3; Columbia Centre on Sustainable Investment et al, *Mapping Mining to the Sustainable Development Goals: An Atlas* (White Paper, Columbia Centre on Sustainable Development, July 2016) 12. https://ccsi.columbia.edu/2016/07/19mapping-mining-to-the-sustainable-development-goals-an-atlas (accessed 26 September 2021).

88 SDGs (n 86); N Yakovleva, J Kotilainen, and M Toivakka, 'Reflections on the Opportunities for Mining Companies to Contribute to the United Nations Sustainable Development Goals in Sub-Saharan Africa' (2017) 4 *The Extractives Industries and Society* 426.

89 Jorden de Haan, Kirsten Dales, and James McQuilken, *Mapping Artisanal and Small-Scale Mining to the Sustainable Development Goals* (Study, University of Delaware's Minerals, Materials and Society, 8 September 2020) i. www.pactworld.org/Mapping%20artisanal%20and%20small-scale%20mining%20to%20the%20sustainable%20development%20goals (accessed 27 September 2021); International Finance Corporation (IFC), 'Working Together – How Large-Scale Mining Can

to eradicate these same challenges and ensure sustainable development through effective mining governance and procedures such as ASM formalization.[90] Mining can help to increase government revenues to finance infrastructural development, create jobs, and drive economic growth. For mining to lead to poverty or prosperity in a given country, the viability or otherwise of its mining governance structure is a critical factor.[91] Thus, it holds that mining can contribute immensely to achieving the SDGs.[92] These positive and negative positions are also true with respect to the potential of ASM as highlighted in many parts of this chapter. Although some authorities posit that only few particular SDGs are mining-related,[93] a critical look at all SDGs and their 169 targets reveals that there are linkages to mining – including ASM.

African governments and international development partners can collaborate to maximize the potential of ASM by encouraging ASM operators to formalize their business structures. They should establish incentives for ASM operators to organize themselves under corporate entities such as mining cooperatives, associations, and enterprises through which they could access the necessary support aimed at assisting them to graduate to commercially viable LSM operators. Examples of such incentives include:

a Access to development grants and other forms of credit. For example, under the now completed Sustainable Management of Mineral Resources Project (SMMRP) implemented by the Ministry of Mines and Steel Development in Nigeria, female-owned ASM businesses that had successfully undergone formalization were given grants to scale up their operations.[94]
b Access to modern mining equipment and cleaner and more efficient technologies. For example, in Ethiopia, lapidaries were developed to support gemstone miners, while in Ghana, the use of induction heating/direct smelting and retorts are being promoted as a safer and more viable alternative to gold amalgamation as part of the implementation of the *Minamata Convention on Mercury*.

Engage With Artisanal and Small-scale Miners' (Guidance Note, IFC, February 2017) 3 and 5. www.eisourcebook.org (accessed 25 September 2021); J Fraser, 'Mining Companies and Communities: Collaborative Approaches to Reduce Social Risk and Advance Sustainable Development' (2018) *Resources Policy* 1, 2; J Caron, S Durand, and H Asselin, 'Principles and Criteria of Sustainable Development for the Mineral Exploration Industry' (2016) 119 *Journal of Cleaner Production* 215; UN Environment and UNDP (n 79) 27.

90 UN Environment and UNDP (n 79) 27; Fraser (n 89); de Haan, Dales and McQuilken (n 89).

91 Columbia Centre on Sustainable Development et al (n 87) 3.

92 *Ibid*; J Caron, S Durand and H Asselin (n 89) 427; UN Environment and UNDP (n 79) 12.

93 Columbia Centre on Sustainable Development et al (n 87) 27.

94 Collins and Lawson (n 11) 17.

c Access to technical support through government extension services which may comprise the provision of geological data and capacity building. For example, under the ACP-EU Development Minerals Project in African, Caribbean, and Pacific countries, PanAfGeo, and other international and local programmes/projects, capacity-building trainings are offered to duly formalized ASM operators.[95]

d Ensuring the security of minerals extraction. With insecurity, the allocation of Cadastre Units to formalized ASM operators can prevent frequent conflicts with LSMs over mining sites and help African governments to effectively track and monitor ASM activities.

2.7 Addressing Contemporary ASM Issues: Ethiopia, Ghana, and Nigeria as Cases Studies

2.7.1 Ethiopia

The exclusive ownership of all natural resources in Ethiopia is vested with the government on behalf of Ethiopians.[96] The *Mining Operations Proclamation*[97] is the chief legal instrument for administering and regulating the mining sector, including legalizing the activities of ASM[98] and requiring artisanal miners to be organized as mining cooperatives.[99] Ethiopia plans to reduce poverty with formalization and support for ASM by 2030.[100] Currently, more than 100,000 miners have been organized into about 500 mining cooperatives, out of at least 1,000,000 artisanal miners in Ethiopia.[101] The Ministry of Mines and Petroleum (MoMP) is the apex body for the general administration of the mining sector. However, regional mining bureaux are specifically in charge of issuing artisanal licences and small-scale mining licences, preferably to Ethiopian citizens, as well as ASM formalization and formation of mining cooperatives/enterprises.[102]

95 European Union (Capacity4dev, 31 March 2017). https://europa.eu/capacity4dev/articles/matter-sdgs-acp-eu-development-minerals-programme (accessed 20 September 2021); EuroGeoSurveys/PanAfGeo. www.eurogeosurveys.org/projects/current-projects/panafgeo (accessed 20 September 2021).

96 Constitution of The Federal Democratic Republic of Ethiopia, art 89(5).

97 Mining Operations Proclamation No. 678/2010 (as amended by Proclamation No. 813/2013).

98 Mining Operations Proclamation, s 9(4) and (5).

99 Mining Operations Proclamation, s 2(2).

100 Office of the Prime Minister, *Home-Grown Economic Reform Agenda: Unlocking Ethiopia's Economic Potential* (White Paper, Government of Ethiopia, August 2019) 13 and 28. www.mofed.gov.et/media/filer_public/0e/4c/0e4c0c11-8262-4e53-b395-0b1c0aa26976/ethiopia-_economic_reform_agenda_.pdf (accessed 29 September 2021).

101 Bekele Assefa, *JDSF Grant to Support to Improve the Economic, Social and Environmental Sustainability of Artisanal Miners Project on ESMF* (Report, Ethiopian Ministry of Mines, 2012) 15.

102 Beyene Tadess, *Artisan Mining Operation – Its Economic Values, Ethiopia* (Report, Extractive Industries Transparency Initiative, December 2015) 15.

Under the *Mining Operations Proclamation*, three categories of miners are recognized: artisanal miners, small-scale miners, and large-scale miners.[103] The first category is that which has gone through different levels of formalization, growth, and development under government supervision and support. The second and third categories are required by law to submit proposed work programmes and environmental impact assessments (EIA).[104] *The Mining Operations Proclamation* places great emphasis on incentivizing mining operations by guaranteeing security of tenure of miners.[105] Unfortunately, there are no existing implementation guidelines to fortify grassroots implementation of the Proclamation, and to create awareness create for miners and government staff. Under Section 63 of the *Mining Operations Proclamation*, royalties of up to 8% are paid by all three categories of miners, and the rates are determined by the relevant laws in existence at the region where the regional bureau is domiciled.[106]

The *Mining Operations Regulations (Amended) No. 124/2006*, the *Transaction of Minerals Proclamation No. 1144/2019*, and the *Cooperative Societies Proclamation No. 985/2016* are also important legal instruments for registering and regulating mining cooperatives through the mandate given to the Federal Cooperatives Agency (FCA). A roadmap called the *ASM National Strategy* has also been developed to foster formalization and responsible governance of ASM.[107] Also, the *Gender Equality Strategy* promotes women's rights in all sectors (including mining) in line with the *Ethiopian National Policy on Women*[108] and other policies for dealing with environmental challenges associated with ASM.[109]

The government has not done much to address infrastructural deficits in Ethiopia, especially in the rural communities. Therefore, there are still challenges regarding mining communities and transporting mined products as roads, railways, bridges, health facilities, schools, and water are acutely inadequate.[110] The government, working with development partners, needs to invest more in bridging these associated infrastructural gaps.

103 Mining Operations Proclamation, ss 26, 28 and 32.
104 Mining Operations Proclamation, ss 26(1)(c) and 28(1)(c).
105 Mining Operations Proclamation, s 4(4).
106 Mining Operations Proclamation, s 37.
107 Ministry of Mines and Petroleum, 'ASM in Ethiopia – Developing the ASM Sector' www.mom.gov.et/index.php/artisanal-and-small-scale-mining/artisanal-mining-asm-in-ethiopia (accessed 20 September 2021).
108 Transitional Government of Ethiopia, *National Policy on Ethiopian Women* (September 1993) 20 and 21. www.abyssinialaw.com/online-resources/policies-and-strategies/policies-and-strategies/national-policy-on-ethiopian-women (accessed 20 September 2021).
109 Environmental Protection Authority, *Environment Policy of Ethiopia* (2 April 1997) 13 and 14. www.informea.org/sites/default/files/imported-documents/UNEP-CHW-NATLEG-NOTIF-Ethiopia01-LAW.English.pdf (accessed 20 September 2021).
110 Ibid 2, 40 and 42.

Regarding international commitments on sustainable mining, Ethiopia has fully complied with EITI requirements as the licensing applications are processed through the Ethiopian Mining Cadastre (EMC) e-portal simply called 'LandFolio'.[111] Ethiopia, alongside other members of the Great Lake Region, in 2006 signed the *Pact on Security, Stability and Development* to curb illegal minerals trading.[112] Ethiopia is also a signatory to the *Minamata Convention on Mercury* but is yet to submit its instrument of ratification.[113]

2.7.2 Ghana

The *Minerals and Mining Act*[114] is the main legal instruments for governing mining activities in Ghana, and it recognizes only small-scale miners[115] and not artisanal miners. Under Section 1 of the Act, ownership of every mineral in its natural state is with the government, while under Section 9, all persons desiring to engage in mining activities require mineral licences granted by the Minister in charge of Mines. Sections 82(1) and 83 of the Act set out conditions for obtaining small-scale licences. Small-scale miners are expected to, among others, pay 5% royalties, application fees, and income tax, and get concessions for import duty for plant, machinery, and equipment. While these requirements are favourable, the major weakness of this law is that the roles of traditional rulers, the custodians of mine host communities, are not adequately recognized by law. Thus, there is need to review these laws.[116]

Specialized institutions such as the Minerals Commission (MC), Environmental Protection Agency (EPA), Precious Metals Marketing Corporation (PMMC), and Water Resources Commission (WRC) are responsible for licensing and regulation, environmental management, official marketing of gold produced by small-scale miners at competitive prices, and management of water resources, respectively. The MC in 2008 provided about $US700,000

111 Ministry of Mines and Petroleum (MoMP), 'E-EITI' (2020). www.mom.gov.et/index.php/international-cooperation/e-eiti (accessed 27 September 2021).

112 International Conference on the Great Lakes Region (ICLGR), *General Overview of the ASM Legislations in the Great Lakes Region (Ethiopia and Ghana as Benchmark Countries)* (Report, ICLGR, July 2017) 27.

113 Convention on Mercury: Parties and Signatories. www.mercuryconvention.org/en/parties (accessed 27 September 2021).

114 Minerals and Mining Act 703 of 2006.

115 Minerals and Mining Act 703 of 2006, s 82.

116 World Bank Group, *Ghana's ASM Policy and Regulatory Regime with Focus on Gold* (PowerPoint Presentation, National Conference on ASM in Ghana: Third World Network- Africa, Accra, Ghana, 14–15 May, 2019) 17. http://twnafrica.org/wp/2017/wp-content/uploads/2019/05/National-Conference-on-ASM-in-Ghana.pdf (accessed 27 September 2021). Note that under section 92(2)(d) of the Minerals and Mining Act, the traditional council in a designated mining area can nominate one person to be a member of the Small-Scale Mining Committee to promote mining in the designated area. This is the farthest that traditional authorities are involved in mining governance.

to assist several mining cooperatives and associations such as Tarkwa Small-scale Mining Association, Talensi-Nadbam Cooperative, Ghana National Association of Small-scale Miners, and Women In Mining, especially in purchasing simple and mechanized equipment.[117] Regarding dialogue among stakeholders, the MC brings together stakeholders on an annual basis for the purpose of policy dialogue. To resolve conflicts, the MC set up a Community and Conflict Resolution Desk to seek ways to mediate problems associated with mining companies (including small-scale companies) and local communities.[118] This unit combines different conflict resolution strategies such as litigation, arbitration, and other alternative dispute resolution mechanisms to resolve land use conflicts.

There are other highly commendable attempts to comply with EITI requirements, especially with the launching of the online data repository called the Mining Cadastre Administration System (MCAS) developed in 2015 to help with better licensing and management of mineral titles and improve access to information, all with a view to promoting accountability and transparency. The *RD Quantum Mobile App* was also developed to equip officers with updated technology for giving quick field reports and to access information from the data repository using handheld devices. The Community Mining Project was initiated by the government to better manage *Galamseys*who were initially outlawed and treated as illegal miners through the failed 'Operation Vanguard'.[119] The project helped with the lifting of the ban on *Galamseys*.

2.7.3 Nigeria

In 2016, the Nigerian government, through its Ministry of Mines and Steel Development, rolled out a *Roadmap for the Growth and Development of the Nigerian Mining Industry* ('the Roadmap')[120] to address the challenges facing the mining sector and proffer a sustainable pathway to optimally harness Nigeria's mineral resources for rapid socio-economic growth and development.[121] The implementation of the Roadmap is leading to some form of resurgence in the Nigerian mining industry (NMI), but the few positive results are yet to make the desired level of impact. Economic development, social inclusion, environmental sustainability, and good governance are the pillars of

117 'Ghana's ASM Profile' (ECA/AMDC). https://knowledge.uneca.org/ASM/Ghana (accessed 27 September 2021).

118 Ibid.

119 Ibid 7.

120 Ministry of Mines and Steel Development, *Roadmap for the Growth and Development of the Nigerian Mining Industry 2016 – On the Road to Shared Mining Prosperity* (2016) ('The Roadmap') <http://www.minesandsteel.gov.ng/roadmap/>.

121 Oxford Business Group, *New Roadmap for Nigeria's Mining Industry* (2018) <https://oxfordbusinessgroup.com/overview/uncovering-potential-new-roadmap-aims-to-help-industry-capitalise-its-substantial-and-diversified>.

sustainable development.[122] This section now attempts to view the Roadmap through the lens of these pillars in a bid to identify and address any sustainable development issues or gaps.

2.7.3.1 ECONOMIC DEVELOPMENT

The Roadmap is a strategic plan to make the NMI globally competitive and contribute to wealth and job creation, drive rapid industrialization, and enhance social security.[123] Although nowhere in the Roadmap was there any allusion to the SDGs, this strategic plan largely reflects the economic development aspirations of the SDGs, particularly SDG 8 (decent jobs and economic growth), SDG 9 (industry, innovation, and infrastructure), and SDG 12 (responsible consumption and production). Mining helps to create jobs for local people in host communities and, through trainings and other capacity-building opportunities, develop their skill sets in various core mining roles as well as associated services. Small- and medium-scale enterprises such as petty traders, food sellers, and stationery service providers also depend on and support mining projects.[124] Mining also helps to develop transport, energy, communications, water, renewables, and waste management infrastructure, which benefit both mining projects and local communities from their shared use.[125]

The Federal Government recognizes that there is an impending global boom in mining investments and regards this boom circle as a good opportunity, using the Roadmap to position Nigeria ahead to benefit economically from its mineral potentials in the coming years.[126] To achieve these aims, the Roadmap sets strategies for attracting local and international investment in the NMI. Emphatic on ensuring a viable tax regime for efficient revenue collection from mining, it advocates the enactment of a *Mining Income Tax Act* for a mining income tax, which will eliminate double taxation and ensure efficient and transparent collection of taxes and royalties from mineral licensing and trading.[127] To this end, the Federal Government has identified and is currently marketing Nigeria's bauxite, bitumen, coal, gold, lead-zinc, limestone, and tin potentials.[128] All these measures are geared towards generating higher revenues for the government.[129]

122 SME (n 80) 2–3.
123 The Roadmap (n 120) 8.
124 Columbia Centre on Sustainable Development et al (n 87) 4; UN Environment and UNDP (n 79) 18.
125 Columbia Centre on Sustainable Development et al (n 87) 4; UN Environment and UNDP (n 79) 18.
126 The Roadmap (n 120) 96.
127 Ibid.
128 Ibid 39 and 51.
129 Ibid 96.

However, the Roadmap is silent on the issue of the minimum share of revenue from minerals that the government will channel towards infrastructural development to support mines and host communities. We submit that it is only fair and equitable for priority to be given to the development of associated mine infrastructure with mineral revenues because both mining projects and host communities will benefit simultaneously.[130] The government would achieve development objectives faster if it continuously leverages mineral revenue to develop infrastructure. Lack of necessary infrastructure can negatively affect investor confidence.

2.7.3.2 SOCIAL INCLUSION

Achieving social development through mining entails the cooperation of stakeholders, particularly the mining companies and the host communities, to identify the positive and adverse impacts of the mining projects in the host communities, and participate and make collective decisions on how to address any issues arising from these impacts.[131] Mining operations should be sensitive to issues of inequality and ensure the inclusion of disadvantaged and vulnerable groups such as women and the physically challenged and resolution of mining-related disputes through corporate social responsibility initiatives. Mining companies should have respect for human rights and indigenous peoples' rights.[132] The Roadmap requires the government to encourage mining companies to negotiate and respect community development agreements (CDAs) with the host communities, seek the free, prior, and informed consent (FPIC) of host communities to embark on mining operations on their lands and give between 7% and 10% equity shares in mining projects to host communities as consideration for their FPIC.[133] These will go a long way in reducing social inequalities and ensuring peace leveraging on the attendant economic benefits of mining.

2.7.3.3 ENVIRONMENTAL SUSTAINABILITY

Mining is responsible for causing negative impacts on land, water, flora, fauna, climate, and environment, and the lives and livelihood of peoples that rely on these natural resources for survival.[134] However, mining can be cultured to reduce its negative impacts and enhance environmental protection and the lives and livelihoods of host communities. These positive impacts are achievable where mining objectives of a country are keyed into the SDGs

130 Centre for Sustainability in Mining and Industry (CSMI), University of the Witswatersrand, 'Good Practice Note for Public Infrastructure and Mining' (EI Sourcebook, CSMI, University of the Witswatersrand, 2010) 6, 11 and 12, www.eisourcebook.org; J Fraser (n 89) 3 and 9.
131 Ibid; UN Environment and UNDP (n 79) 19.
132 Columbia Centre on Sustainable Development et al (n 87) 4.
133 Ibid 59.
134 Columbia Centre on Sustainable Development et al (n 87) 4; R Eggert (n 82) 225.

aimed at protecting the environment.[135] Particularly, SDG 6 (clean water and sanitation), SDG 15 (life on land), SDG 7 (affordable and clean energy), and SDG 13 (climate action) have direct bearings on environmental protection and sustainability.[136] Mining requires land, water resources, and energy, and these activities come with their negative environmental impacts, yet they can be leveraged to reduce their negative impacts and ensure greater efficient and sustainable natural resource and energy use.[137]

The Roadmap contains far-reaching strategies to enhance environmental sustainability, including improving statutory and regulatory enforcements of extant environmental laws and liaising with other relevant agencies to enforce them;[138] establish and implement an Environmental Protection and Rehabilitation Fund;[139] and set up a Mines Field Police and Special Mines Surveillance Taskforce to effectively police mines sites, check illegal mining activities, and prevent under-reporting and tax avoidance.[140] The Roadmap provides a general statement to work with other government ministries to ensure that necessary support infrastructure to aid the growth of the NMI will be included into relevant national plans and budgets.[141]

Going forward, the Ministry of Mines and Steel Development should seek to promote renewable energies as part of efforts to reduce carbon emissions from mining. Relevant government ministries, departments, and agencies in charge of national planning, energy, petroleum, and environment would be key stakeholders in this regard.[142]

2.7.3.4 GOOD GOVERNANCE

Achieving socio-economic development and environmental sustainability through mining will be a Herculean task, if not a Sisyphean task, without strong governance structures.[143] These include sound regulatory and policy frameworks; competitive tax regimes and incentives; institutional transparency and accountability; adequate checks and balances; all-inclusive participation in resource governance; and access to information and justice.[144] In order to reduce bureaucracy and enhance efficiency, the Roadmap proffers the creation of a new 'super' regulatory agency for the NMI by merging the Mining Cadastral Office (MCO), Mines Inspectorate Department (MID), Mines Environmental

135 International Council on Mining and Metals (ICMM), *Role of Mining in National Economies* (Report, ICMM, 3rd ed, 2016) 60, www.icmm.com/romine/index (accessed 26 September 2021).
136 Columbia Centre on Sustainable Development et al (n 87) 4.
137 Ibid.
138 Ibid 56.
139 Ibid.
140 Ibid 57.
141 Ibid 67.
142 Ibid 49–50 and 67.
143 SME (n 100) 2–3.
144 Ibid.

Compliance Department (MECD), and Artisanal and Small-Scale Mining Department (ASMD) within a two-year timeline; establishing a Solid Minerals Development Fund to finance mining regulatory or joint venture activities; establishing a clear separation of powers between federal and state governments regarding mining regulation and taxes; ensuring continuity and consistency of programmes; enhancing Nigeria's geological survey databank; and formalizing the participation of ASMs in the mining sector.[145] Since the launch of the Roadmap in 2016, this 'super' agency is yet to be established. It is a major setback for the NMI because investors who want to see a strong institutional governance framework will be dissuaded from investing their capital in an uncertain or weak regulatory space. It is essential that this proposed new regulatory agency is quickly established. The mining sector needs strong institutions to be viable and sustainable, and this proposed regulatory agency is key to achieving this. The Roadmap recognizes this position but it is now a question of implementation.

3 The Need for Proper Organizational Structures in ASM Operations

ASM operations are small-scale businesses and are confronted with several challenges revolving around human capital, technological, legal, finance, marketing, and organizational issues.[146] The plurality of ASM operations do not have the benefit of local workers with the modern skills to support mining operations.[147] Small-scale businesses offer socio-economic benefits to their host communities as they are a source of income and employment. However, ASM operations are vulnerable small-scale businesses because of the lack of strong managerial skills and a lack of a sustainability strategy.[148] For any growing business, the organizational structure is an essential way of providing direction and clarification on concerns of organizational functions such as production, marketing and human resources. Finding the appropriate organizational structure for ASM operators can be very challenging as they require proper guidance to get it right. The ultimate consequence of setting up a proper business organizational structure in ASM ventures is to formalize their operations and integrate them into the global best practices.

3.1 Proposed Functional Organizational Structure for ASM Operations

Figure 5.2 depicts three levels of hierarchy in the business organizational structure: Level 1 –owners/partners; Level 2 – managers; and Level 3 – unit workers in sales, public relations, budget, financial accounts, production, maintenance,

145 The Roadmap (n 120) 43–46.

146 Thomas Hentschel, Felix Hruschka, and Michael Priester, *Artisanal and Small-Scale Mining, Challenges and Opportunities* (Working Paper, International Institute for Environment and Development, 2003) 7.

147 Pontsho Ledwaba Pontsho and Nellie Mutemeri Nellia, *Preliminary Study on Artisanal and Small-Scale Mining in South Africa* (Report, Open Society Foundation of South Africa/Centre for Sustainability in Mining and Industry, 2017) 14.

148 Ibid 26.

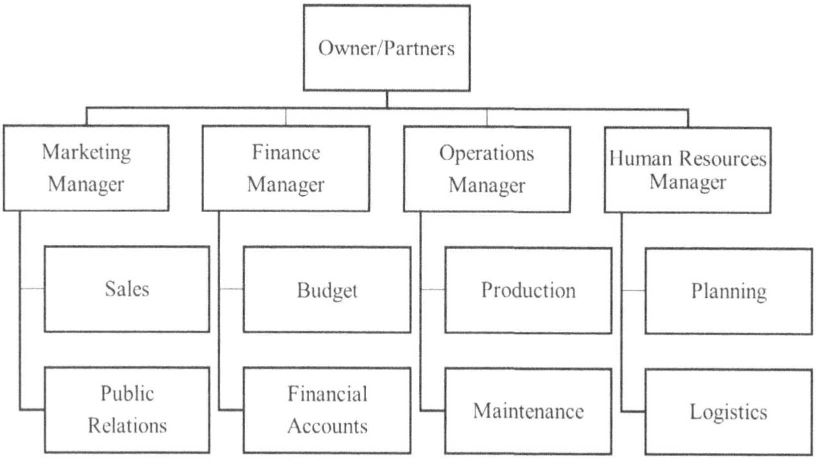

Figure 5.2 Proposed simple model organogram for ASM Businesses

planning, and logistics. The line of communication in the functional organizational structure is vertical; information travels from the top to the bottom of the structure. Managers communicate with the owner or partners of the organization and pass down the information to the workers. The operations department ensures that all the equipment and workers are functional. They also have to enforce quality measurements on the product that they are processing. The human resources department will provide the staff and the necessary training required for the tasks. The finance department will provide financial data analysis of the production unit, determine whether there are enough funds to hire and pay new staff members, purchase new equipment, and cater for contingencies. The marketing department must fully understand the prospective players and customers in market for the mineral ores being mined by the company before embarking on strategic marketing and advertising actions. Thus, it is clear from the foregoing that all departments heavily depend on one another for a successful ASM organizational and business output.

3.2 Government Support for ASM Organizational Structures

The African Mineral Development Centre (AMDC)[149] developed a guidebook on domesticating the AMV,[150] including aspects related to ASM,[151]

149 A technical partnership between the African Union (AU), African Development Bank (AfDB), and United Nations Economic Commission for Africa (UNECA).
150 African Minerals Development Centre (AMDC), *A Country Mining Vision Guidebook: Domesticating the Africa Mining Vision* (Guidebook, 2014), www.academia.edu/31058042/A_COUNTRY_MINING_VISION_GUIDEBOOK_Domesticating_the_Africa_Mining_Vision_African_Minerals_Development_Centre?email_work_card=view-paper (accessed 22 September 2021).
151 Ibid 107–126.

and, following suit, certain African countries have developed their own training handbooks for artisanal and small-scale miners within their jurisdictions. These training manuals generally give guidance on economic, social, and environmental considerations while harnessing the potential of ASM in their respective countries. These training handbooks also contains information for the education of ASM operators about more productive and safer mineral processing techniques, and guidelines for securing mining and operating licences.

Government support for ASM operations, aside from organizational structure modelling, has been made in the areas of training and capacity building as well as enforcement, monitoring, and evaluation. Many mineral-rich countries in Africa have enacted laws or created policies that encourage or mandate the training and education of stakeholders in the ASM sub-sector. For example, the Nigerian MMA provides that the Solid Minerals Development Fund under the ministry in charge of regulating matters relating to the exploration and development of mineral resources shall be responsible for extension services including the teaching of adequate skills in mining to small-scale and artisanal miners and regularly introducing them to new mining technology.[152] While these trainings usually border on the economic potentials of ASM, they are increasingly emphasizing the importance of environment, health, and safety in ASM. ASM operators are increasingly educated about life-threatening health risks and environmental degradation that can have dire consequences for food and water security among others.[153] Governments also try to set up enforcement, monitoring, and evaluation mechanisms for the ASM sub-sector.

Formalization is one of the issues that tops the ASM development agenda. In our view, the creation of organizational business structure models for ASM operations is integral to the formalization process. The Ghanaian government created an agency to drive the formalization of ASM operations and have recorded commendable successes in its implementation drive to the extent that over 1,300 local ASM operators have been formalized, creating direct and indirect employment for about a million Ghanaians.[154]

3.3 Organizational Business Structure for ASM Operators: Challenges and Recommendations

ASM businesses are mostly poorly and informally organized. They do not have a formal structure in place that will guide their engagement with external parties such as the government and host communities. It is recommended that ASM operators be assisted to formalize and then draw up and implement management plans which will make them more attractive for direct

152 MMA, ss 34(1)(e) and 91(e).
153 AMDC (n 151) 113–114.
154 Ibid 121–122.

investment, instil more business operation certainty, and attain higher productivity in mineral exploration driven by a synergy of departments and units under the ASM operators.

ASM operators do not usually have access to major markets or required marketing skills for trade in mineral commodities compared to LSM operators. It is advised here that ASM operators can access funding from creditors such as angel investors, venture capitalists, and crowd funders, but this would only become feasible and necessary upon undergoing the process of formalization. This is because an organized business structure instils confidence in investors to risk their capital for corporate growth and feasible returns on investment.

The directors of small-scale operations are often not trained or inadequately trained in the area of business management and administration skills within the context of global best practices. It is recommended that governments continue to assist ASM operators to build administration and management capacity and improve financial and economic literacy by linking them with funding for trainings and other extension exercises necessary to maximize their socio-economic potential.

Due to the often-subsistent nature of ASM, capital and revenues generated from such operations are meagre and not enough to expand mining operations into large-scale operations towards optimal mineral extraction. Local and international banks and investors are often loath to lend to ASM operators due to non-formalization and non-integration into the mainstream of the mining industry. It is recommended that to reverse this trend, ASM operations would need to be formalized in order to access available assistance. It is clear that the challenges of ASM across African countries and the solutions thereto are heavily linked to the extent of formalization of the sub-sector within their respective jurisdictions.

4 The Way Forward

Perhaps the argument for ASM as a socio-economic empowerment opportunity in developing countries is now hackneyed. Therefore, it is undesirable and potentially counterproductive for countries to completely force its closure. Recognizing its importance, governments, development partners, and civil society organisations (CSOs)/non-governmental organizations (NGOs) need to increase the support they offer, especially in the areas of capacity building in minerals development and improved health, safety, and environment (HSE) practices in the sector. However, the effectiveness of such interventions relies heavily on the formalization of the ASM sector.

This chapter has iterated the widely agreed position that the key to advancing ASM in Africa is that African governments recognize it as a crucial subsector bearing huge potentials for the continent. It observed that there is a gradual but steady shift towards formalization of ASM in Africa as many African governments are beginning to see the benefits that the sub-sector offers. However, there is still a lot of work to be done in this regard because the formalization of ASM goes beyond recognition and regulation. To turn this prospect into prosperity, it requires effective strategies to drive the integration of ASM into the formal mining sector and the larger economy. For this to happen, the following recommendations are pertinent.

National governments, sub-national governments, and development partners must genuinely shift their perception of ASM from being illegal and a nuisance to a sector that makes legitimate and meaningful contributions to livelihoods and to local economies. This requires them to be determined and focused on solving its problems, rather than treating it with levity. One way to do this effectively is for African governments and development partners to create, review/update, and be committed to implementing workable national legislation and policies geared towards addressing ASM issues, especially those around formalization and how its barriers could be overcome.

Robust policies and guidelines for managing conflicts arising from or connected with ASM activities must also be put in place and ombudsmen appointed to address grievances and resolve disputes. Incentives need to be provided to encourage ASM formalization through the establishment of mining cooperatives, associations, and enterprises. This can be in the form of grants, loans, and platforms for mineral trainings that exist already in countries such as Nigeria, Ghana, and Ethiopia to ensure ASM operators are able to sell their products to private companies and the government at competitive prices. The importance of this is that it helps address one of the biggest known challenges ASM faces – access to markets.

In addition, all African governments engaged in ASM must be resolute in making adequate budgetary allocations to ensure the delivery of quality extension services, especially with respect to identifying geological formations that are rich in mineral resources, assessing their economic viability, and delineating them into Cadastre Units for allocation to ASM operators. There must be renewed commitment to training and retraining of professionals involved in formalizing and regulating ASM, and adequate logistical support provided to ensure their jobs are performed at optimal levels. Institutional capacities have to be improved and the challenge of corruption in public institutions and regulators in the mining sector properly addressed.

ASM operators need to receive adequate training, especially on HSE, safer mining techniques, use of PPE, gender equity, child labour, entrepreneurship, value addition, etc. through the extension services programme of the relevant government agencies. There is also the need to increase investments in baseline studies and data collection, storage, and management to ensure proper planning for ASM, and also in research and development, especially in terms of creating alternative technologies that are more environmentally friendly.

African governments and development partners can also support ASM operations through the promotion and distribution of alternative and safer mining technologies, enabling access to market with initiatives such as mineral buying and processing centres, equipped lapidaries, training centres, delineation of exclusive ASM zones and their allocation to formalized miners, and the provision of bankable geodata following geological and resource estimation efforts free of charge or with an opportunity to make repayment over a reasonably wide window to the duly formalized miners. In such special ASM exclusion zones, there is need for the provision of infrastructure (especially water, power, housing, and roads) and creation of enabling environments for service providers to exist.

African governments need to invest massively in ensuring the protection of local communities where ASM activities take place, similarly to the attention paid to LSM areas – for example, a mines police (those saddled with the responsibility of providing security and monitoring of mining areas) – and supporting them with appropriate modern technologies such as drones and detective equipment to ensure ASM operations are effectively tracked and managed. These personnel must be trained in such a way that they are able to aggressively dominate their areas of responsibility and check any disruptive activities that could scare the investor away, while also staying focused on protecting genuine ASM operations.

For formalization efforts to succeed, national governments and their development counterparts must realize that the best approach to success lies with broadening and levelling the field to allow enhanced communication and information sharing via conferences, forums, workshops, and routine networking events where the miners and the local host communities actively participate in discussions. To this end, local knowledge on resource management must be respected and sought to incentivize the effort. Forums for networking, data gathering and management, and information sharing between countries must be set up and well funded to ensure ready access to information on current issues and sharing of effective strategies to manage the mining sector. Such platforms would also allow stakeholders to share strategies on ASM formalization which can benefit any government authority saddled with the responsibility of administering and/or regulating the ASM sector in a given country.

African countries and institutions must collaborate to create, where necessary, and push for the implementation of continental mineral governance frameworks such as the AMV, sector-specific standards, certification, and protocols relevant to the African mining ecosystem. African countries and institutions must begin to collaborate in designing and managing continental systems for mining, processing, storing, beneficiating/value addition, and sales of mineral resources derived from ASM, as these must not be limited to LSM only. Where possible, these systems can be modelled around international systems such as Mines to Markets, the Kimberly Process, and the BGR Certified Trading Chains Scheme to expand the sub-sector.

All African countries should also endeavour to expand continental platforms that allow for peer review. ASM formalization is a ripe developmental issue for the African Peer Review Mechanism (APRM), an African Union platform where African countries can gain access to understudy each other and, learning from these experiences, improve on processes and policies for a given developmental issue.[155]

Private consultants, NGOs, development partners, institutions (e.g. UNECA) and those focused on research (e.g. African universities and peak bodies) need to be allowed and empowered to participate by providing technical support towards the overall management of the ASM sub-sector. African countries that have not done so already can follow the examples set by Ghana by mandating mining companies and cooperatives to set up special funds for land reclamation which can be used for mitigating the environmental impacts of ASM. In a country such as Nigeria, which set up funds such as the Natural Resources Development Fund and the Solid Minerals Development Fund to help with optimization of economic contributions of the sector to the national economy and training, robust mechanisms need to be put in place to make them more effective. ASM needs to receive special priority when making budgetary allocations to such funds.

Public participation in the mining sector is key to the formalization process. Avenues must be created for advocacy and awareness campaigns to effectively educate local communities and ASM operators on the management of issues relating to ASM. These avenues need to be independent and transparent to enable the voicing of grievances, and redress when wrongs are perceived to have been done.

African toolkits containing methods for data collection, analysis, reporting, and management need to be designed and made fit for purpose. This means that such toolkits would made taking into cognizance the cultural and social practices of local host communities. The importance of this is that the drivers of formalization efforts would deploy strategies that recognize the protocols associated with the 'unshakeable' hierarchical structures when visiting African elders to inform them of a team's mission aimed at addressing ASM challenges. This would be in accordance with the practices of most Africans and would help the locals welcome the efforts at formalization. The advantage of local systems such as this is that they speak directly to the hearts and minds of the people affected. ASM can best be managed when context and area-specific frameworks are put in place to help address its challenges. Formalization can work best when systems that are African or country-specific are in place. The miners can equally recognize the 'language' these systems speak. To this end, appropriate government authorities in African countries must work to domesticate and implement the AMV, and push aggressively for formalization using strategies outlined therein, as well as other relevant guides designed by other organizations.

155 African Union, 'The African Peer Review Mechanism', https://au.int/en/aprm#:~:text=APRM%20is%20a%20voluntary%20arrangement,economic%20growth%20and%20sustainable%20development.&text=democracy%20and%20political%20governance (accessed 22 September 2021).

6 Artisanal and Illegal Mining in Nigeria's Ungoverned Spaces

New Frontiers of Social Risks

David Yibakuo Amakiri and Solomon Ugochukwu Ezike

1 Introduction

Nigeria is rich in mining and mineral resources. These include a host of metallic and non-metallic minerals such as baryte, bitumen, gold, iron ore, limestone, lead/zinc, coal, tin, silver, granite, marble, topaz, sapphire, lithium, etc.[1] There are currently many states in Nigeria where these mining activities are taking place, but the majority of these are found in the northern states of Kebbi, Kogi, Nasarawa, Niger, Plateau, and Zamfara, and the southern states of Osun, Oyo, and Enugu.

However, the absence of policy clarity, legal and regulatory conflicts, inaccurate geological data, lack of infrastructure, illegal and artisanal mining, corruption, diverse forms of banditry, and kidnappings have loomed over the mining landscape in Africa, and Nigeria's in particular. Despite ambitious developmental plans to have 50 mines in operation by 2023,[2] agitations by host communities for inclusion and environmental protection, among others, brew on. Specifically, environmental concerns and social acceptability of the exploitation of minerals are issues that have remained ill-addressed in Nigeria. The sector is also vastly underdeveloped, informal, and non-standardised. These issues contribute significantly to the negative social impacts of mining in Nigeria, leading to the loss of royalties and taxation revenues for the government[3] and means of livelihood for the host communities.

1 For a full list of the available mineral resources in Nigeria, see publication of the Nigerian Investment Commission on page 2. Available at https://nctig.org/wp-content/uploads/2020/01/Mining-in-Nigeria.pdf, accessed 2 April 2021.

2 Kayode Fayemi and Chi Onwurah, 'Nigeria's Solid Minerals Sector: Alternative Investment Opportunities' [2016] Chatham House (Royal Institute of International Affairs) 1; Paul Carsten, 'Nigeria Says Can Tap Mineral Wealth despite COVID, Unrest' (Reuters, 2020), www.reuters.com/article/us-nigeria-mining/nigeria-says-can-tap-mineral-wealth-despite-covid-unrest-idUSKBN25T1K4, accessed 2 April 2021.

3 Oxford Business Group, 'Nigeria's Untapped Mineral Reserves Could Aid the Goal of Economic Diversification' (2016), https://oxfordbusinessgroup.com/overview/digging-deep-significant-and-largely-untapped-mineral-reserves-could-aid-goal-economic, accessed 2 February 2021.

DOI: 10.4324/9781003284437-8

This chapter tackles these issues, highlighting the role of government and steps that need to be taken to protect communities and legitimate miners against these highlighted social conditions. It also addresses the government's critical role in attending to the interest of host communities, the interest of legitimate miners, and national interests against illegal mining activities which damage the environment and economy.

We use the concept of ungoverned spaces to demonstrate the link between ineffective and limited government presence in host communities, and illegal mining activities in those areas. We also theorise that socio-environmental risks – which pertain to the negative impacts of social risks on the environment – play a central role in the escalating insecurity in many mining communities. In doing this, our research is an attempt to answer the following questions:

- i What are social tensions created by Artisanal and Small-Scale Mining (ASM) in Nigeria?
- ii What is the nexus between social risks and insecurity in the areas where mining is prevalent in Nigeria?
- iii How are the challenges associated with ASM and social risks related to the problems in the legal framework for mining in Nigeria?

The chapter addresses the above questions in six sections. After this introduction, Section 2 reviews existing literature on the concepts that will underpin our discourse, which include Nigerian mining investment risks, social risks of the mining industry, and ASM. Section 3 draws the causal link between ASM, social risks, and insecurity in mining communities in Nigeria. Section 4 examines the legal and licensing framework of Nigeria's mining sector and its adequacy or otherwise in protecting the industry against social risks. Section 5 proposes various policy options that could assist the government in protecting the mining communities on the one hand and the mining industry on the other. Section 6 concludes the chapter and suggests areas where further research is necessary.

2 Analysis of Underlying Concepts

To appreciate how the social impacts of ASM and illicit mining affect the security situation of host communities, we first lay down the conceptual basis upon which this chapter uses the terms 'ASM' and 'social risks' as risks to the mining industry. In other words, what are ASMs? Why do most mining activities carried out in Nigeria fall under ASM and how do these activities portend risks for the larger investment environment? What are these social risks, and how do these risks manifest in mining activities? Additionally, what is the current social situation in key mining areas in Nigeria and how does this impact mining investments?

Artisanal and Small-Scale Mining

ASM refers to mining operators who use their own equipment and facilities to conduct mining activities on a small scale. These miners are usually subsistent or self-employed.[4] Illegal mining, however, implies mining activities that are undertaken without state permission, land rights, mining licences, or mineral transportation permits. It can be subsistent in nature (like ASM) but it can also manifest in large-scale activities linked to organised crime.[5]

ASM activities play an important role in the economy of developing countries, supporting an estimated 100 million people globally.[6] Both Nalule[7] and Hilson and Maconachie[8] estimate that there are around 500,000 artisanal and small-scale miners in Nigeria dealing in gold and other precious metals, who support over two million dependants (these dependants include women who are sometimes forced to smuggle metals and children who are coerced into child labour[9]). According to Adeyemi and Olagunju,[10] 95 per cent of mining activities in Nigeria are artisanal in nature, while a further 95 per cent of these ASMs are illegal. In a study funded by the European Union, this latter figure is put at 80 per cent of all mining activities.[11] That is, eight out of every ten mines in Nigeria are potentially illegal due to the fact that they are undertaken without state permission, land rights, mining licences, and transportation permits.

This shows that there is a sizeable body of ASM activities in Nigeria that potentially pose risks to the social structure of local communities in states where ASM and illicit mining is prevalent. In contrast to the challenges of social risks, ASMs are not inherently unlawful except where they are mixed with illegal activities. But their existence and prevalence, if unchecked, could lead to social

4 Victoria R Nalule, 'Social and Environmental Impacts of Mining' in *Mining and the Law in Africa* (Springer International Publishing 2020), http://link.springer.com/10.1007/978-3-030-33008-8
5 Maurice Ogbonnaya, 'Illegal Mining and Rural Banditry in North West Nigeria' (ENACT Observer 2021), http://enact-africa.s3.amazonaws.com/site/uploads/2020-11-19-illegal-mining-policy-brief.pdf, accessed 20 April 2021.
6 ICMM, *Minerals Taxation Regimes a Review of Issues and Challenges in Their Design and Application.* (Commonwealth Secretariat and the International Council on Mining and Metals 2009)
7 Nalule (n 4).
8 Gavin Hilson and Roy Maconachie, 'Formalising Artisanal and Small-scale Mining: Insights, Contestations and Clarifications' (2017) 49 *Area* 443.
9 YM Ahmed and ED Oruonye, 'Socioeconomic Impact of Artisanal and Small Scale Mining on the Mambilla Plateau of Taraba State, Nigeria' (2016) 3 *World*; Adriana Eftimie and others, 'Gender Dimensions of Artisanal and Small-Scale Mining: A Rapid Assessment Toolkit'. See also Maurice Ogbonnaya, 'Women and Girls in Illegal Mining in Northern Nigeria' (ENACT Observer 2020), https://enactafrica.org/enact-observer/women-and-girls-in-illegal-mining-in-northern-nigeria, accessed 20 April 2020.
10 Sesan Adeyemi and Ayodele Olagunju, 'Putting Small-Scale Mining in Perspective: An Analysis of Risk Perception of a Southwestern Nigerian Community' (2017) 2 *Review of Social Sciences.*
11 Ogbonnaya (n 5).

mishaps, illegal mining, and a breakdown of law. With most mining occurring illegally, there is a strong case for the causal link between ASM, illegal mining, and insecurity in Nigeria. This link will be addressed in the next section.

Investment Risks in Nigerian Mining

Investment risks in Nigeria's mining sector relate to, among others, (i) the negative effects of political interference in the regulatory process or with the mining activity unilaterally by state actors; (ii) lack of administrative capacity to effectively regulate the industry; (iii) lack of sufficient technical and geological data to attract bankable mining projects; (iv) environmental concerns and the risks of negative social impact as a result of mining activities. The chapter focuses on this last category of risk – that is, the social risks to and of mining activities in Nigeria and the role and responsibility of the national government to ensure that these risks do not continue to fester creating a negative environment for stakeholders.

Social Risks of Mining

The concept of social risks in the extractives industry began to gain traction in the literature from the beginning of the last decade.[12] It has developed from merely concerns about the social acceptability of mining projects to more complex issues of a varied range.[13] Academic and grey literature see social risks from two broad perspectives: risks from the communities to the mining investment, and risks from the mining activity to the communities.

The first view is that social risks are the potential negative impacts on a project which result in communities' resistance, challenges by stakeholders, violence, or external third-party actions.[14] It has been suggested that this approach serves to link social risks with business risk and, therefore, the investor's financial bottom line.[15] As a result, this viewpoint ignores vital community interests.

12 Susan Joyce and Ian Thomson, 'Earning a Social Licence to Operate: Social Acceptability and Resource Development in Latin America' (2000) 93 *CIM Bulletin* 49.

13 Ibid. Joyce and Thompson said: 'Social risk is now an aspect of all mining activities which needs to be calculated into project assessments, and to be managed throughout the life of a project. While social risk has a number of aspects at the corporate level, such as company image and shareholder relations, at the level of the individual project, its primary component is social acceptability, and the principal tool for managing it is the Social Licence to Operate.'

14 Rio Tinto, 'Social Risk Analysis Guidance Note' [2011] Rio Tinto, Melbourne; Mary Anne Barclay, Daniel M Franks, and Catherine Pattenden, 'Risk Communication: A Framework for Technology Development and Implementation in the Mining and Minerals Processing Industries'; Barrick Gold, 'Community Relations Standard' [2012] Unpublished corporate document.

15 Geordan Graetz and Daniel M Franks, 'Conceptualising Social Risk and Business Risk Associated with Private Sector Development Projects' (2016) 19 *Journal of Risk Research* 581.

A second perspective sees social risks as suspected or anticipated challenges and the unwanted social effects on communities resulting from the social transition(s) precipitated by the actions of mining companies, financiers of mining projects, government entities, and non-governmental organisations.[16] This perspective focuses on activity-to-community impacts and it prioritises community interests over those of the business. A number of researchers[17] have conceptualised social risk in this way.

In addition, there exists other literature that conceptualises social risks to encompass both perspectives: from the mining activity to the local communities as well as acts of resistance or violence from these communities to the mining activity. Bekefi et al.[18] in a paper for Harvard University's Corporate Social Responsibility Initiative, conceived social risk as 'challenges by stakeholders to companies' business practices due to real or perceived business impacts on a broad range of issues related to human welfare – for example, working conditions, environmental quality, health, or economic opportunity'.[19] In other words, social risks are risks to individuals and groups arising from development interventions, as well as those actions of individual and group actors that challenge business practices.[20]

These conceptualisations show that social risks could manifest as resistance, violent attacks,[21] kidnappings, legal challenges from host communities or groups, etc. But they could also be human rights infringements, destruction of sacred sites and displacement of aboriginal people,[22] distortion of local economies, inequitable distribution of benefits, etc. Graetz and Franks[23]

16 Ibid.

17 Ibid; Deanna Kemp, Sandy Worden, and John R Owen, 'Differentiated Social Risk: Rebound Dynamics and Sustainability Performance in Mining' (2016) 50 *Resources Policy* 19; David Brereton and Joni Parmenter, 'Water, Communities and Mineral Resource Development – Understanding the Risks and Opportunities', Water in Mining Conference, Brsibane, Australia (2006).

18 Tamara Bekefi, Beth Jenkins and Beth Kytle, 'Social Risk as Strategic Risk' (2006) 30 Corporate Social Responsibility Initiative Working Paper.

19 See also Lise Aurore Lapalme, *The Social Dimension of Sustainable Development and the Mining Industry: A Background Paper* (Mineral and Metal Policy Branch Natural Resources Canada 2003); Hidayat Hasan, 'Managing Social and Environmental Risk' [2011] industrial minerals; Sidney Schafrik and Vassilios Kazakidis, 'Due Diligence in Mine Feasibility Studies for the Assessment of Social Risk' (2011) 25 *International Journal of Mining, Reclamation and Environment* 86; Daniel Franks and others, 'Technology Futures Discussion Paper: Technology Assessment and the CSIRO Minerals Downunder National Research Flagship' [2010] Prepared for CSIRO Minerals Down Under Flagship, Minerals Futures Cluster Collaboration by the Centre for Social Responsibility in Mining, Sustainable Minerals Institute, University of Queensland: Brisbane.

20 For a criticism of this approach, see Graetz and Franks (n 15).

21 Kemp, Worden, and Owen (n 17).

22 Gail Whiteman, 'All My Relations: Understanding Perceptions of Justice and Conflict between Companies and Indigenous Peoples' (2009) 30 *Organization Studies* 101, http://journals.sagepub.com/doi/10.1177/0170840608100518

23 Graetz and Franks (n 15).

divide these examples of social risks into six categories as show in Table 6.1 below.

3 Theorising the Linkage Between ASM, Social Risk, Insecurity and Mining Investment

The north-central part of Nigeria is rich in mineral resources such as gold, tin, and lead. The mining of these contributes directly and significantly to socio-environmental challenges,[24] security imbalance, and economic

Table 6.1 Types of social risk and examples

Types	*Examples*
Socio-economic risk	Threat of future loss of income Risk of social inequality Inequitable distribution of goods, services, and monies Dependence on resource rents/royalties for subsistence (in the case of mining developments)
Socio-environmental risk	Threat of environmental contamination and lost access to lands Potential future loss of land use Potential land use change
Cultural risk	Threats to cultural heritage Potential desecration of sacred sites Threats to traditional ways of life Loss of language Threat of assimilation Community fragmentation
Risk of human rights transgressions	Threats to property Denial of free association and expression Denial of right to consent and to control development(s)
Health and safety risk	Threats to human health, including stress, cancers, respiratory illness, and substance abuse Risk of injury or death
Livelihood risk	Threat to one's livelihood, including employment, access to goods and services, and housing Threats to traditional way of life (also a cultural risk)

Source: Graetz and Franks

24 Ikenna Mike Alumona and Kingsley Maduka Onwuanabile, 'Environmental Insecurity in Nigeria' in Oshita O. Oshita, Ikenna Mike Alumona and Freedom Chukwudi Onuoha (eds), *Internal Security Management in Nigeria* (Palgrave Macmillan 2019).

inequality[25] in those parts of the country. To put these in a clearer perspective, the key manifestations of social risks in Nigeria are in its socio-environmental impacts, socio-economic risks, risk of human rights transgressions, and birthing of ungovernable spaces.

Socio-environmental risk. The first prism through which social risks are viewed in this section is the impact mining has on the environment – which has been referred to as socio-environmental risk.[26] The environmental impacts of illegal mining have led to deforestation and soil damage, among others, especially in the northern parts of Nigeria – where the majority of mining activities take place. Consequently, nomadic cattle herders (usually marauding in the north) no longer find enough green vegetation in the north and out of necessity move southwards in search of areas for open grazing. This has led to violent clashes between herders from the north and farmers in the south – who are acultural to the practice of open cattle grazing. This means that the socio-environmental consequences of mining in Nigeria (such as deforestation and soil damage) influence the security dynamics in the host communities and beyond. This manifested in the southern part of the country where the environmental impacts of the petroleum industry resulting from air and water pollution contributed immensely to the rise of militancy and consequent destruction of oil and gas assets in the Niger Delta region of the country. This is also currently manifesting in northern Nigeria, where, for instance, because of artisanal and small-scale gold mining, children are being recruited into mining and eventually crime. Beyond this, the use of mercury in (artisanal) gold mining in Zamfara State led to an outbreak of lead poisoning that the Human Rights Watch estimated to have affected more than 3,500 children.[27]

Socio-economic risk. Another angle from which to examine the impact of mining on the security of the host communities and investments is the socio-economic risk it poses.[28] Despite the abundance of gold in Nigeria, there was no commercial-scale gold-mining entity in the country until August 2016 when Thor Explorations, a Toronto Venture Exchange listed company, acquired a 100 per cent interest in developing a gold-mining project in Osun State in southwest Nigeria. Thor Exploration was expected to produce its first gold in the second quarter of 2021. However, there are already concerns that this plan could be undermined by 'structural challenges pertaining to insecurity and infrastructure deficiencies' surrounding the mining landscape in the

25 John Sunday Ojo, 'Governing "Ungoverned Spaces" in the Foliage of Conspiracy: Toward (Re)Ordering Terrorism, from Boko Haram Insurgency, Fulani Militancy to Banditry in Northern Nigeria' (2020) 13 *African Security* 77, https://doi.org/10.1080/19392206.2020.1731109

26 Graetz and Franks (n 15).

27 Human Rights Watch, 'A Heavy Price: Lead Poisoning and Gold Mining in Nigeria's Zamfara State' (2011).

28 Graetz and Franks (n 15).

country.[29] Also, the potential for violence grows, as mining both contributes to and competes with the subsistence of the local populations. In addition to this is the role poverty plays in proliferating mining conflicts.[30]

In-country security remains a decisive factor in investment decisions. Disruptions to mining investments because of social challenges (such as organised crime) can be very severe for investors. For example, community conflicts in a Latin American community have caused projects to lose as much as 20 million dollars per week.[31] While the loss of revenue to the industry is unlikely to be as high in Nigeria, such losses create further socio-economic risks and imbalances, exacerbating indigence among members of the mining communities in the northern parts of the country.

Ungoverned spaces. The lack of an effective government presence in areas prone to illegal and artisanal mining is a risk to these local communities as the regulating and policing role of the state, which is taken for granted in developed countries or mature mining provinces, is absent. As noted earlier, most of the mining activities in Nigeria are artisanal in nature and therefore susceptible to illegal activities.[32] These ASMs and illegal mining activities often occur in isolated areas that are most appropriately described as ungoverned spaces.[33] These are areas in the country where the state institutions cannot or do not exercise full authority to regulate law and order or lack legitimacy to control.[34] These spaces often provide a haven for criminal networks, terrorists/terrorist groups, and bandits. This is because such places

29 'Nigeria Aims to Have 50 Operational Mines by 2023 – Mines Minister' (Africa Inc. 2020), https://africaincmag.com/nigeria-aims-to-have-50-operational-mines-by-2023-mines-minister-says, accessed 2 April 2021.

30 Marta Conde and Philippe Le Billon, 'Why Do Some Communities Resist Mining Projects While Others Do Not?' (2017) 4 *The Extractive Industries and Society* 681, https://linkinghub.elsevier.com/retrieve/pii/S2214790X17300035

31 DM Franks and others, 'Conflict Translates Environmental and Social Risk into Business Costs' (2014) 111 *Proceedings of the National Academy of Sciences* 7576, www.pnas.org/cgi/doi/10.1073/pnas.1405135111

32 Muhammad Rabiu Musa, 'Understanding Foreign Investment Regulations in Nigerian Mining Industry' (2020) 2 *International Review of Law and Jurisprudence (IRLJ)*.

33 Literature has described ungoverned spaces as where 'the state or central government is unable or unwilling to extend control, effectively govern, or influence the local population and where a provincial, local, tribal, or autonomous government does not fully or effectively govern, due to inadequate governance capacity, insufficient political will, gaps in legitimacy, the preserves of conflict, or restrictive norms of behaviour. Ungoverned areas should be assumed to include under-governed, ill-governed, contested, and exploitable areas.' See Jennifer Keister, 'The Illusion of Chaos: Why Ungoverned Spaces Aren't Ungoverned, and Why That Matters' [2014] Cato Institute Policy Analysis. See also Ogbonnaya (n 5).

34 Ojo (n 25). See also Richard Olaniyan and Olukayode A Faleye, 'Civil Society and Terrorism in Africa: Rethinking Borderland Security in Northeastern Nigeria' in RA Olaniyan and RT Akinyele (eds) Nigeria's Ungoverned Spaces (Awolowo University Press, 2016)

necessitate the emergence and control of non-state armed actors, and the lack of an effective state presence means that laws do not have the potency required to maintain stability and/or deter non-state elements.

In Nigeria, therefore, the link between ungoverned spaces and insecurity is clear. Evidence of this is currently prevalent in many villages in the northern parts of Nigeria, which are also littered with many small mines where government presence is scarce or non-existent.[35] There is also a lack of government-backed geological data on ASM activities in such places.

Countries with significant swathes of ungoverned areas are often vulnerable to the pervasiveness of (un)organised crime, as experienced in some African and Middle Eastern countries where militias and criminal groups collaborate to mine or loot metals and minerals.[36] These hazards have led mining communities in Nigeria into a crossfire of human rights violations and instability fuelled by crimes such as kidnapping, banditry, and human rights violations.

To attract sustainable and much-needed investment in Nigeria's mining sector, therefore, there need to be holistic efforts directed towards addressing the negative impacts of (in)security on local communities on one hand and private investments into the sector on the other hand.

4 Is the Current Legal Framework Adequate?

The peculiarities of the issues raised above present no easy route for the Nigerian government to address, and in such areas resort is usually had to the use of legal tools to change individual and group behaviour. But are the available legal structures (including bilateral and multilateral obligations) enough to protect local communities on one hand and secure investment on the other hand?

The Nigerian Constitution[37] and the Nigerian Minerals and Mining Act 2007 (the 'Mining Act') create the basic legislative framework for mining activities in the country.[38] Although there exists much other legislation[39] that

35 Ogbonnaya (n 5).
36 Ojo (n. 25).
37 Constitution of the Federal Republic of Nigeria, 1999 (as amended).
38 There is also a regulation, the Minerals and Mining Regulations of 2011 made pursuant to the Mining Act.
39 These include the Land Use Act, 1978; the Environmental Impact Assessment Act, 1992; the Mines and Quarries (Control of Building etc.) Act, 1969; the Explosives Act, 1964; the Nuclear Safety and Radiation Protection Act, 1995; the Companies Income Tax Act, 1961; the Nigerian Investment Protection Act, 1989; the Nigerian (Export Incentives and Miscellaneous Provisions) Act, 1992; the National Geological Survey Agency (Establishment) Act 2006; the National Environmental Standards and Regulations Enforcement Agency (Establishment) Act, 2007; the Companies and Allied Matters Act, 2020; and the Finance Act, 2020. There are also policy documents such as National Minerals and Metals Policy 2008 and Roadmap for the Development of Solid Minerals and Metals 2012.

impacts the mining sector, these two are fundamental and create the basis upon which mineral rights are owned and controlled.

Ownership and Control of Minerals in Local Areas

In consonance with the Constitution,[40] the Mining Act creates state ownership and control over of mineral resources and prohibits unauthorised exploration or exploitation of minerals whether in federal-owned, state-owned, or privately held land.[41] The Mining Act further mandates that all lands in which minerals are found in commercial quantities must be acquired by the federal government in accordance with the Land Use Act 1978.[42] Property in mineral resources is divested from the government to any person through whom the mineral resources are lawfully won, upon their recovery in accordance with provisions of the Mining Act.[43] But exercise of rights over the use of land, simpliciter, is controlled by the state governments according to the Land Use Act.

It has been argued that this structure (whereby federal authorities provide mineral rights while state authorities provide rights over the land on which such minerals are found) creates disharmony between the different levels of government (federal and state), and between the government entities and the communities or landowners.[44] This is because there is a perception that the government wrestles ownership and control of land from communities or indigenous people once there is economic value to it. However, where no such economic value is apparent, the government shirks responsibility in such areas to the people.

Priority of Mining for Use of Land

The above-highlighted disharmony is exacerbated by the fact that mining, under the Mining Act, is a priority in respect of the use of land to the exclusion of any other purpose.[45] The overriding public interest in the use of land for mining means that where there are minerals and metals discovered in sacred land or places that are being preserved by local communities, such social capital (for land) is relegated due to public interest in mining such land.

40 Section 44(3) of the Constitution provides that the entire property in and control of all minerals, mineral oils and natural gas in, under or upon any land in Nigeria or in, under or upon the territorial waters and the Exclusive Economic Zone of Nigeria shall vest in the Government of the Federation and shall be managed in such manner as may be prescribed by the National Assembly.

41 Section 1 of the Mining Act.

42 A federal law that regulates ownership and use in Nigeria and which vests all lands in each state of Nigeria in the Governor of the state.

43 Section 1 of the Mining Act.

44 Ogbonnaya (n 5).

45 Section 22 of the Mining Act.

The public in this context is usually not the local community but the government being represented by the presence of mining investors or contractors. This has the potential of creating resistance by communities since the mining operation is to be carried out by contractors who may not be locals. This then creates the impression that precious community land is being acquired by mining contractors or investors.

Although the law does prohibit mining in sacred areas or desecration of objects of veneration, the power to determine what qualifies as sacred is vested solely in the Mining Cadastral Office.[46] Again, this creates the impression that state authorities determine the culture of the people with outcomes that are usually favourable to the state or deep-pocketed miners, rather than the community.

Licensing Regime for Miners and ASM

The licensing regime created by the Mining Act provides for entry-level certification, prospecting licence, reconnaissance permit, exploration licence, mining lease and/or a small-scale mining lease. The latter is issued for small-scale mining activities – carried out in areas of no more than three square kilometres. The holders of such ASM leases are required to perform certain environmental and social responsibilities such as the effective rehabilitation of the mined-out areas to the 'satisfaction of the government'. Likewise, holders of an exploration licence, which is a pre-cursor to a mining lease, are required to obtain landowners' consent (pre-grant) and to (post-grant): provide an undertaking to conduct an environmental impact assessment (EIA) and submit an EIA report; compensate all users of land within the mining area; and provide a mine closure and rehabilitation plan.[47] These conditions attached to the issuance of a mining lease are a good step, but where mining operators have already carried out exploration activities before meeting these conditions, communities could already have been subjected to the likely occurrence of risks before the conduct of an EIA or concluding a community development agreement. The receipt of a mining lease, however, is intended to serve as an incentive to the lessee to address community social concerns. In respect of ASM, the government ministry has the responsibility of providing EIAs and mine designing and planning for such operators.

These requirements are intended to address the environmental and social risks of mining activities in Nigeria. But the obligations on mining lease holders and ASM lessees are not far-reaching enough. The requirement to obtain the consent of landowners is at best pedantic. This is because

46 Section 98 of the Mining Act. The Mining Cadastral Office administers, grants, and processes mineral titles and maintains a cadastral registry which collates registers of all granted mining-related permits and licences.

47 Section 71 of the Mining Act.

landowners, for instance, in the communities are usually uninformed as to the full extent of the impacts of the activities on the environment and society. When governments at the state level issue formal rights over land use, they do so taking property rights and obligations into account, rather than socio-environmental rights, socio-cultural rights, and the socio-economic impacts of mining. This is further compounded by the fact that governments at state and local levels are only empowered to regulate property rights. Federal authorities t provide mineral rights are far removed from the activities in the local regions.

The Mining Act is a useful piece of legislation. However, it does not appear to have foreseen and looks incapable of solving the myriad of social issues that have hampered the sector and local communities.

5 Towards the Protection of Communities and Investors in the Nigerian Mining Industry

In the foregoing sections, we have shown that there is a causal link between the prevalence of ASM in parts of Nigeria and social risks of insecurity suffered in those areas. In addition, the insecurity challenge has led to the erosion of legitimate large-scale mining investments in the country. Therefore, this section focuses on steps and policy tools that could be used in the short and long term to address these issues.

Procedural justice. Established principles of energy justice require that communities need to have a voice in the exploitation of their natural resources and to use their local decision-making mechanisms to express this voice. This, however, contrasts with the cursory and inconsequential participatory processes still frequently used by mining companies and governments during the EIA. As a result, it is important that government actions support the ability of local communities to build alliances with extra-local actors to aid the communities' desire for self-organisation. To achieve this, government regulations/policies need to introduce processes that constantly evaluate the relationship between miners and the local communities, beyond mere issuance of regulations and permits.

Community engagement agreements. The mining policy climate needs to (i) promote effective evaluation and management of the social impacts of ASM and (ii) support community engagement and community decision making all through the various stages of mining activities. The standardisation of such interactions into agreements between government, miners, and native communities – with localised grievance-handling approaches – is vital in maintaining effective management of social threats and conflict resolution.

Regional partnerships. Considering that social conflicts often involve cross-border elements, it is imperative that the Nigerian government partners with governments in West and Central Africa to strengthen the protection of local communities in the region and promote sustainable mining practices.

Inspiration can be drawn from the Aarhus Convention Framework[48] which is applicable to European countries, and which spells out public consultation processes for cross-border projects. This framework serves as a regional tool to tackle any disregard for social engagement in building transboundary projects. There is no similar framework across West and Central Africa yet, but such a framework could also benefit local communities as is the case with the Aarhus Convention.

Support for ASM. Beyond licensing small-scale miners, government policies must not ignore the need to strengthen ASM miners so that they are not continually used as a cover to perpetuate illegal mining activities across the northern parts of Nigeria. Apart from licensing, the ASMs require support in improving their operations as well as guidance in structuring their relationships with the local population to avoid social risks.[49]

Good governance. Fifth, good governance that transcends ethnic idealisation and interests is a veritable weapon to triumph over the manifold security challenges experienced by mining communities in Nigeria. The disharmony (identified above) between state and federal authorities, the neglect of rural areas manifested in an abject lack of basic amenities, and the weakness of law enforcement in large portions of the northern parts of Nigeria are all conditions that create the enabling environment for organised crime groups.[50] These groups take advantage of the gap in effective governance to radicalise locals and recruit them for illicit mining activities.[51] Therefore, government at federal, state, and local levels needs to ensure the maintenance of effective law and order by: first, ensuring community participation in governance and law reforms so that mining laws will reflect the spirit and interest of the communities; second, by frequently interfacing with communities with a view to ensuring that laws and regulations can evolve to meet the different dynamics of mining's social impacts; and third, and importantly, enforcing laws effectively in a manner that involves the buy-in of the local communities. This latter step will invariably involve the training and re-training of law

48 United Nations ECE Convention on Access to Information, Public Participation in Decision-making and Access to Justice in Environmental Matters, known as the Aarhus Convention; see Sean T McAllister, 'The Convention on Access to Information, Public Participation in Decision-Making, and Access to Justice in Environmental Matters' (1999) 10 Colorado Journal of International Environmental Law and Policy 187.

49 Olga Sidorenko, Rauno Sairinen, and Kathryn Moore, 'Rethinking the Concept of Small-Scale Mining for Technologically Advanced Raw Materials Production' (2020) 68 *Resources Policy* 101712, www.sciencedirect.com/science/article/pii/S0301420719307871; T Hentschel, F Hruschka, and M Priester, 'Artisanal and Small-Scale Mining: Challenges and Opportunities. Iied' [2003] WBCSD, London; Maria Laura Barreto and others, 'The Impact of Small-Scale Mining Operations on Economies and Livelihoods in Low-to Middle-Income Countries' [2018] UK Department for International Development, London.

50 Kemp, Worden and Owen (n 17).

51 Keister (n 33).

enforcement agents. Ancillary to the role of good governance is the need to stamp out corruption from those within law enforcement authorities and government departments[52] who collude with both legal and illegal miners to render the rule of law ineffective.

Behavioural regulation. The role of behavioural regulation of corporate entities is also imperative – that is, regulating corporate conduct through social norms and international best practices.[53] The government has a role in helping miners to take steps that are meaningful for the communities. For instance:

i through meaningful engagement and dialogue with affected communities in a bid to obtain and maintain the miners' social licence to operate;
ii by creating processes that track and respond to claims, complaints, and grievances by communities; and commitments made by the mine operators; and
iii by incentivising ASM miners to continue formalising their activities – one vital element of this is by constantly removing regulatory bottlenecks in the formalisation process for ASM miners.

Finally, although questions around the amendment of the country's Constitution are usually controversial and politicised, the legal framework relating to mining and property rights accruable to landowners/communities and the government is in need of recalibration.

6 Discussion and Conclusion

A dominant deduction from this chapter is that the presence of a multiverse of ASMs and illegal miners has created social tensions in mining communities, and they are a crucial and frequent source of social risks to the mining industry and local communities in the mining-prevalent regions of the country. Therefore, where large proportions of ASM remain informal, they portend risk for legitimate investors/operators and host communities. These risks then create a crossfire of human rights abuses and insecurity in (a large proportion of) mining communities in Nigeria.

Another pivotal deduction is that the same issues giving rise to social risks from mining activities are also creating the plinth for criminality to thrive in northern Nigeria, affecting both the local population and legitimate investors. Therefore, the government has a responsibility to protect all stakeholders. Blanket bans on mining activities, as pronounced in Nigeria in 2019 and

52 Sheelagh Brady, 'Crude Oil's Ugly Sister: The Political-Criminal Nexus and Corruption Inside Nigeria's Solid Minerals and Mining Sector' in Y Zabyelina and D van Uhm (eds) *Illegal Mining* (Springer International Publishing 2020), http://link.springer.com/10.1007/978-3-030-46327-4_14

53 Franks and others (n 31).

2021,[54] have not addressed the social conflicts that have defined the daily experiences of mining host communities; nor have they helped to protect the activities of legitimate miners (who in many cases are ASMs).

As the basic piece of legislation for the industry, the Nigerian Mining Act remains useful. However, the regulatory and administrative framework it creates is not enough to protect stakeholders against social risks. It does not appear to have foreseen and looks incapable of solving the myriad of social issues that have hampered the sector since it was enacted. For example, the formalisation regime created for ASM under the Mining Act in Nigeria is not innovative enough and it focuses almost entirely on providing licence terms for ASMs. The lack of political will and governance necessary to complement the regime of the Mining Act make the issues complex and persistent.

Six policy steps available to the government in the short and long term to address the issues raised in this chapter have been recommended. These recommendations are not exhaustive and would require testing to see how much impact they can make in Nigeria. Admittedly, more research is required to show the comprehensive implications of legislative and structural disharmonies for the plight of local economies and the legitimate mining industry. Also, more attention needs to be paid to the effects of insecurity on the socio-cultural fabric of mining communities, especially for young people and children who have to accept spiralling social conflicts as part of their daily lives.

References

Adeyemi S and Olagunju A, 'Putting Small-Scale Mining in Perspective: An Analysis of Risk Perception of a Southwestern Nigerian Community' (2017) 2*Review of Social Sciences*

Ahmed YM and Oruonye ED, 'Socioeconomic Impact of Artisanal and Small Scale Mining on the Mambilla Plateau of Taraba State, Nigeria' (2016) 3*World*

Alumona IM and Onwuanabile KM, 'Environmental Insecurity in Nigeria' in Oshita O Oshita, Ikenna Mike Alumona, and Freedom Chukwudi Onuoha (eds), *Internal Security Management in Nigeria* (Palgrave Macmillan 2019)

Barclay MA, Franks DM, and Pattenden C, 'Risk Communication: A Framework for Technology Development and Implementation in the Mining and Minerals Processing Industries'

Barreto ML and others, 'The Impact of Small-Scale Mining Operations on Economies and Livelihoods in Low-to Middle-Income Countries' [2018] UK Department for International Development, London

Bekefi T, Jenkins B and Kytle B, 'Social Risk as Strategic Risk' (2006) 30 Corporate Social Responsibility Initiative, Working Paper

54 Brady (n 53); Abdur Shaban, 'Nigeria Bans Mining in Zamfara State after Surge in Banditry' (Africa News,7 April 2019), www.africanews.com/2019/04/07/buhari-decries-ridiculous-claims-of-apathy-to-insecurity-in-zamfara

Brady S, 'Crude Oil's Ugly Sister: The Political-Criminal Nexus and Corruption Inside Nigeria's Solid Minerals and Mining Sector' in Y Zabyelina and D van Uhm (eds) *Illegal Mining* (Springer International Publishing 2020), http://link.springer.com/10.1007/978-3-030-46327-4_14.

Brereton D and Parmenter J, *Water, Communities and Mineral Resource Development—Understanding the Risks and Opportunities*, Water in Mining Conference, Brisbane, Australia (2006)

Carsten P, 'Nigeria Says Can Tap Mineral Wealth despite COVID, Unrest' (Reuters, 2020), www.reuters.com/article/us-nigeria-mining/nigeria-says-can-tap-mineral-wealth-despite-covid-unrest-idUSKBN25T1K4, accessed 2 April 2021

Conde M and Le Billon P, 'Why Do Some Communities Resist Mining Projects While Others Do Not?' (2017) 4*The Extractive Industries and Society* 681, https://linkinghub.elsevier.com/retrieve/pii/S2214790X17300035.

Eftimie A and others, 'Gender Dimensions of Artisanal and Small-Scale Mining: A Rapid Assessment Toolkit'

Fayemi K and Onwurah C, 'Nigeria's Solid Minerals Sector: Alternative Investment Opportunities' [2016] Chatham House (Royal Institute of International Affairs)1

Franks D and others, 'Technology Futures Discussion Paper: Technology Assessment and the CSIRO Minerals Downunder National Research Flagship' [2010] Prepared for CSIRO Minerals Down Under Flagship, Minerals Futures Cluster Collaboration by the Centre for Social Responsibility in Mining, Sustainable Minerals Institute, The University of Queensland, Brisbane

Franks DM and others, 'Conflict Translates Environmental and Social Risk into Business Costs' (2014) 111*Proceedings of the National Academy of Sciences* 7576, www.pnas.org/cgi/doi/10.1073/pnas.1405135111

Gold B, 'Community Relations Standard' [2012] Unpublished corporate document

Graetz G and Franks DM, 'Conceptualising Social Risk and Business Risk Associated with Private Sector Development Projects' (2016) 19*Journal of Risk Research* 581

Hasan H, 'Managing Social & Environmental Risk' [2011] *Industrial Minerals*.

Hentschel T, Hruschka F and Priester M, 'Artisanal and Small-Scale Mining: Challenges and Opportunities. Iied' [2003] WBCSD, London

Hilson G and Maconachie R, 'Formalising Artisanal and Small-scale Mining: Insights, Contestations and Clarifications' (2017) 49*Area* 443

Human Rights Watch, 'A Heavy Price: Lead Poisoning and Gold Mining in Nigeria's Zamfara State' (2011)

ICMM, *Minerals Taxation Regimes a Review of Issues and Challenges in Their Design and Application.* (Commonwealth Secretariat and the International Council on Mining and Metals 2009)

Joyce S and Thomson I, 'Earning a Social Licence to Operate: Social Acceptability and Resource Development in Latin America' (2000) 93*CIM Bulletin* 49

Keister J, 'The Illusion of Chaos: Why Ungoverned Spaces Aren't Ungoverned, and Why That Matters' [2014] Cato Institute Policy Analysis

Kemp D, Worden S and Owen JR, 'Differentiated Social Risk: Rebound Dynamics and Sustainability Performance in Mining' (2016) 50*Resources Policy* 19

Lapalme LA, *The Social Dimension of Sustainable Development and the Mining Industry: A Background Paper* (Mineral and Metal Policy Branch Natural Resources Canada 2003)

McAllister ST, 'The Convention on Access to Information, Public Participation in Decision-Making, and Access to Justice in Environmental Matters' (1999) 10 *Colorado Journal of International Environmental Law and Policy* 187

Musa MR, 'Understanding Foreign Investment Regulations in Nigerian Mining Industry' (2020) 2*International Review of Law and Jurisprudence (IRLJ)*

Nalule VR, 'Social and Environmental Impacts of Mining' in *Mining and the Law in Africa* (Springer International Publishing 2020), http://link.springer.com/10.1007/978-3-030-33008-8

'Nigeria Aims to Have 50 Operational Mines by 2023 – Mines Minister' (Africa Inc. 2020), https://africaincmag.com/nigeria-aims-to-have-50-operational-mines-by-2023-mines-minister-says, accessed 2 April 2021

Ogbonnaya M, 'Women and Girls in Illegal Mining in Northern Nigeria', (ENACT Observer 2020), https://enactafrica.org/enact-observer/women-and-girls-in-illegal-mining-in-northern-nigeria, accessed 20 April 2020

Ogbonnaya M, 'Illegal Mining and Rural Banditry in North West Nigeria' (ENACT Observer 2021), http://enact-africa.s3.amazonaws.com/site/uploads/2020-11-19-illegal-mining-policy-brief.pdf, accessed 20 April 2021

Ojo JS, 'Governing "Ungoverned Spaces" in the Foliage of Conspiracy: Toward (Re) Ordering Terrorism, from Boko Haram Insurgency, Fulani Militancy to Banditry in Northern Nigeria' (2020) 13*African Security* 77, https://doi.org/10.1080/19392206.2020.1731109

Olaniyan R and Faleye OA, 'Civil Society and Terrorism in Africa: Rethinking Borderland Security in Northeastern Nigeria' in RA Olaniyan and RT Akinyele (eds) *Nigeria's Ungoverned Spaces* (Awolowo University Press, 2016)

Oxford Business Group, 'Nigeria's Untapped Mineral Reserves Could Aid the Goal of Economic Diversification' (2016), https://oxfordbusinessgroup.com/overview/digging-deep-significant-and-largely-untapped-mineral-reserves-could-aid-goal-economic, accessed 2 February 2021

Schafrik S and Kazakidis V, 'Due Diligence in Mine Feasibility Studies for the Assessment of Social Risk' (2011) 25*International Journal of Mining, Reclamation and Environment* 86

Shaban A, 'Nigeria Bans Mining in Zamfara State after Surge in Banditry' (Africa News, 7 April 2019), www.africanews.com/2019/04/07/buhari-decries-ridiculous-claims-of-apathy-to-insecurity-in-zamfara

Sidorenko O, Sairinen R, and Moore K, 'Rethinking the Concept of Small-Scale Mining for Technologically Advanced Raw Materials Production' (2020) 68*Resources Policy* 101712, www.sciencedirect.com/science/article/pii/S0301420719307871

Rio Tinto, 'Social Risk Analysis Guidance Note' [2011] Rio Tinto, Melbourne

Whiteman G, 'All My Relations: Understanding Perceptions of Justice and Conflict between Companies and Indigenous Peoples' (2009) 30*Organization Studies* 101, http://journals.sagepub.com/doi/10.1177/0170840608100518

7 Administration of Fiscal Regimes and Youth Empowerment

A Case for Stability in Zambia's Royalty for Gold Mining

Kangwa-Musole George Chisanga[1], *Mwaba Chileya Mulenga*[2], *and Chileshe Mwaba*[3]

1 Introduction

Zambia is a Southern African country blessed with many natural resources. For instance, the country's savanna climate and the wildlife found in it are renowned and serve as the basis of the tourism sector.[4] The agriculture sector recorded bumper harvests in the years leading to the COVID-19 pandemic.[5] Yet tourism and agriculture are not what the country is known for; Zambia is known as a copper-producing country. Barrick Gold, Vedanta, and China Non-Ferrous metals are among the international mining companies operating in the country.

The mining industry has been the social backbone of Zambia since the 1930s, when early exploration works were undertaken, which subsequently led to large-scale mining operations of a number of minerals.[6] The building blocks for mining are laws and implementing regulations. Often the laws are sector-specific but intermesh with other domestic laws dealing with tax, investment, and environment matters, for example. Any legal framework for mining has to be consistent with the country's constitutional framework and

1 Lecturer, National Institute of Public Administration, Lusaka, Zambia.
2 Associate, Christopher Russell Cook & Co., Lusaka, Zambia.
3 Chairperson, Twikatane Imisepela Multi-Purpose Society, Lusaka, Zambia.
4 Office of the Auditor-General, *Report of the Auditor General on the Performance of the Tourism Sector in Ensuring an Increase in the Length of Stay of International Tourist* (Office of the Auditor-General 2020); Binyi Liu, Floyd M Mwanza, 'Towards Sustainable Tourism Development in Zambia: Advancing Tourism Planning and Natural Resource Management in Livingstone (Mosi-oa-Tunya) Area' (2014) *Journal of Service Science and Management* (7) 30–45; World Bank, *Zambia: Economic and Poverty Impact of Nature-based Tourism* (World Bank 2007)
5 Ministry of Finance, 'Treasury Releases K1 Billion for Crop Purchases', www.mof.gov.zm/?p=6184, accessed on 20 November 2021.
6 Muna Ndulo, 'Mining Legislation and Mineral Development in Zambia' (1986) 19 (1) *Cornell Law Faculty Publications* 1, 4.

DOI: 10.4324/9781003284437-9

must operate within the wider legal system in the host state shaped by civil law or common law.[7]

The economic impact of the pandemic, however, hit all sectors, prompting the country to develop an economic recovery programme spanning the years 2020 to 2023. Mining features prominently in the plan, proving that the sector will play a key role in post-pandemic economic recovery.[8] The road to economic recovery is riddled with challenges such as a lack of economic diversification due to the country's reliance on copper mining and high levels of unemployment among youths. While the mining sector cannot be expected to solve all such ills, there is a high expectation placed on the sector in general, and the recent discovery of gold deposits is set to increase such expectations. It is in this regard that the fiscal regime and the challenges arising from its implementation form a suitable topic for this chapter.

Although Zambia is known for being a copper-producing country, other prominent minerals including cement, cobalt, and gold deposits are being mined in the country. Academic discourse about the mining sector spans various disciplines including history, town planning, medicine, and law. In discourse about Zambia's fiscal regime in particular, there is a theme that stability is lacking and that this is undermining the impact of mining on the country's development agenda.

Zambia's government discovered that deposits of gold were being mined by artisanal miners in the Eastern Province of the country.[9] The government, through its mining investment company Zambia Consolidated Copper Mines Investment-Holdings Limited ('ZCCM-IH') entered the market by establishing a subsidiary specially designed for gold mining.[10] The potential of emerging out of the pandemic as a gold producer that can compete with other producing countries, such as South Africa, makes the development of gold mining in Zambia an inevitable priority for post-COVID-19 economic recovery.

High on the agenda is economic diversification. This is logical given the relatively wide base of natural resources the country possesses. That, coupled with the massive investment in infrastructure development, sets the country on to the path to industrialization in the post-COVID-19 era. In this regard, gold mining has a lot of potential to play a part in development. This chapter contributes to the academic discourse on the issue by focusing on the fiscal regime for gold mining in Zambia.

7 Erik Richer La Fleche (ed.) *The Mining Law Review* 8th edition (Law Business Research 2019).

8 Ministry of Finance & National Planning, *2022 Budget Address delivered to the National Assembly* (National Budget, 2022) paras 50–51.

9 Carlos Williams, 'Zambia's Mining Industry Strikes Gold', https://borgenproject.org/zambias-mining-industry, accessed 14 May 2021.

10 ZCCM-IH, 'Press Statement: ZCCM-IH and Its Investment Strategy in the Gold Sector in Zambia' (24 May 2020), www.zccm-ih.com.zm/2020/05/26/zccm-ih-and-its-investment-strategy-in-the-gold-sector-in-zambia, accessed 21 May 2021.

A fiscal regime is a critical component of any nation's mining regulatory framework due in part to the fact that government revenues are made up of earnings from the mining sector.[11] The mining industry also has some unique characteristics which create challenges in designing a fiscal regime. High potential rents and the obsolescing bargain inherent in extractives projects are some of the characteristics that can present a challenge for governments in this regard.[12] Despite the conventional method of revenue collection through taxes, if companies are perceived as not paying their fair share, governments tend to impose other extraneous changes or abruptly revise taxes as a corrective measure, and this has been identified as a manifestation of resource nationalism.[13] This chapter was developed on the argument that stability in the fiscal regime is a prerequisite for the achievement of positive impacts of mining in the country.[14]

Zambia's fiscal regime for mining is repeatedly criticized by analysts as being unstable. By one estimate, the country has changed its fiscal regime once every 18 months since 2001.[15] In 2019, for instance, the national budget directed over half of the state's proposed revenue measures towards the mining sector, representing another seismic shift in mining taxation policy. This instability in the fiscal regime is symptomatic of constrained capacity in the tax administration whereby changes in tax policy are made to address challenges in revenue collection. This is a challenge for the country; hence, this chapter attempts to address this problem.

Using a qualitative content analysis methodology, the chapter constructs a literature review on stability in tax administration for the mining sector and

11 La Fleche n 7; James Otto and John Cordes, *The Regulation of Mineral Enterprises: A Global Perspective on Economics, Law and Policy* (Rocky Mountain Mineral Law Foundation 2002).

12 Jack Calder, *Administering Fiscal Regimes for Extractive Industries* (International Monetary Fund 2015) 2, 8; Peter D Cameron and Michael C Stanley, *Oil, Gas, and Mining: A Sourcebook for Understanding the Extractive Industries* (The World Bank 2017) 40–42.

13 Ian Bremmer and Robert Johnston, 'The rise and fall of resource nationalism' (2009) 51(2) *Survival* 149; David R Mares, *Working Paper: Resource Nationalism and Energy Security in Latin America: Implications for Global Oil Supplies* (James A. Baker III Institute for Public Policy 2010); Francisco J Monaldi, 'Is resource nationalism fading in Latin America? The case of the oil industry' (2014) Issue Brief, www.bakerinstitute.org/research/resource-nationalism-fading-latin-america-case-oil-industry; Jeffrey D Wilson, 'Understanding resource nationalism: Economic dynamics and political institutions' (2015) 21(4) *Contemporary Politics* 399. Andreasson, Stefan. 'Varieties of resource nationalism in sub-Saharan Africa's energy and minerals markets' (2015) 2(2) *The Extractive Industries and Society* 310–319; Natalie Koch and Tom Perreault, 'Resource nationalism' (2019) 43 *Progress in Human Geography* 611.

14 James Otto, *Mining Royalties: A Global Study of Their Impact on Investors, Government, and Civil Society* (World Bank Publications 2006) 12, 14.

15 Twivwe Siwale and Benjamin Chibuye, 'Mining taxation policy in Zambia: The tyranny of indecision' 13 June 2019, www.theigc.org/blog/mining-taxation-policy-in-zambia-the-tyranny-of-indecision, accessed 19 November 2021.

presents the Zambian regime as an example of what can be deemed an unstable fiscal regime. Analysis of historical changes in the mining fiscal regime is used to forecast the likelihood of change in the mineral royalty tax charged on gold.

A case study of the Zambian small-scale and artisanal mining company Migodi-Auric Limited is presented as an example of youth empowerment through its Rural Upliftment Programme ('RUP'). The argument in this chapter is that the tax authorities should be encouraged to take a long-term view for stability in the design of their revenue collection and management systems if the proposed support for gold mining through the RUP is to achieve the aim of youth empowerment.

2 Mineral Tax Administration: A Literature Review

Literature under this section was selected based on the respective authors' discussion of the core ideas of designing fiscal regimes and mineral tax administration. The literature reveals some lived realities in Zambia in contrast to the ideals found in numerous texts on the subject of mineral tax administration. Literature was also gathered on the basis of discussion of the problem of youth employment as a problem not only faced in Zambia at a national level but at the continental and global level.

The Organisation for Economic Co-operation and Development (OECD), in a guide to addressing tax challenges, restates effectiveness and fairness (i.e. collecting the right tax at the right time) among the core principles of taxation.[16] Other principles included are certainty and simplicity (i.e. enabling taxpayers to know where they stand).[17] This chapter takes this view and applies it in assessing the fiscal regime for gold mining in Zambia. Commentary on Latin American fiscal regimes is also instructive for the Zambian reality as proper fiscal regimes are a key to development.[18] Collier identifies the lack of long-term commitment on the part of mining companies as a challenge faced by low-income countries.[19] To overcome this challenge, one of the proposed principles for such countries to adopt is to have permanent income and future generation funds. Such funds can only be created by a consistent fiscal regime. These ideas helped shape the research focus regarding youth empowerment and the effectiveness of fiscal regimes regulating gold mining.

16 OECD, 'Fundamental principles of taxation' in *Addressing the Tax Challenges of the Digital Economy* (OECD Publishing, 2014) 30.
17 Ibid at 34.
18 Christian Aid, *Undermining the Poor: Mineral Taxation Reforms in Latin America* (Christian Aid, 2009).
19 P Collier, 'Resource taxation for low-income countries' in P Daniel, M Keen, and C McPherson (eds), *The Taxation of Petroleum and Minerals: Principles, Problems and Practice* (Routledge 2010) 76.

Tom and Nalule,[20] in their discussion on management of petroleum revenue, opine that tax planning and administration of revenue is one of the key internal challenges faced by various governments in resource-rich developing countries. This chapter dissects the specific aspects of tax planning and tax administration with which the Zambian government grapples – namely, revenue collection, administration, and management.

Manley[21] discusses the substratum of Zambia's tax policy; stating that the categories of tax most relevant to the mining sector are indirect taxes, such as value added tax ('VAT') or customs and import duty, and direct taxes such as corporate income tax, excess or variable profit tax, mineral royalty tax ('MRT'), and windfall tax. Direct taxes are revenue-based taxes, although one could make the argument that royalties are a form of ownership transfer tax rather than profit-based tax, given that they are a fixed percentage of the total minerals sold. Taking this argument in the Zambian context, it is then possible to assert that the revision to the tax policy in 2018 to scrap variable profit tax and focus on a sliding scale for royalties is reflective of the government's efforts to focus on ownership transfer tax as opposed to revenue-based tax. In its attempt to ensure that it retains a greater share of the economic value in the mining sphere, the government has, however, neglected to design a consistent tax policy that proactively responds to the changes in prices, by opting to focus on varying royalties and leaving the problem of price fluctuation in minerals unaddressed.

Manley, in his analysis of amendments to Zambia's fiscal regime, also comments that the amendments have not introduced progressivity in the tax regime.[22] Using Manley's research to formulate a hypothesis, this chapter explores the idea that constant revisions to the tax administration in Zambia are reflective of the complications that concern the government.

Manley emphasizes that the tax regime is problematic in two respects.[23] First, the price thresholds are much too low to sufficiently cover the rent when there is an increase in the price of copper. Second, the price-based royalty is not reflective of fluctuating mining costs which are subject to changes in copper prices. Tracing the trend of the tax policy in recent years evidences the manner in which the government has benefitted from and simultaneously been harmed by the fluctuating prices of minerals, particularly copper.

20 Ayebare Tom and Victoria Nalule, 'Key Issues in Petroleum Revenue Management', http://afrienergyminerals.org/wp-content/uploads/2020/01/REVENUE-MANAGEMENT-AND-KEYISSUES22222222222.pdf, accessed 30 February 2021.

21 Ibid.

22 David Manley, 'Ninth time lucky: Is Zambia's Mining Tax the Best Approach to an Uncertain Future?' in Briefing October 2014 (Natural Resource Governance Institute).

23 Ibid.

Prior to the amendment in 2018, the tax policy had been changed a total of eight times in 15 years. This is typically reflective of policy instability.[24] Policy instability matters for these three reasons. First, investors are afraid of a government that would expropriate the investment or disproportionately raise taxes abruptly after the investment. Second, policy instability affects the collection of rent. Finally, the changes in policy allow for the government to make mistakes and for other companies and stakeholders to lobby for incentives, which might lead to conflict in the relationships between natural resource corporations and other stakeholders such as civil society organizations.

Tied to the problem of failing to design a consistent tax regime is the growing concern of risk assessment. In this regard, risk assessment signifies an effective administration of revenue legislation through which a proper strategy to identify, understand, and manage the weaknesses in the law is developed.[25] The aim, then, is to understand the legislative provisions and the possible risks they present. Arguably, the government does make strides in this area, as the main collector of taxes is empowered with the authority to issue administrative guidelines to resolve uncertainties arising from gaps in legislation. However, even with these strides, the government still comes up short, because in some instances the tax authorities are left to decide how to apply unclear legislation. Arguably, the government's failure to understand and address the weaknesses of the tax policy as regards fluctuation in mineral prices falls short of internationally accepted standards for risk assessment and management.

Calder,[26] in his chapter on the implementation and design of natural resource taxes, states that governments design their tax regime to suit their particular economic objectives. He states further that ease of administration of taxes must be secondary to these objectives, meaning that governments must consider the factors that influence the administration and implementation of the tax regime. The author asserts that developing countries rely on output-based taxes because these can be administered more effectively in relation to profit-based taxes. To the extent that the author underscores the complexity of a predominantly royalty-based tax system, this chapter agrees with Calder's hypothesis.

In this regard, the OECD has also compiled a list of the dangers of certain tax incentives aimed at attracting capital for mining.[27] Among these pitfalls is that tax incentives erode the amount of revenue that a government would collect, and they enable mining companies involved in aggressive tax planning to avoid tax wherever the law allows,[28] potentially

24 Ibid.
25 Calder (n 12) 22.
26 Ibid at 23.
27 OECD and IGF, *The Hidden Costs of Tax Incentives* (OECD Publishing, 2018) 14.
28 Ibid at 11.

creating an industry that is dissociated from the local community hosting the mine. These problems arising as a result of a government's focus on output-based taxes may be increased in a tax regime with poorly designed legislative provisions. For this reason, Calder emphasizes the importance of identifying administrative difficulties arising from badly designed legislative provisions. This chapter analyses these administrative difficulties which are made manifest in the gaps or ambiguities that make the law unworkable in practice.

In assessing the factors that affect stability in tax administration in Zambia, this chapter also analyses the inadequacies of in-built mechanisms in the tax regimen to deal with changes in revenue collection. Calder discusses the variation in the base for different taxes and explains further that unnecessary complexity in tax administration is worsened by the variation in cost calculation for the purposes of different profit taxes.[29] It calls into question whether such variations are motivated and backed by a change in policy or whether a clear reason exists for the variation. Calder also asserts that differences in revenue and cost calculation should have a clear policy justification and should not be mere accidents of history. This chapter further deconstructs the history of variation in cost calculation for different minerals, with a view to assessing the efficiency of existing mechanisms by which costs of different profit taxes are varied.

The Zambia Chamber of Mines[30] argues that to develop an effective tax regime, there is a need for a sound understanding of the science of mining. Notably, publications from the World Bank[31] and the International Monetary Fund[32] recommend that the effectiveness of a tax regime is linked to its standard consistency with general tax law and adapted to reflect the distinctive characteristics of mining. Using gold mining as a case example, this chapter further explores the notion that a generic tax regime that doesn't cater specifically for each stage of a mine's life cycle is problematic, and this is bound to manifest itself in several inconsistent changes to the country's tax policy.

The argument in this chapter is that mineral tax administration should yield both tangible and intangible benefits.[33] Youth empowerment and economic diversification encompass both forms of benefits directed at a specific

29 Calder (n 12) 21.

30 Zambia Chamber of Mines, *Taxing the Mining Sector* (Mining for Zambia 2019).

31 For example, Cameron and Stanley (n 12).

32 International Monetary Fund, *Revenue Mobilization in Developing Countries* (IMF 2011) 6; International Monetary Fund, *Fiscal Regimes for Extractive Industries: Design and Implementation* (IMF 2012) 6.

33 N Woodroffe, 'Force Majeure and Other Coronavirus-Era Legal Challenges: Lessons for Resource-Dependent Countries from the Glencore-Zambia Dispute' (2020), https://resourcegovernance.org/blog/force-majeure-coronavirus-legal-challenges-resource-dependent-zambia-glencore; KMG Chisanga, *Exploring a wider*

group – namely, youths in Zambia. The prerequisite for this is a stable, predictable fiscal regime. The next section describes the legal framework for mineral royalty tax, the changes made, and the potential impact continued changes would have on gold mining in Zambia.

3 Zambia's Legal Framework for Gold Mineral Royalty Tax

Most (arguably all) governments have an underlying policy that drives their agenda for national economic development.[34] The policy of the Zambian government, particularly, is the 'massive investment and infrastructure development'.[35] The basis of state economic policy, as provided for in Article 10 of the Constitution of the Republic of Zambia,[36] is to maximize government revenue, while simultaneously promoting investment in the mining sector. This requires a sound tax administration system.

The potential for Zambia to contribute to the global supply of gold is truly attractive for the nation aiming to recover its economy in record time after the COVID-19 pandemic. Zambia uses a tax-royalty fiscal regime as provided for in Part 7 of the Mines and Minerals Development Act ('MMDA')[37] which governs mining in the country from large-scale to small-scale and artisanal mining. This section compares the regime in Zambia with that of the more established gold mining jurisdiction, South Africa. This aids in understanding the challenges Zambia may face in revenue collection and evaluating the progressivity of the current tax administration system in Zambia. This section narrows the discussion of the law to provisions governing mineral royalty tax, insofar as these relate to gold mining. For purposes of this section, a mineral royalty is understood as a payment received as consideration for the extraction of minerals.[38] This section also makes reference to the role of administrative bodies (i.e. the Commissioner-General of Zambia Revenue Authority ('ZRA'), Ministry of Mines, and Ministry of Finance) in administering mineral royalty tax.

conceptualization of value in the mining sector: Analysis of the impact of automation on Zambia's labour-related local content (Unpublished LLM Dissertation, University of Dundee 2020); SE Cook, 'Community management of mineral resources: the case of the Royal Bafokeng Nation' (2013) 113 *Journal of the Southern African Institute of Mining and Metallurgy* 61, 67.

34 Ash Amin, 'An Institutionalist Perspective on Regional Economic Development' (1999) Events and Debates 365, 375.

35 Ministry of National Development Planning, *Zambia's 7th National Development Plan 2017–2021* (2017) paras 1.3–1.4.

36 The Constitution of the Republic of Zambia, Chapter 1 of the Laws of Zambia (as Amended by Act No.2 of 2016).

37 Mines and Minerals Development Act No. 11 of 2015.

38 P Daniel, M Keen, and C McPherson, *The Taxation of Petroleum and Minerals: Principles, Problems and Practice* (Routledge, 2010).

3.1. Fiscal Regime for Gold Mining

In Zambia, the calculations of mineral toyalty are based on either the norm[39] or gross value[40] of the minerals. The gross value is used when the royalty is charged on industrial minerals, energy minerals, and gemstones.[41] The norm value is applicable when calculating the royalty charged for base metals (including copper) and precious metals (including gold). This method differs from the South African method of calculating royalties as provided in the Mineral and Petroleum Resources Royalty Act ('MPRRA').[42]

The South African method is far more complex as it factors into the royalty calculation key data such as the mining company's earnings before interest and taxes ('EBIT') and gross sales[43] before capping the maximum payable royalty at 5% for refined minerals[44] and 7% for unrefined minerals.[45] This presents a key difference in the calculations for royalties for minerals such as gold.

South Africa's framework does not classify minerals by their forms as the Zambian framework does but classifies minerals according to their levels of refinement. This is in line with the country's policy priority for local beneficiation.[46] Zambia risks promoting the export of unrefined gold if the current wording is applied. However, the South African regime applies a complex method which increases the sophistication of skills needed to administer its royalty system. Zambia has been known for copper mining that is largely exported in cathode form. The level of local beneficiation has been so low that it became a policy objective in the Mineral Resources Development Policy.[47] If the fiscal regime for gold is to yield economic diversification, the South African example of charging higher royalties for unrefined forms of

39 Defined in s 89(5) of the MMDA as:

'(a) the monthly average London Metal Exchange cash price per tonne multiplied by the quantity of the metal or recoverable metal sold;

(b) the monthly average Metal Bulletin cash price per tonne multiplied by the quantity of metal sold or recoverable metal sold to the extent that the metal price is not quoted on the London Metal Exchange; or

(c) the monthly average cash price per tonne, at any other exchange market approved by the Commissioner-General, multiplied by the quantity of the metal or recoverable metal sold to the extent that the metal price is not quoted on the London Metal Exchange or Metal Bulletin.'

40 Defined in s 89(5) of the MMDA as 'the realised price for a sale free on board at the point of export from Zambia or point of delivery within Zambia'.

41 MMDA, 89(1)(b), (2).

42 No. 28 of 2008.

43 Defined in s 6(1)(a) of the Minerals and Petroleum Resources Revenue Act ('MPRRA') as 'the amount received or accrued during the year of assessment in respect of the transfer of that mineral resource'.

44 MPRRA, s 4(3)(a).

45 MPRRA, s 4(3)(b).

46 Broad-Based Black Socio-Economic Empowerment Charter for the South African Mining and Minerals Industry, 2017 (published June 2017) para 1.

47 Zambia Mineral Development Policy, 2013 10.

minerals can be followed. However, the use of complex methods of calculating royalties may prove a challenge given the level of capacity building and training necessary to administer the gold-mining framework.[48] The future choices made by the Zambian government can be predicted based on the historical changes made to the mining fiscal regime. These are discussed in the next section.

3.2. Changes in the Fiscal Regime

In the October 2014 Budget Speech, the Minister of Finance announced that mineral royalties on the norm value of base metals produced would increase from 6% to 20% on open-cast mining and to 8% on underground mining. *Ad valorem* royalty rates for copper vary between countries, generally ranging between 0% and 8%. The new rate in Zambia for open mines was way above this range. The corporate income tax rate applicable on mining operations, with the exception of mineral processing, was revised from 30% to 0%. Variable profits tax of up to 15% when the taxable income exceeded 8% of gross sales was abolished. These changes, effective from 1 January 2015, changed the mining tax regime from a profit-based tax system to a revenue-based tax system. The royalty regime aimed to close the loopholes companies had used to evade corporate income tax.

In 2018, the MMDA was amended to introduce a semi-progressive system of taxation, in which royalties now follow an accelerating schedule. The effect of the amendment was to remove the 9% royalty on copper, introduce a price-based royalty, and remove the variable profit tax. The argument for the amendment was that the price-based royalty would enhance tax collection, while some stakeholders argue that the proposition was retrogressive, and would negatively impact the collection of tax revenue.

In 2019, the mineral royalty tax increased and now ranges from 5.5% (previously 4%) when prices are low, to 7.5 % (previously 6%) and 10% depending on whether the price goes higher. Mineral royalty increased by 1.55 across the sliding mineral bands makes Zambia uncompetitive in attracting capital by pushing out break-even times for mining projects.[49] Making MRT non-deducible means that in a high price environment at 10% on the new increased scale, Zambia metals and mining operators will on average remain with 2.4% before tax (1.6% after) in a high copper price environment.

The provisions discussed above present a picture of a clear regime for mineral royalty tax payable for gold mining and the relevant administrative bodies responsible for tax administration.

48 Calder (n 12) 64.
49 Zambia Chamber of Mines (n 30).

3.3 Potential Effect of Changes to the Mineral Royalty Tax Regime on the Administration of Tax in Gold Mining

This section will examine the administration of tax in artisanal and small-scale gold mining in light of the changes to the tax regime discussed in the section above. The frequency of these changes gives the impression that Zambia's fiscal regime is unstable.[50]

The 2020 budget from the mining industry made two major proposals. The first proposal was to set a maximum of 7.5% mineral royalty tax payable across the mining sector. The industry players proposed that a 1.5% increment on each band of the sliding scale has a maximum of 7.5% royalty applying when copper prices equal or exceed an average price of US$7,500/tonne. The second was to make mineral royalty tax deductible for purposes of computing corporate income tax as the current regime raised the tax burden on mining companies to unsustainable levels.

Setting mineral royalty tax in this way (i.e. through annual budget speeches) sets a precedent where the royalty rate is set is too low and will not cover rents when there is an increase in commodity price. Additionally, these changes to the mineral royalty tax regime do not account for fluctuating costs such as the cost of inputs for mineral processing which must be considered when setting mineral royalty tax. Further, because the taxes paid by a mining company are not limited to mineral royalty tax, it is necessary to have a regime that affords stability in the collection of royalties if gold mining is to achieve aims such as youth empowerment.

It also appears that there is no proper strategy to identify, understand, and manage the weaknesses in the law is developed.[51] Each consecutive budget indicates a change that the industry players reject, and this calls into question whether there was consultation with the aim of increasing understanding of the legislative provisions and the possible risks they present to the mining sector.

Developing an effective fiscal regime requires sound understanding of the science of mining,[52] a critical issue when artisanal and small-scale gold mining is concerned. The net effect of designing the fiscal regime with a focus on ownership of minerals is that the government has neglected to design a consistent tax policy. As will be seen, consistency will be necessary if the benefits from gold mining are to reach the youth affected by gold mining activity.

Additionally, the changes targeted at large-scale mining companies also have an impact on artisanal and small-scale mining in ways that the government may not always envision. When one considers the roles of government agencies governing mineral royalty tax (the Commissioner-General of

50 Manley (n 22).
51 Calder (n 12) 22.
52 Zambia Chamber of Mines (n 30).

ZRA, Ministry of Mines, Ministry of Finance), these difficulties are put into context.

The ZRA is established pursuant to the Zambia Revenue Authority Act[53] as a body corporate and an agency of government under the direction of the Minister of Finance. ZRA is led by the Commissioner-General appointed by the President. The Commissioner-General is the chief executive of ZRA and is responsible for the execution of functions. The role of ZRA is to assess, charge, levy, and collect all revenue due to the government and ensure that all revenue collected is, as soon as reasonably practicable, credited to the treasury.[54] The government has the discretion to grant tax relief to mining companies in the form of various allowances.[55] The pitfalls of such tax reliefs have previously been mentioned.[56] The main pitfall is that the tax reliefs effectively reduce the revenue collectable by the government, which would otherwise be channelled towards youth empowerment. Moreover, such reliefs are granted subject to administrative discretion, and there is thus no mechanism to challenge the merits of the decision to grant these reliefs.

There are two ministries that affect the administration of mineral royalty tax. The Ministry of Mines and Minerals Development provides administrative oversight over mining and is responsible for the development and management of mineral resources in a sustainable manner for the benefit of the people of Zambia.

The Ministry of Finance manages government financial assets and proposes economic and financial policy. The Ministry monitors how the mining sector is operating annually. It also prepares the annual fiscal budget and issues adequate regulations for its execution. These regulations and policy statements affect the distribution of revenue collected through mineral royalties. Notably, there are no specific policies or directives in place to regulate the administration of royalties in a way that would directly impact youth economic empowerment.

This shows that there are unique but related roles for the government agencies governing mineral royalty tax. However, these agencies do not make provision for tax administration mechanisms that are specific to artisanal and small-scale miners ('ASMs'), who may not have the requisite knowledge of all the procedures involved with the various government agencies. As discussed in Section 1 above, administrative difficulties arising from badly designed legislative provisions are also problematic,[57] as the process is not streamlined to cater for ASMs whose formalization into the sector is being undertaken gradually. It is important to identify administrative difficulties arising from badly designed legislative provisions, but this is outside the scope of this chapter.

53 Chapter 321 of the Laws of Zambia ('ZRA Act').
54 ZRA Act, s 11(b).
55 MMDA, s 93.
56 Text to n 27.
57 Calder (n 12) 22

The issue of principal concern in this chapter is youth empowerment. The design of mining laws does not allow for express reference to specific issues such as youth empowerment.[58] However, after considering the Mineral Resources Development Policy, no reference is made to youth empowerment. Issues pertaining to youth unemployment are found in the National Development Plan designed to be pursued from 2017 to 2022.[59] The design of the fiscal regime must take into account national development goals as this will guide how revenues accrued from mining will be applied by the government once collected. The next section describes two potential applications for the revenue collected from gold mining that can be achieved only if the fiscal regime stays stable.

4 Potential Application of Gold Revenues

This section of the research discusses the potential application of gold revenues towards youth employment and empowerment and diversification beyond mining. The section gives an example of a company whose activities indicate what can be achieved if there is a collaborative effort between private and public sector stakeholders responsible for gold mining.

4.1. Achieving Youth Empowerment and Employment Through Gold Mining

The COVID-19 pandemic led African economies into recession,[60] exacerbating existing problems such as unemployment.[61] Africa Development Bank reports highlight unemployment and lack of empowerment[62] among the

58 James Otto and others, *Mining Royalties: A Global Study of Their Impact on Investors, Government, and Civil Society* (World Bank, 2006).

59 Government of the Republic of Zambia, *7th National Development Plan* (Ministry of National Development and Planning, 2017) ('Zambia NDP') 20.

60 Albert G Zeufack, Cesar Calderon, Gerard Kambou, Calvin ZDjiofack, Megumi Kubota, Vijdan Korman, Catalina Cantu Canales, *Africa's Pulse, No. 21, Spring 2020: An Analysis of Issues Shaping Africa's Economic Future* (World Bank, 2020) 1.

61 African Development Bank (AfDB) *Ghana – Country Strategy Paper 2019–2023* (AfDB 2019); *Enhancing Capacity for Youth Employment in Africa Some Emerging Lessons* (AfDB, 2011); World Bank, *Republic of the Gambia: Overcoming a No-Growth Legacy – Systematic Country Diagnostic* (World Bank, 2020) 62; World Bank, *Central African Republic: Priorities For Ending Poverty and Boosting Shared Prosperity – Systematic Country Diagnostic* (World Bank, 2019) 60; World Bank, *Nigeria on the Move: A Journey to Inclusive Growth – Systematic Country Diagnostic* (World Bank, 2019) 9.

62 For current purposes, the term 'empowerment' is taken by analogy from the definition of 'women empowerment' which means gaining access to opportunities and developing a person's capacities to shape that person's life or that of the person's community, in all spheres of life. See SP Ng'ambi and KMG Chisanga, *International Investment Law and Gender Equality: Stabilization Clauses and Foreign Investment* (2020, Routledge).

youth.[63] Addressing youth unemployment requires considering issues in the informal sector which accounts, by some estimates, for 90% of Zambia's labour force.[64] The main sector currently giving opportunities to informal artisans is the construction sector for jobs such as bricklaying.[65] Despite the annual economic growth rate of more than 5% in the years preceding 2011, African countries still face high levels of youth unemployment.[66] This is lamentable given that Africa is the youngest continent in the world with a growing youth labour force at 37% in 2011.[67]

A related area of concern is the level of skills transfer to the youth population surrounding mines. In this regard, the United Nations developed an atlas to show how mining can contribute to the achievement of the Sustainable Development Goals (SDGs).[68] Mining companies are encouraged to integrate the fourth SDG (SDG 4) into their core business by training their workforce and collaborate with universities and schools in this regard. For example, the World Bank established the African Centre of Excellence in Sustainable Mining at Zambia's Copperbelt University.[69] Further, BHP Billiton financed the creation of the International Energy Policy Institute in Adelaide, and the Institute for Sustainable Resources in London at the Bartlett Faculty of the Built Environment, University College London.[70] SDG 4 is particularly important regarding introduction of technologies in gold mining.[71]

Achieving youth empowerment requires applying gold revenues to address societal inequality.[72] Further, the level of distribution of benefits to local

63 AfDB (n 61) 1. For present purposes, 'youth' means any person between the ages of 15 and 24. See Secretary-General's Report to the General Assembly, A/36/215, 1981.

64 M Shah, *The Informal Sector in Zambia: Can It Disappear? Should It Disappear?* (International Growth Centre, 2012)

65 J Sutton and G Langmead, *An Enterprise Map of Zambia* (International Growth Centre, 2014) 134.

66 AfDB (n 61) 1.

67 OECD, *African Economic Outlook 2011: Africa and its Emerging Partner* (2011, OECD Publishing).

68 United Nations Development Programme (UNDP), *Mapping Mining to the Sustainable Development Goals: An Atlas* (World Economic Forum, 2016).

69 Jameson Mbale, *Interrogating the Constructs, Scope and Prospects for Achieving the National Vision 2030, the SDGs and the Africa Union Agenda 2063* (International Conference on Geology, Mining, mineral and Groundwater Resources of Sub-Saharan Africa, 2017).

70 Jane Rendell, 'Giving an account of oneself: Architecturally' (2016) 15 *Journal of Visual Culture* 334.

71 Ricardo Vinuesa and others, 'The role of artificial intelligence in achieving the Sustainable Development Goals' (2020) 11 *Nature Communications* 1.

72 KL Sokoloff and EM Zolt, 'Inequality and taxation: Evidence from the Americas on how inequality may influence tax institutions'(2005) 59 *Tax Law Review* 167.

communities is a key concern in gold mining.[73] Another application of revenues can be economic diversification.

4.2. Economic Diversification Beyond Gold Mining

The post-COVID-19 economy will present both challenges and opportunities for gold mining in Zambia. Governments designing and redesigning their tax regime will have to consider particular economic objectives such as economic diversification. Companies operating in the gold mining sector will also have to consider ways to contribute to diversification.

One mining project showing promise for contributing to diversification beyond gold mining is that of Migodi-Auric Limited ('Migodi'). Migodi is an investment company that facilitates formalization of artisanal and small-scale gold mining in rural areas.[74] The government of the Republic of Zambia has signed an Investment Promotion and Protection Agreement with Migodi to facilitate Migodi's RUP.[75] This section describes the details of the RUP. The information on the project was gathered through a questionnaire[76] sent to the company and a report[77] compiled by a youth-led cooperative working closely with the company. The RUP is designed around the different dimensions of mining as illustrated in Figure 7.1 below.

The RUP Industrialization Training Centre was set up to conduct extensive research to understand artisanal science, technology, and innovation, as such research will spur the RUP objective of facilitating Inclusive Sustainable Economic Development through Rural Industrialization.[78] At the core of RUP are four interlinked programmes. Each of these is described below.

4.2.1. The Small-Scale Mining Mechanization Project

RUP facilitates the improvement skills in mining and in the requisite legal procedures associated in this sector.[79] Further, RUP provides financial assistance to improve productivity in mining and capitalize on value-adding opportunities through the implementation of a Small-Scale Mining Mechanization Project ('SSMMP').[80] This programme provides machinery to artisanal and small-scale gold miners at Migodi's cost. In this way, skills are transferred to the community and mining is done in a more sustainable manner.

73 E Dietsche, T Ruffer, and O Cisse, *Mineral Taxation Regimes: A Review of Issues and Challenges in Their Design and Application* (London: International Council on Mining and Metals, 2009).
74 Migodi-Auric Limited *Extended Profile* (Migodi-Auric Limited, 2017) 1.
75 Government of the Republic of Zambia ('GRZ') *Investment Promotion and Protection Agreement with Migodi-Auric Limited* (2018).
76 K Chisanga, Migodi-Auric RUP Questionnaire, 25 January 2021.
77 Twikatane Imisepela Multi-Purpose Society (TIMS), *Report on the Migodi-Auric Project* (TIMS, 2018).
78 Chisanga (n 76) para 3.
79 Chisanga (n 76) para 1.
80 GRZ (n 75) 1.

Figure 7.1 Dimensions of value drawn from mining
Source: World Economic Forum (2013)

4.2.2. The Small-Scale Farming Mechanization Program

Zambia's government, through the Farmer Input Support Programme ('FISP') provides agricultural inputs such as fertilizer and seed. However, productivity remains low as a result of many factors including low mechanization of farming activities.[81] Seeing an opportunity to assist, Migodi aligned its activities to the Second National Agricultural Policy ('SNAP').[82] The policy's main aim is to increase production and productivity, promoting the commercialization of agriculture among various other measures.

4.2.3. The Small-Scale Tourism Project

This part of the RUP is premised on the notion that all chiefdoms have extraordinary potential for a vibrant tourism industry. Evidence of such potential is witnessed by the celebration of Zambian traditional ceremonies such as the Eastern Province's *Ncwala* ceremony celebrated by many artisanal and small-scale gold miners.[83] RUP through its Small-Scale Tourism Project facilitates tourism-related skill sets and infrastructure to contribute to the

81 Chisanga (n 76) para 3.
82 Ibid. See also Ministry of Agriculture and Ministry of Fisheries and Livestock, *Second National Agricultural Policy* (GRZ, 2016).
83 TIMS (n 77) 2.

social and economic development of chiefdoms. In this regard, Migodi aligned itself with the 2015 National Tourism Policy[84] wherein Zambia's Government is guided by the principles of sustainable development as it plans and develops tourism. Migodi, in its efforts to expand tourism in the chiefdoms, will be guided by these same principles.

4.2.4. The Chiefdom Infrastructure Development Trust

The critical RUP intervention at chiefdom level is development of infrastructure, as chiefdoms require adequate and high-quality social, technological, and economic infrastructure.[85] Many of Zambia's chiefdoms are endowed with various minerals, but most mining activities in chiefdoms are small scale and/or illegal and poorly coordinated.[86] This significantly limits the benefits from mining that may accrue to the chiefdoms, their people, and ultimately the Zambian economy. Migodi's RUP facilitates improvement in the coordination and regulation of small-scale mining activities in the chiefdoms where it operates (i.e. in the Lusaka, Eastern, and Central Provinces) in order to enable the chiefdoms and their citizenry to benefit through inclusive participation in mining activities.

The above are some of the ways the revenues from gold mining can address economic diversification and youth empowerment. The next section will relate the success of these proposed programmes, in light of the changes to the mineral royalty tax regime and the potential effect of these changes on the administration of tax in gold mining.

5 The Potential Effect of Changes to the Mineral Royalty Tax Regime in Gold Mining on the Sustainability of Economic Diversification Based on the RUP

So far, this chapter has shown the changes to the mineral royalty tax regime over the years. The frequency of these changes gives the impression that Zambia's fiscal regime is unstable.[87] This section will appraise the sustainability of the RUP based on revenue from ASM gold mining considering the changes to the tax regime considered in the sections above. As will be discussed, consistency in the tax regime is necessary if the benefits from gold mining are to be distributed to the youth as part of the government's empowerment programme.

84 Ministry of Tourism and Arts, *Tourism Policy for Zambia* (Ministry of Tourism 2015). See also Policy Monitoring and Research Centre, *National Tourism Policy Presentation* (Policy Monitoring and Research Centre, 2017); Policy Monitoring and Research Centre, *PMRC Policy Analysis of the National Tourism Policy: Repositioning the Tourism Sector as One of the Country's Major Economic Contributors* (Policy Monitoring and Research Centre, 2018).

85 Chisanga (n 76) para 3; GRZ (n 75) 1.

86 TIMS (n 77) 1.

87 Chisanga (n 76).

The instability of the regime will undoubtedly impact on the success of the RUP. First, the manner in which mineral royalty tax has been set up in the preceding years (i.e. through annual budget speeches) gives the impression that there is no proper strategy to identify, understand, and manage the weaknesses in the way the law is developed, which involves multi-stakeholder consultation. Each consecutive budget indicates a change that the industry players reject, and this calls into question whether there was consultation with the aim of increasing understanding of the legislative provisions and the possible risks they present to the mining sector. This in essence means there is a lack of effective risk assessment in the development of the fiscal regime.

Second, the royalty rate is restrictively defined and does not cover rents when there is an increase in commodity price. These changes to the mineral royalty tax regime do not account for fluctuating costs such as the cost of inputs for mineral processing which must be considered when setting mineral royalty tax. In relation to the effectiveness of tax administration and gold mining, this is problematic because it interposes the risk of instability in the collection of royalties. The success of the RUP and youth empowerment is in this regard jeopardized, as the revenue collected in this manner does not represent the fullest extent of revenue collectable when there is an increase in commodity price.

Developing an effective fiscal regime requires a sound understanding of the science of mining,[88] and this is particularly critical where artisanal and small-scale gold mining is concerned. The changes discussed have been predominantly targeted at large-scale mining companies, and this has an impact on artisanal and small-scale mining in ways that the government may not always envision.

This shows that there are unique but related roles for the government agencies governing mineral royalty tax. However, these agencies do not make provision for tax administration mechanisms which are specific to ASMs, who may not have the requisite knowledge of all procedures involved with the various government agencies. As discussed in Section 1 above, administrative difficulties arising from badly designed legislative provisions are also problematic, as the process is not streamlined to cater for ASMs whose formalization into the sector is being undertaken gradually. It is important to identify administrative difficulties arising from badly designed legislative provisions but this is outside the scope of this chapter.

All in all, the viability of the RUP is impaired by the fact that, by design, the principal legislative laws do not expressly reference youth empowerment. The principal policy, the Mineral Resources Development Policy, also does not make any reference to youth empowerment. Issues pertaining to youth unemployment are found in the National Development Plan designed to be pursued from 2017 to 2022. The net effect is that the fiscal regime does not provide for the collection and administration of revenue in a manner that can be consistently applied towards youth empowerment.

88 Zambia Chamber of Mines (n 30).

6 Conclusion

In conclusion, this chapter has shown that there is a clear framework for determining mineral royalty tax for gold mining. It has also been spotlighted that there are various organs of the government, mainly the Ministry of Mines, the Ministry of Finance, and the Zambia Revenue Authority, which have complementary roles in tax administration. The Ministry of Mines is tasked with the responsibility of administrative oversight of the Mines and Minerals Development Act, which makes provision for mineral royalties payable for base metals such as gold. A historical analysis of the current framework has shown that the various changes to the fiscal regime are unlikely to affect stability in the mineral royalty tax charged on gold. This is because the changes are primarily targeted to enhance revenue from large-scale mining, and do not take into account the different costs and nuances of artisanal and small-scale gold mining.

This chapter has also explained the potential for revenues from gold mining to be applied towards youth empowerment and economic diversification. The example of the RUP by Migodi-Auric aimed at promoting rural economic empowerment through gold mining has shown that the main goal set by the government and Migodi-Auric is to assist in the formalization of artisanal and small-scale gold mining so as to enhance revenue generation and collection. This is in line with the emerging recognition that ASM is of significant social and economic importance and should be brought within efforts to regulate the mining industry so as to maximize mineral resource development.

The chapter has shown that the risks and flaws connected with tax administration are exacerbated by the current mineral royalty regime which does not adequately cater for all the nuances of artisanal and small-scale gold mining. The RUP does have the potential to achieve youth empowerment by introducing sustainability in artisanal and small-scale gold mining. The progression of the RUP can, however, be hindered by the instability of the fiscal regime and the current challenges affecting tax administration, which are largely attributable to the absence of strategies to identify, understand, and manage the weaknesses inherent in the mineral royalty system.

As most artisanal and small-scale activity is driven by poverty, the challenges in tax administration could potentially force the miners to opt out of the framework, making them vulnerable to exploitation on the market. This could jeopardize the success of the RUP, given the critical role of gold mining in the overall scheme of youth empowerment. There is therefore need to focus on implementing programmes that focus on multi-stakeholder collaboration in order to streamline the process of developing a tax regime and/or framework which addresses the inherent peculiarities of artisanal and small-scale mining.

References

Amin A. 1999. 'An Institutionalist Perspective on Regional Economic Development.' *Events and Debates* 365, 375.

African Development Bank (AfDB)2011. *Enhancing Capacity for Youth Employment in Africa Some Emerging Lessons.*

African Development Bank (AfDB)2019. *Ghana – Country Strategy Paper 2019–2023.*

Andreasson, S. 2015. 'Varieties of resource nationalism in sub-Saharan Africa's energy and minerals markets.' *The Extractive Industries and Society* 2, no. 2: 310–319.

Bremmer I and Johnston R. 2009. 'The rise and fall of resource nationalism.' *Survival* 51, no. 2, 149–158.

Calder J. 2015. *Administering Fiscal Regimes for Extractive Industries.* International Monetary Fund.

Cameron PD and Stanley MC. 2017. *Oil, Gas, and Mining: A Sourcebook for Understanding the Extractive Industries.* World Bank.

Chisanga KMG. 2020. *Exploring a wider conceptualization of value in the mining sector: Analysis of the impact of automation on Zambia's labour-related local content.* Unpublished LLM Dissertation, University of Dundee.

Chisanga KMG. 2021. Migodi-Auric RUP Questionnaire 25 January.

Christian Aid. 2009. *Undermining the Poor: Mineral Taxation Reforms in Latin America.* Christian Aid.

Collier P. 2010. 'Resource taxation for low-income countries' in Daniel P, Keen M, and McPherson C (eds), *The Taxation of Petroleum and Minerals: Principles, Problems and Practice.* Routledge.

Cook SE. 2013. 'Community management of mineral resources: the case of the Royal Bafokeng Nation.' *Journal of the Southern African Institute of Mining and Metallurgy* 113: 61–67.

Daniel P, Keen M, and McPherson C. 2010. *The Taxation of Petroleum and Minerals: Principles, Problems and Practice.* Routledge.

Dietsche E, Ruffer T, and Cisse O. 2009. *Mineral Taxation Regimes: A Review of Issues and Challenges in Their Design and Application.* London: International Council on Mining and Metals.

Government of the Republic of South Africa. 2017. *Broad-Based Black Socio-Economic Empowerment Charter for the South African Mining and Minerals Industry, 2017* (published June 2017).

Government of the Republic of Zambia. 2013. *Zambia Mineral Development Policy.* Ministry of Mines.

Government of the Republic of Zambia. 2015. *Tourism Policy for Zambia.* Ministry of Tourism and Arts.

Government of the Republic of Zambia. 2017*7th National Development Plan.* Ministry of National Development and Planning.

Government of the Republic of Zambia. 2018. *Investment Promotion and Protection Agreement with Migodi-Auric Limited.*

Government of the Republic of Zambia. 2016. *Second National Agricultural Policy.* Ministry of Agriculture and Ministry of Fisheries and Livestock.

International Monetary Fund. 2011. *Revenue Mobilization in Developing Countries.* IMF.

International Monetary Fund. 2012. *Fiscal Regimes for Extractive Industries: Design and Implementation.* IMF.

Koch N and Perreault P. 2019. 'Resource nationalism.' *Progress in Human Geography* 43, no. 4: 611.

La Fleche, ER (ed.). 2019*The Mining Law Review 8th edition.* Law Business Research.

Liu B and Mwanza FM. 2014. 'Towards Sustainable Tourism Development in Zambia: Advancing Tourism Planning and Natural Resource Management in Livingstone (Mosi-oa-Tunya) Area.' *Journal of Service Science and Management*7: 30–45.

Manley D. 2017. 'Ninth time lucky: Is Zambia's Mining Tax the Best Approach to an Uncertain Future?' in *Briefing October 2017*. Natural Resource Governance Institute.

Mares, DR. 2010. *Working Paper: Resource Nationalism and Energy Security in Latin America: Implications for Global Oil Supplies*. James A. Baker III Institute for Public Policy.

Mbale J. 2017. *Interrogating the Constructs, Scope and Prospects for Achieving the National Vision 2030, the SDGs and the Africa Union Agenda 2063*. (International Conference on Geology, Mining, mineral and Groundwater Resources of Sub-Saharan Africa, 2017).

Migodi-Auric Limited. 2017. *Extended Profile*. Migodi-Auric Limited.

Ministry of Finance. 2020. 'Treasury Releases K1 Billion for Crop Purchases.' 19 June 2020, www.mof.gov.zm/?p=6184

Monaldi, FJ. 2014. 'Is resource nationalism fading in Latin America? The case of the oil industry.' Issue Brief. www.bakerinstitute.org/research/resource-nationalism-fading-latin-america-case-oil-industry

Ndulo M. 1986. 'Mining Legislation and Mineral Development in Zambia.' *Cornell Law Faculty Publications* 19, no. 1: 1.

Ng'ambi SP and Chisanga KMG. 2020. *International Investment Law and Gender Equality: Stabilization Clauses and Foreign Investment*. Routledge.

OECD and IGF. 2018. *The Hidden Costs of Tax Incentives*. Paris: OECD Publishing.

OECD. 2011. *African Economic Outlook 2011: Africa and its Emerging Partner*. Paris: OECD Publishing.

OECD. 2014. 'Fundamental principles of taxation' in *Addressing the Tax Challenges of the Digital Economy*. Paris: OECD Publishing.

Office of the Auditor-General. 2020. *Report of the Auditor General on the Performance of the Tourism Sector in Ensuring an Increase in the Length of Stay of International Tourist*. Lusaka: Office of the Auditor-General.

Otto J and Cordes J. 2002. *The Regulation of Mineral Enterprises: A Global Perspective on Economics, Law and Policy*. Westminster, CO: Rocky Mountain Mineral Law Foundation.

Otto J. 2006. *Mining Royalties: A Global Study of Their Impact on Investors, Government, and Civil Society*. World Bank Publications.

Otto J, Andrews CB, Cawood F, Doggett M, Guj P, Stermole F, Stermole J and Tilton J. 2006. *Mining Royalties: A Global Study of Their Impact on Investors, Government, and Civil Society*. The World Bank.

Policy Monitoring and Research Centre. 2017. *National Tourism Policy Presentation*. Policy Monitoring and Research Centre.

Policy Monitoring and Research Centre. 2018. *PMRC Policy Analysis of the National Tourism Policy: Repositioning the Tourism Sector as One of the Country's Major Economic Contributors*. Policy Monitoring and Research Centre.

Rendell J. 2016. 'Giving an account of oneself: Architecturally.' *Journal of Visual Culture* 15, no. 3, 334.

Shah M. 2012. *The Informal Sector in Zambia: Can It Disappear? Should It Disappear?*International Growth Centre.

Siwale T and Chibuye B. 2019. 'Mining taxation policy in Zambia: The tyranny of indecision.' 13 June. www.theigc.org/blog/mining-taxation-policy-in-zambia-the-tyranny-of-indecision

Sokoloff KL and Zolt EM. 2005. 'Inequality and taxation: Evidence from the Americas on how inequality may influence tax institutions.' *Tax Law Review* 59, no. 2: 167.

Sutton J and Langmead G. 2014. *An Enterprise Map of Zambia*. International Growth Centre.

Tom A and Nalule A. 2020. 'Key Issues in Petroleum Revenue Management.' http://afrienergyminerals.org/wp-content/uploads/2020/01/REVENUE-MANAGEMENT-AND-KEY-ISSUES2222222222.pdf

Twikatane Imisepela Multi-Purpose Society (TIMS). 2021. *Report on the Migodi-Auric Project*.

United Nations (UN). 1981. *Secretary-General's Report to the General Assembly, A/36/215*.

United Nations Development Programme. 2016. *Mapping Mining to the Sustainable Development Goals: An Atlas*. World Economic Forum.

Vinuesa R, Azizpour H, Leite I, Balaam M, Dignum V, Domisch S, Felländer A, Langhans SD, Tegmark M and Nerini FF. 2020. 'The role of artificial intelligence in achieving the Sustainable Development Goals.' *Nature Communications* 11, no. 1: 1–10.

Williams C. 2020 'Zambia's Mining Industry Strikes Gold.' https://borgenproject.org/zambias-mining-industry

Wilson JD. 2015. 'Understanding resource nationalism: Economic dynamics and political institutions.' *Contemporary Politics* 21, no. 4: 399–416.

Woodroffe N. 2020. 'Force Majeure and Other Coronavirus-Era Legal Challenges: Lessons for Resource-Dependent Countries from the Glencore-Zambia Dispute.' https://resourcegovernance.org/blog/force-majeure-coronavirus-legal-challenges-resource-dependent-zambia-glencore.

World Bank. 2019a. *Central African Republic: Priorities for Ending Poverty and Boosting Shared Prosperity - Systematic Country Diagnostic*. Washington, DC: World Bank.

World Bank. 2019b. *Nigeria on the Move: A Journey to Inclusive Growth – Systematic Country Diagnostic*. Washington, DC: World Bank.

World Bank. 2020. *Republic of the Gambia: Overcoming a No-Growth Legacy – Systematic Country Diagnostic*. Washington, DC: World Bank.

Zambia Chamber of Mines. 2019. *Taxing the Mining Sector*. Mining for Zambia.

ZCCM-IH. 2020. 'Press Statement: ZCCM-IH and Its Investment Strategy in the Gold Sector in Zambia.' www.zccm-ih.com.zm/2020/05/26/zccm-ih-and-its-investment-strategy-in-the-gold-sector-in-zambia

Zeufack, AG, Calderon C, Kambou G, Djiofack CZ, KubotaM; Korman V and Cantu Canales C. 2020. *Africa's Pulse, No. 21, Spring 2020: An Analysis of Issues Shaping Africa's Economic Future*. Washington, DC: World Bank.

8 Regional Cooperation and Mining Governance in Africa

An Assessment of The Africa Mining Vision 2009

Solomon N. Obulor and Eddy L. Wifa

1 Introduction

The realisation that decades of energy and natural resource mining in Africa has not manifestly promoted economic prosperity and sustainable development to catapult the continent to modernisation as seen in the Global North in the eighteenth and nineteenth centuries is debatably one of the reasons that informed the African Union's (AU) decision to formulate the Africa Mining Vision (AMV) in 2009 with the lofty aim of ensuring a 'transparent, equitable and optimal exploitation of mineral resources to underpin broad-based sustainable growth and socio-economic development' in the continent.[1]

This chapter, therefore, broadly assesses the AMV potency in promoting cooperation in African-rich resource countries' mining governance framework that will in turn facilitate resource-based sustainable socio-economic growth and development in Africa. It will be demonstrated that while the AMV may be the silver bullet (as it 'makes some headway in reflecting global norms for the governance of mineral sectors') that AU resource-rich member states can leverage to enact laws that will ensure that energy and natural resources exploitation within their border facilitates socio-economic growth, sustainable development, and industrialisation to catapult Africa to modernisation, the AMV like previous AU resource-based development strategies is ostensibly deficient.[2] The argument here is not that the AMV is a completely failed attempt in restructuring resource exploitation regulation to facilitate broad-based sustainable development in the continent, but that the current form in which the AMV exists lacks the mechanisms to ensure its domestication and implementation in AU member states and, therefore, may not achieve the anticipated uniform resource development regulation in the continent that can facilitate socio-economic growth that will lead to industrialisation.

1 Africa Union, Africa Mining Vision 2009, https://docs.igihe.com/IMG/pdf/africa_mining_vision_english.pdf, accessed 14/03/2022.

2 United Nations Economic Commission for Africa, *A Country Mining Vision Guidebook: Domesticating the Africa Mining Vision* (2014), https://repository.uneca.org/handle/10855/22836, accessed 25/05/2022.

DOI: 10.4324/9781003284437-10

In the first instance, the AMV did not address the trajectories including but not limited to human capital, political institutional quality, and commodity price volatility that interact with resource development to facilitate socio-economic growth and sustainable development.[3] Worthy of note in this regard is the fact that the AMV tasks the heads of AU resource-rich member states (most of whom have rent-seeking practices) to implement the AMV without first addressing the internal issues of poor governance, institutional weaknesses, and corruption. Second, issues relating to environmental justice, host communities' engagement, and energy transition were not properly addressed. To address issues relating to host communities' engagement in decision making and benefit sharing in relation to the development of subsoil resources on host communities' land, the chapter recommends the adoption of an impact and benefit agreement (IBA) from a regional point of view to create a tripartite partnership between the host country, resource developers, and host communities in order to promote local citizens' participation in resource development on their land and facilitate a holistic rather than just national economic growth and sustainable development.

Finally, the domestication of the AMV, and translation of the underpinning philosophy of the AMV to effective country mining visions (CMV) or even resource development policies across Africa have been very slow and challenging owing to the AMV soft law (non-binding) nature. Mixed results have been witnessed in countries that have managed to domesticate the AMV owing to different approaches in implementation informed by differences in socio-economic development priorities.[4] 'Questions, therefore, still linger as to whether the AMV can indeed become a catalyst for inclusive economic development.'[5] This chapter proffers solutions to these questions.

Following the brief introduction, the chapter is divided as follows. Section 2 briefly analyses resource development and the resource curse thesis in Africa using only Nigerian experience due to word-limit constraints. Section 3 analyses the essentials of the AMV as a regional mining governance regime. Section 4 assesses the potency of the AMV in promoting cooperation in African countries' mining regimes in order to ensure that mining in Africa catalyses diversified industrial development that is inclusive. It also discusses some shortcomings of the AMV.

Section 5 recommends the adoption of IBA on a regional basis that will create a tripartite partnership needed for resource development to facilitate

3 Jean-Philippe C Stijns, 'Natural Resource Abundance and Economic Growth Revisited' (2005) 30 *Resources Policy* 111; Christopher Blattman et al, 'Winners and Losers in the Commodity Lottery: The Impact of Terms of Trade Growth and Volatility in the Periphery 1870–1939' (2007) 82(1) *Journal of Development Economics* 156–179.

4 Paul Justice Kamlongera, 'Domesticating the Africa Mining Vision in Malawi: Approaches and Experiences' (2020) *Canadian Journal of Development Studies* 470.

5 Ibid.

holistic socio-economic growth and development. Finally, Section 6 concludes that the AMV's ability to promote resource-based development in Africa as a region is dependent on (1) the AMV becoming mandatory to have a binding effect on member states and (2) AU member states' cooperation in developing requisite human capital and political institutions conducive for resource development to facilitate socio-economic sustainable development. This is because 'no successful implementation of a robust compliance framework can succeed unless the regulator has the capacity and capability to monitor and enforce compliance'.[6]

2 Subsoil Mineral Development and the Concept of the Resource Curse in Africa: Myth or Reality

The importance of subsoil mineral resources[7] development in fuelling socio-economic growth and sustainable development of countries abundantly endowed has been a subject of much scholarship.[8] Energy and natural resource endowment usually form part of a country's natural stock of capital and, when mined, generate wealth for the country. The wealth generated (when properly managed) promotes socio-economic prosperity for the present and future generations.[9] It is common knowledge that the majority of the developed world's (Global North) socio-economic growth and industrialisation were largely hinged on energy and natural resources mining.[10]

Britain, France, Belgium, Germany, and the United States of America are examples of countries that industrialised and attained economic development during the eighteenth and nineteenth centuries based chiefly on energy and natural resource wealth.[11] In particular, abundant coal deposits in Britain are

6 Otsile Matlou et al, *African Mining Legislation Atlas: Toolkit on Local Development & Community Engagement in Mining Projects* (The World Bank Group 2022) 25.
7 That is, energy and natural resources such as diamond, silver, gold, tin, petroleum, and iron ore.
8 Jim Cust and David Manley, 'The Natural Resource Charter Second Edition' (2014) *Natural Resource Governance Institute* 4, http://citeseerx.ist.psu.edu/viewdoc/download?doi=10.1.1.468.8759&rep=rep1&type=pdf, accessed 10/01/2022.
9 Syed Rahim et al, 'Do Natural Resources Abundance and Human Capital Development Promote Economic Growth? A Study on the Resource Curse Hypothesis in Next Eleven Countries' (2021) 4 *Resources, Environment and Sustainability* 1–2.
10 Thorvaldur Gylfason et al, 'A Mixed Blessing: Natural Resources and Economic Growth' (1997) Centre for Economic Policy Research Discussion Paper No 1668 1–5, https://papers.ssrn.com/sol3/papers.cfm?abstract_id=49193, accessed on 08/03/2022
11 Gavin Wright, 'The Origin of American Industrial Success, 1878–1949' (1990) 80 (4) *American Economic Review* 635; Alan Fernihough and Kevin Hjortshøj O'Rourke, 'Coal and the European Industrial Revolution' (2014) NBER Working Paper No. 19802, www.nber.org/system/files/working_papers/w19802/w19802.pdf, accessed 10/01/2020, accessed 19/01/2023.

said to have been fundamental for its industrialisation and economic development.[12] As Pollard opines, 'the map of the British Industrial Revolution, it is well known, is simply the map of the coalfields ... and Britain's good fortune in being well endowed with coal has often been noted'.[13] This historical conventional insight has been referred to as a resource blessing: the idea that abundant mining promotes the overall economic growth of a country.[14]

However, scholarship in development and political economics and, recently, international economic law has since the Second World War defied the conventional insight of mining being a blessing that promotes the economic prosperity of the countries where the mining takes place.[15] There is a growing body of literature with empirical findings of, arguably, a negative correlation between mining and economic growth in many energy- and natural resource-rich countries of the Global South.[16] The central claim of the literature tends to the fact that when resource rent is mismanaged due to endogenous and exogenous factors such as lack of institutional quality, bad fiscal policies, postcolonial legacy, etc., it clogs economic growth and brings about poverty amid plenty.[17]

The findings in the literature are the basis for the natural resource curse thesis or paradox of plenty – a catchphrase for the paradox that energy and natural resources abundance (mining) tend to clog economic growth and development.[18] Unfortunately, this has usually been the case, particularly in African natural resource-rich countries, except for Botswana.[19] Among

12 Edward Anthony Wrigley, *Energy and the English Industrial Revolution* (Cambridge University Press 2010) 23.
13 Sidney Pollard, *Peaceful Conquest: The Industrialisation of Europe 1760–1970* (Oxford University Press 1981) 4 ff, 40.
14 Stewart M Patrick, 'Why Natural Resources Are a Curse on Developing Countries and How to Fix It' (2012) the Atlantic, www.theatlantic.com/international/archive/2012/04/why-natural-resources-are-a-curse-on-developing-countries-and-how-to-fix-it/256508, accessed 19/01/2023.
15 Pr Atangana Ondoa Henri, 'Natural Resources Curse: A Reality in Africa' (2019) 63 *Resource Policy* 1.
16 Jeffrey A Frankel, 'The Natural Resource Curse: A Survey' (2010) National Bureau of Economic Research Working Paper No 15836, 1–38, www.nber.org/papers/w15836, accessed 10/01/2022.
17 Manuella Appiah and Ting Zhang, 'Escaping the Resource Curse in Sub-Saharan Africa' (2013) The Hague Institute for Global Justice Policy Brief 4, https://thehagueinstituteforglobaljustice.org/wp-content/uploads/2015/10/PB4-Escaping-Resource-Curse-Sub-Saharan-Africa.pdf, accessed 19/07/2022.
18 Eddy Lenusira Wifa, 'The Paradox of Plenty: An Analysis of the Environmental and Socio-Economic Symptoms of the Resource Curse in the Niger Delta Region of Nigeria' in Ibibia Worika and Sylvester Popnen (eds) *The Challenge of Justice: Contemporary Legal Essays in Honour of B.M. Wifa, OFR, SAN, DSSRS, KJW* (Princeton and Associates Publishing 2018) 576–601; Terry Lynn Karl, 'Ensuring Fairness: The Case for Transparent Fiscal Social Contract' in Macartan Humphreys et al (eds), *Escaping the Resource Curse* (Columbia University Press 2007) 258–293.
19 C Mlambo and Gabriel Borz, 'Politics and the Natural Resource Curse: Evidence from Selected African Countries' (2022) 8(1) *Cogent Social Sciences* 1–17.

several reasons for this is the impact of colonialism that left the majority of African countries existentially fragile.[20] Such existential fragility remains largely responsible for the rent-seeking, corrupt, and weak governance institutions that have plagued the continent. While there are several suggestions of both legal and political ways to cure or perhaps mitigate the resource curse within the African continent, this chapter, as mentioned in the introduction, recommends making the AMV a hard law and also explores a relational and contractual model known as IBA from the perspective of an organic regional framework.[21]

Journalistically, African energy and natural resource-rich countries have been represented as prime examples of resource curse countries.[22] This is because decades of energy and natural resources mining in almost all resource-rich African countries except Botswana have not translated to economic prosperity and sustainable development as seen in the Global North but rather clogged the socio-economic growth and development of the continent; while enriching a few elites, this has led to greater poverty of the masses, environmental degradation, and civil war.[23] Readily cited examples include but are not limited to Angola, the Democratic Republic of Congo, Nigeria, and Sudan which are endowed with petroleum, diamond, and other excellent minerals, yet decades of mining these resources have not promoted socio-economic growth and sustainable development but rather fuelled civil war, political instability, environmental degradation, and poverty.[24]

20 Mark Igiehon and Mariam Fawole-Masini, 'Problematising the Persistence of Resource Curse –Propounding New Third Explanation of `Existentially Fragile States Disorder' (2020) Working Paper Series, Oil and Minerals for Good Global Forum, https://oilandminerals.com/wp-content/uploads/2020/07/Research-Paper-Final-22July2020.pdf, accessed 15/06/2022.

21 Open Society Justice Initiative, *Legal Remedies for the Resource Curse: A Digest of Experience in Using Law to Combat Natural Resource Corruption* (Open Society Institute 2005); Todd Moss et al, *Oil to Cash: Fighting the Resource Curse through Cash Transfers* (Center for Global Development 2015); Thiemo Fetzer and Stephan Kyburz, 'How Can Countries Escape the Natural Resource Curse? Answer: Democracy' (2019) Centre for Global Development, www.cgdev.org/blog/how-can-countries-escape-natural-resource-curse-answer-democracy, accessed 19/07/2022.

22 Kwamina Panford, *Africa's Natural Resources and Underdevelopment: How Ghana's Petroleum Can Create Sustainable Economic Prosperity* (Palgrave Macmillan 2017) 33.

23 Annegret Mähler, 'Nigeria: A Prime Example of the Resource Curse? Revisiting the Oil-Violence Link in the Niger Delta' (2010) German Institutes of Global Area Studies Working Paper No 120, 5–37, www.giga-hamburg.de/en/system/files/publications/wp120_maehler.pdf, accessed 06/02/2022; Rhoks Ako and Nilopar Udin, 'Good Governance and Resource Management in Africa' in Francis N Botchway (ed) *Natural Resource Investment and Africa's Development* (Edward Elgar Publishing 2011) 35.

24 Ingrid Samset, 'Conflict of Interests or Interests in Conflict? Diamonds and war in the DRC' (2002) 29(93–94) *Review of African Political Economy*; Adrian Gonzalez, 'Petroleum and Its Impact on Three Wars in Africa: Angola, Nigeria and Sudan' (2010) 16 *Journal of Peace, Conflict and Development*; Mähler (n 23).

The journalistic representation of Africa's energy- and natural resource-rich countries as prime examples of resource curse countries may, on a cursory examination, appeal to common sense. A brief analysis of Nigeria's experience with the development of hydrocarbons will reinforce this appeal.[25]

Petroleum exploitation in Nigeria since 1959, when commercial production commenced, has generated billions of dollars in rent to the Nigerian government. It is estimated that Nigeria earns over $28 billion annually from oil revenues. These earnings constitute over 90% of the country's aggregate revenues.[26] In this context, petroleum is considered the life-wire of the Nigerian economy.[27] However, these earnings have not increased Nigeria's per capita gross domestic product (GDP), which has remained the same since the 1960s. Nigeria's GDP was estimated to be $1,113 in the 1970s and decreased to $1,084 in the 2000s.[28] Nor has it improved the well-being of common Nigerians; rather, the earnings have enriched political elites and their cronies. According to the 2022 World Bank assessment, more than 82.9 million Nigerians live in poverty and below the national poverty line of 137,430.[29] Wenar summarises this position thus: 'from 1970 to 2000 the Nigerian government received very large revenues (around $300 billion) from oil sales. Yet during this period, the percentage of Nigerians living in extreme poverty ($1 per day) increased from 36 per cent to almost 70 per cent'.[30] Furthermore, petroleum exploitation has become a source of environmental degradation and polluted the environment including arable land and fresh waterways. The resultant effect of this is that most Niger Delta indigenes are deprived of a means of livelihood as they are mainly farmers and fishermen who depend on the environment for sustenance.[31] It is reported that Nigeria has experienced

25 This is a generalised statement using just one country and should be taken with a pinch of salt; an examination of the developmental outcome of all the African resource-rich countries is not feasible here due to space constraints.

26 Nelson E Ojukwu-Ogba, 'Legislating Development in Nigeria's Oil Producing Region: The NDDC Act Seven Years On' (2009) 17 *African Journal of International and Comparative Law* 136.

27 Olubunmi Awoyemi, *The Impact of Oil Pollution on the Environment of the Oil Producing Communities of Nigeria: A Critical Analysis of the Statutory and Regulatory Provisions in Nigeria* (MO Awoyemi and Co Independent Publisher 2014) 3.

28 Xavier Sala-i-Martin and Arvind Subramanian, 'Addressing the Natural Resource Curse: An Illustration from Nigeria' (2013) 22(1) *Journal of African Economies* 570–615.

29 Shubham Chaudhuri et al, 'Nigeria Poverty Assessment 2022: A Better Future for All Nigerians' (2022) International Bank for Reconstruction and Development/The World Bank 13, https://documents1.worldbank.org/curated/en/099730003152232753/pdf/P17630107476630fa09c990da780535511c.pdf, accessed 21/07/2022.

30 Leif Wenar, 'Rights and the Resource Curse' (2008) 36(1) *Philosophy and Public Affairs* 4–5.

31 Adeoye O Akinola, *Globalization, Democracy and Oil Sector Reform in Nigeria: African Histories and Modernities* (Palgrave Macmillan 2018) 5.

over 4,000 oil spills amounting to 9–13 million barrels of oil, and gas flaring continues unabated.[32]

However, it must be noted that African resource-rich countries may not have experienced economic growth and sustainable development despite decades of continued exploitation of the subsoil mineral resources and the rent generated, not necessarily due to the abundance of the resources but because the countries either lack the requisite institutional quality or appropriate fiscal and regulatory frameworks that are the 'crucial links between resource endowments … on the one hand and economic outcomes on the other hand'.[33] Institutional quality connotes a country's bureaucratic ability to 'competently and continuously' deliver policies and services even in the face of changes in political leadership.[34] The lack of institutional quality in Africa and other colonised nations is generally attributable to colonial baggage.[35]

3 The African Mining Vision (AMV) 2009: The Perceived Silver Bullet for Resource-Based Sustainable Development

In recognition of the fragmented nature of energy and natural resource regimes in the African continent that cannot deliver resource-based sustainable development, the AU birthed the idea of a regional cooperation regime in the form of the AMV.

32 Commission on Environmental, Economic, and Social Policy, 'Niger Delta Natural Resource Damage Assessment and Restoration Project' (Phase I Scoping Report) (2006) 3, www.bing.com/search?q=iucn%2Fceesp+2006%29&go=Submit+Query&qs=bs&form=QBRE, accessed 22/01/2022; Eddy Wifa Leusira, 'The Role of Environmental Impact Assessment (EIA) in the Nigerian Oil and Gas Industry Using the United Nation's Programme EIA on Ogoni as a Case Study – Lessons from Some International Good Practices' (2014) *IELR* 112; Sylvester O Nliam, 'International Oil and Gas Environmental Legal Framework and the Precautionary Principle: The Implications for the Niger Delta' (2014) 22 *African Journal of International and Comparative Law* 23; Oluwaṣeun S ⊠k⊠, 'Gas Flaring in Nigeria and the Flexed Muscles of the 2018 Regulations: Key Implications and Investment Considerations' (2019) 17(1) *OGEL* 1–15.

33 Mansoob S Murshed, 'When Does Natural Resource Abundance Lead to a Resource Curse?' (2004) Environmental Economics Programme/International Institute for Environment and Development Discussion Paper 04-01, http://pubs.iied.org/pdfs/9250IIED.pdf, accessed 10/03/2022; Frederick van der Ploeg, 'Natural Resources: Curse or Blessing?' (2011) 49(2) *Journal of Economic Literature* 269–370.

34 Mark C Thurber et al, 'The Limits of Institutional Design in Oil Sector Governance: Exporting the "Norwegian Model"' (2010) ISA Annual Convention 2, https://resourcegovernance.org/sites/default/files/Thurber_Hults_%20and_Heller_ISA2010paper_11Feb10.pdf, accessed 10/01/2022; Mark C Thurber et al, 'Exporting the "Norwegian Model": The Effect of Administrative Design on Oil Sector Performance' (2011) 39(9) *Energy Policy* 5367.

35 Paul Cleary, 'Poles Apart: Comparative Resource Sector Governance in Australia and Norway' (2016) 51(1) *Australian Journal of Political Science* 150–162.

The AMV is one of the several Pan-African policy frameworks that the AU created in 2009 under Agenda 2063 to address the connection between energy and natural resource development and the industrialisation of the continent: economic growth and sustainable development based on resource development.[36] This was informed by the casual observation that resource-rich African countries rank among the poorest in the world in terms of economic growth and sustainable development in all ramifications including economic, social, political, and human development indexes.[37]

Agenda 2063 is Africa's long-term development strategy that emphasises, among other things, the need to harness Africa's energy and natural resources efficiently and effectively to ensure that, first and foremost, African citizens and African countries benefit from Africa's natural endowments by implementing policies, laws, and other regulatory frameworks to guide the extraction and use of these natural resource endowments for economic growth and sustainable development.[38]

The AMV broadly focuses on the structural transformation of Africa's energy and natural resource sector through 'enhanced linkages with the local economy, increased value addition, promotion of local citizen participation and human resources development'.[39] Its primary objective is to provide a regional framework geared towards the 'transparent, equitable and optimal exploitation of mineral resources to underpin broad-based sustainable growth and socio-economic development'.[40]

The AMV primarily seeks to address the resource curse paradox in Africa; a resource-rich continent that continues to see enduring poverty.[41] It is seen to have a holistic approach that moves beyond a mere focus on economic growth to socio-economic and sustainable development considerations. In this

36 African Union, 'The African Mining Vision: Transparent, Equitable and Optimal Exploitation of Africa's Mineral Resources' (2021), https://au.int/en/articles/africa n-mining-vision, accessed 23/05/2022.

37 Economic Commission for Africa, 'Africa Mining Vision: African Minerals Governance Framework' (2018) Economic Commission for Africa 1, www.kimberleyp rocess.com/en/system/files/documents/the_africa_mining_vision_amv_african_m ineral_governance_framework_amgf_rev.pdf, accessed 25/05/2022.

38 African Union Commission, 'Agenda 2063 the Africa We Want: First Ten-Year Implementation Plan 2013–2023' (2015), https://wedocs.unep.org/bitstream/ha ndle/20.500.11822/20823/Agenda%202063%20-%20FIRST%20TEN%20YEAR %20PLAN%20%20%20September%20%202015.pdf?sequence=1&isAllowed=y, accessed 31/05/2022.

39 Amani Mustafa Mhinda, 'The Africa Mining Vision in Tanzania: Is it a Step Down or a Way Forwards?' in Andreas Bohne and Ibrahima Thiam (eds) *The Africa Mining Vision: Can Business as Usual Be Visionary? Perspectives From Four Regions* (Rosa-Luxemburg-Stiftung 2018) 25.

40 African Union, 'African Mining Vision' (2009) 5, https://au.int/sites/default/files/ documents/30995-doc-africa_mining_vision_english_1.pdf, accessed 26/05/2022.

41 Nadège W R Compaoré, 'The Africa Mining Vision: Prospects and Challenges for Implementing Countries' (2017) 8(10) Institute for Peace and Security Studies Policy Brief 1–2.

respect, Dauda opines that the AMV 'signified the onset of a regionally driven paradigm shift, away from "a resource development model anchored on extractivism toward one in which mineral resources are harnessed to accelerate broad-based development and build resilient, diversified and competitive economies"'.[42] The AMV is considered an aspirational and transformational statement of goals and principles and represents a development blueprint that transcends the confines of the extractive sector. It entails a multi/cross-sectoral approach to mineral development policy, especially the link with industrial and trade policy, infrastructure development, empowerment of local enterprises, and skills formation.[43]

3.1 The African Mining Vision 2009 in Context: A Move in the Right Direction or a Futile Mechanism?

The AMV envisions an African mining sector that is:

- Knowledge-driven and contributes to growth and development which is fully integrated into a single African market;
- Sustainable and well-governed and effectively garners and deploys resource rents, is safe, healthy, gender and ethnically inclusive, environmentally friendly, socially responsible, and appreciated by surrounding communities;
- A key component of a diversified, vibrant and globally competitive industrialising African economy;
- Helping to establish a competitive African infrastructure platform, through the maximisation of its propulsive local and regional economic linkages;
- Optimising Africa's finite mineral resource endowments that is diversified, incorporating both high-value metals and lower-value industrial minerals at both commercial and small-scale levels;
- Harnessing the potential of artisanal and small-scale mining to stimulate local/national entrepreneurship, improve livelihoods and advance integrated rural social and economic development;
- A major player in vibrant and competitive national, continental, and international capital and commodity markets.[44]

42 Suleman Dauda, 'Operationalising the "Africa Mining Vision": Critical reflections from Ghana' (2020) 41(3) *Canadian Journal of Development Studies* 505–506.

43 Antonio MA Pedro, 'The Country Mining Vision: Towards a New Deal' in *Österreichische Entwicklungspolitik* (Rohstoffe Und Entwicklung 2015) 28, www.oefse.at/fileadmin/content/Downloads/Publikationen/Oepol/Artikel2015/Teil1_02_Country_Mining_Pedro.pdf, accessed 26/05/2022.

44 African Union, 'African Mining Vision' (2009), https://au.int/sites/default/files/documents/30995-doc-africa_mining_vision_english_1.pdf, accessed 30/05/2022.

However, it has been greeted with both enthusiasm and apathy. This is not only because it tries to reinforce capacity and bolster governance and institution quality or make existing institutions more efficient as if the problem were essentially technical or managerial.[45] It also requires changes in paradigm which include fundamental reorientation of past strategies to develop the mining sector in Africa and new goals and instruments, based on new norms that seek a renewal of relations among the actors concerned as it envisions cooperation among African countries in the management of resource exploitation and the wealth generated therefrom in the African continent.[46] This vision is well captured by Dauda who states that the AMV 'seeks to create conditions for sustainable growth in, and to facilitate the broad-based development of, (sub-Saharan) Africa through its mineral wealth'.[47] Nevertheless, the AMV is potentially deficient in terms of requiring domestication by member states to have a uniform resource development regime that will facilitate resource-based sustainable development. The next section assesses the AMV in creating and promoting the conditions necessary for energy and natural resource exploitation to facilitate broad-based socio-economic growth and sustainable development of the African continent.

4 An Assessment of the African Mining Vision as a Regional Cooperation Framework that Can Transform the African Subsoil Mineral Sector to Facilitate Broad-Based Sustainable Development

Some analysts have expressed optimism that the AMV has the potential to transform Africa's energy and natural resource sector in a way that will underpin broad-based sustainable growth and socio-economic development of the continent and, subsequently, tackle the resource curse in the continent.[48] According to these analysts, the AMV is nothing but the pivoting silver bullet that African resource-rich countries could hinge on to eliminate or at least mitigate the resource curse that has bedevilled the continent. According to Dauda, the AMV 'signified the onset of a regionally driven paradigm shift' in resource exploitation management in Africa.[49]

For Pedro, the AMV is

> Africa's efforts to transform the mining sector away from its colonially-created enclave features that have so far met with very limited success.

45 Daniel Béland, Bonnie Campbell, Mylène Coderre, and Privilege Haang'andu, 'Policy Change and Paradigm Shifts in Sub-Saharan Africa: Implementing the Africa Mining Vision' (2022) 56(1) *Canadian Journal of African Studies* 79.

46 Ibid.

47 Dauda (n 42) 505

48 Gavin Hilson, 'The Africa Mining Vision: A Manifesto for More Inclusive Extractive Industry-Led Development?' (2020) 41(3) *Canadian Journal of Development Studies* 420.

49 Dauda (n 42) 505.

> The Africa Mining Vision offers a framework for integrating the sector more coherently and firmly into the continent's economy and society.[50]

Others have qualified this optimism to mean that if the AMV is effectively implemented by the AU member states, it could serve as Africa's 'continental framework to promote resource-driven broad-based development and structural transformation'.[51] It is our candid view that qualifying the optimism becomes necessary because the AMV will be no more than a toothless bulldog or perhaps a worthless piece of paper hanging on the AU shelf if, first, AU member states are not under any obligation to translate it into national mineral development policies and laws in the form of CMV and, second, there are no institutions in the AU member states to ensure the enforcement of the CMV. This is because, as Harvey and many resource curse sceptics have rightly pointed out, natural resource wealth only tends to be a curse in countries that lack or have weak institutions.[52]

4.1 Key Strengths of the African Mining Vision 2009 in Linking the Extractive Sector with Wider Economic Sectors: A Course for a Regulatory Paradigm Shift

The optimistic view that the AMV is the silver bullet that African resource-rich countries need (either in its raw or qualified form) is premised on several grounds. First and foremost are the perceived key strengths of the AMV in bringing about a shift in paradigm regarding the role of the mining sector in the sustainable socio-economic growth and development of mineral-rich AU member states.[53]

The AMV tends to pursue harmonised regulations in AU member states' extractive sector in order to change the narrative of resource-producing African countries (from resource curse to resource blessing) being mere quarries for developers (i.e. exporters of raw minerals) with no value addition to the economy by creating resource-based economic linkages locally.[54] The essence of these linkages is for the extractive sector to positively impact other sectors of the economy with a view towards facilitating broad-based and inclusive local socio-economic sustainable development. In this regard, the AMV is said to have a wider focus than any prior strategy that has been proposed or adopted in the continent. This

50 Pedro (n 43) 1.
51 Ibid 27.
52 Ross Harvey, 'Why Being Endowed with Oil is Not Always a Boon: The Case of Nigeria and Angola' (2021) The Conversation, https://theconversation.com/why-being-endowed-with-oil-is-not-always-a-boon-the-case-of-nigeria-and-angola-153892, accessed 30/05/2022.
53 Béland et al (n 45)..
54 Patricia Ackah-Baidoo, 'Implementing Local Content Under the Africa Mining Vision: An Achievable Outcome?' (2020) 41(3) *Canadian Journal of Development Studies* 488.

is because it 'focuses on balancing transparency and accountability with the need to integrate mining into countries' long-term development plans'.[55]

Mining sector linkages with wider economic sectors can only be achieved through building integrated resource industrial clusters: forward or down-stream linkages into mineral beneficiation and manufacturing, backward or upstream linkages into mining capital goods, consumables, and services industries, and lateral or side-stream linkages into infrastructure (power, logistics, communications, water) and skills and technology development, investment into human resource development, and research and development.[56] These linkages are best achieved through local content policies and, according to the AMV document, can only be realised if the resource sector is unbundled to identify entry points for:

i down-stream linkages into mineral beneficiation, value addition, and manufacturing;
ii increasing local upstream linkages into mining inputs in the form of capital goods, consumables, and services;
iii catalysing resource-based infrastructure to open up opportunities in other sectors of the economy;
iv spearheading side-stream linkages into skills, research, and technology development;
v facilitating lateral migration of mining technologies to other industries;
vi promoting the development of sustainable livelihoods in mining communities, and
vii creating small- and medium-sized enterprises and a more balanced and diversified economy with greater multiplier effects and potential to create employment.[57]

Second, the AMV covers the entire mineral value chain, promotes a fiscal regime that is capable of 'haemorrhaging the continent's resources through tax avoidance, evasion, and illicit financial flows from the mineral sector', upholds the principle of free, prior, and informed consent (FPIC) for host communities, advocates for linkages of resource sector with other economic sectors, which in turn will promote local value addition and economic diversification, addresses negative environmental and social impacts of resource development, and finally provides a means for countries to adapt it to local circumstances – country mining vision (CMV).[58]

55 Paul Justice Kamlongera, 'Domesticating the Africa Mining Vision in Malawi: Approaches and Experiences' (2020) 41(7) *Canadian Journal of Development Studies* 470.
56 African Union, 'African Mining Vision' (2009) 5, https://au.int/sites/default/files/documents/30995-doc-africa_mining_vision_english_1.pdf, accessed 26/05/2022.
57 Ibid.
58 Oxfam, 'From Aspiration to Reality Unpacking the Africa Mining Vision' (2017) Oxfam International 23, www-cdn.oxfam.org/s3fs-public/bp-africa-mining-vision-090317-en.pdf, accessed 25/05/2022.

Additionally, the AMV 'legitimised' moves by some AU member state governments to adopt the tenets of the AMV into national mining policies and laws and renegotiate mining contracts. The new laws and renegotiated contracts terminated the generous tax breaks and other economic incentives granted to industry proponents that exploit the resources.[59]

4.2 Challenges and Shortcomings of the Africa Mining Vision 2009

Since the establishment of the AMV in 2009, its implementation by the AU member states has been slow, and energy and natural resource development has not engendered any reasonable and positive difference in the economic growth and sustainable development of African resource-rich countries.[60] Nigeria, Angola, Algeria, Chad, Libya, Tanzania, etc., major energy and natural resource producers in Africa, are prime examples.[61] This has caused many questions to be asked by civil societies, scholars, and policy analysts concerning the AMV as Africa's regional cooperation instrument in ensuring that mining engenders socio-economic sustainable development and leads to the industrialisation of the continent. These questions fall mostly within the purview of political economy, environmental justice, corporate social responsibility, sustainable development, and sustainability.[62] Although some limb of the questions has legal implications, especially regarding the binding nature of the AMV, curiously, it seems no one has contextually questioned the soft law nature of the AMV and why its domestication into national law is voluntary rather than mandatory.

This section, therefore, attempts to question this and examines one of the prominent legal questions relating to whether the AMV is fit for purpose, especially in supporting cooperation in mining governance among AU member states to ensure that energy and natural resource development in Africa facilitates industrialisation and modernisation of the continent. This question is prominent primarily because several years after its establishment, the AMV is yet to be domesticated by all member states and, arguably, seems not to have catalysed the strategic harnessing of Africa's energy and mineral resources for broad-based sustainable development despite having the

59 Hilson (n 48) 421.

60 Kamlongera (n 55) 470.

61 Ivailo Izvorski et al, 'Reinvigorating Growth in Resource-Rich Sub-Saharan Africa' (2018) World Bank 41–43, https://documents1.worldbank.org/curated/en/617451536237967588/pdf/5-9-2018-17-9-2-SSAGrowthforweb.pdf, accessed 26/05/2022; Robert I Rotberg and John Campbell, 'Nigeria Is a Failed State' (2021) *Foreign Policy*, https://foreignpolicy.com/2021/05/27/nigeria-is-a-failed-state, accessed 26/05/2022.

62 Hilson (n 48) 420 has asked the following questions: (1) Is the AMV still seen as a blueprint for development in mineral-rich Africa, and if so, by whom? (2) What impact, developmentally, has the AMV had thus far? (3) Are individual countries still responding to the recommendations enshrined in the AMV, and if so, how? (4) What does the future hold for the AMV?

aspiration of 'a knowledge-driven African mining sector that catalyses and contributes to the broad-based growth and development of, and is fully integrated into, a single African market' through back, forward, and lateral linkages.[63]

The section, therefore, assesses the effectiveness of the AMV in ensuring cooperation in mining governance in Africa's resource-rich countries that will support energy and mineral development, facilitating economic growth and sustainable development leading to the industrialisation of the continent. It is important to note that the section will not assess the implementation of the AMV in the AU member states, primarily because not all member states have implemented it and scholarship exists that evaluates the AMV implementation in member states that have completely or partially domesticated the AMV.[64] Rather, the section examines the 'legal reason' behind the non-implementation of the AMV despite the overwhelming support by AU member states' heads of state during its launch in 2009 and its implications. As such, some generic (legal) features of the AMV are examined against some known benchmarks of international and regional cooperation frameworks to evaluate its effectiveness in commandeering cooperation in mining governance in Africa that will lead to resource-based socio-economic development. This does not detract from reference to country-specific laws that have domesticated the tenets of the AMV or bear some marks of the AMV where necessary to clarify any point or support an argument.

4.2.1 The African Mining Vision 2009 as a Soft Law Instrument

We argue that the non- or partial implementation of the AMV by most AU member states is attributable to the voluntary or non-binding nature of the policy framework, which is akin to a soft law instrument in international law. Soft law in the international order has a range of possible meanings. However, from the perspective of law making, which aligns with our argument, it

63 Economic Commission for Africa, 'Africa Mining Vision: African Minerals Governance Framework' (2018) Economic Commission for Africa 1, www.kimberleyprocess.com/en/system/files/documents/the_africa_mining_vision_amv_african_mineral_governance_framework_amgf_rev.pdf, accessed 25/05/2022.

64 For an evaluation of the implementation of the AMV in Ghana, Kenya, Malawi, Namibia, Zambia, and Zimbabwe, see David van Wyk et al, 'Attainable Vision or Fairy Tale?' in Andreas Bohne and Ibrahima Thiam (eds) *The Africa Mining Vision: Can Business as Usual Be Visionary? Perspectives From Four Regions* (Rosa-Luxemburg-Stiftung 2018) 19–22. For the implementation of the AMV in Tanzania, see Amani Mustafa Mhinda, 'The Africa Mining Vision in Tanzania: Is It a Step Down or a Way Forwards?' in Andreas Bohne and Ibrahima Thiam (eds) *The Africa Mining Vision: Can Business as Usual Be Visionary? Perspectives from Four Regions* (Rosa-Luxemburg-Stiftung 2018) 25–36; for the domestication of the AMV in Malawi, see Paul Justice Kamlongera, 'Domesticating the Africa Mining Vision in Malawi: Approaches and Experiences' (2020) 41(7) *Canadian Journal of Development Studies* 467–485.

is a 'convenient description for a variety of non-binding legal instruments used in contemporary international relations by States and international organisations'.[65] In this respect, Guzman and Meyer opine that soft law is a 'continuum, or spectrum, running between fully binding treaties and fully political positions'.[66] From the above definitions, the primary characteristic of soft law is that it is legally non-binding on the parties and, as such, limited in relevance. This characteristic notwithstanding, soft law often generates widespread and consistent state practice and often sets the stage for a treaty.[67]

It is understandable why the AU, in dealing with energy and natural resource management in the continent, pursued soft legalisation rather than hard legalisation in the form of a treaty such as the *African Charter on Human and Peoples' Rights*.[68] Hard legalisation would arguably infringe the international doctrine of permanent sovereignty over natural resources[69] which has become a mantra for many resource-rich African countries and is upheld by the *African Charter on Human and Peoples' Rights*.[70]

The AU member states are not obligated or bound to domesticate the AMV in national policies and laws.[71] Member states that have implemented the AMV do so merely on grounds of political commitment as non-implementation does not attract any sanction from the AU.

The 'AMV envisages that all African governments establish and lead a multi-stakeholder CMV task force' to 'identify gaps between the existing mining regime and the key tenets of the AMV' in order to domesticate the AMV nationally.[72] However, as a result of the non-binding nature of the AMV, only Lesotho has fully implemented it, while 25 other AU member states including Congo, Eritrea, Ghana, Kenya, Malawi, Namibia, Niger,

65 Alan Boyle, 'Soft Law in International Law-Making' in Malcom Evans (ed) International Law (4th edn, Oxford University Press 2014) 120.

66 Andrew T Guzman and Timothy L Meyer, 'International Soft Law' (2010) 2(1) *Journal of Legal Analysis* 173.

67 Boyle (n 65) 122.

68 *African (Banjul) Charter on Human and* Peoples' *Rights* (Adopted 27 June 1981, OAU Doc. CAB/LEG/67/3 rev. 5, 21 I.L.M. 58 (1982), entered into force 21 October 1986).

69 *United Nations General Assembly Resolution 1803 (XVII) on Permanent Sovereignty over Natural Resources.*

70 *African Charter on Human and Peoples' Rights* article 21(1); Sangwani Patrick Ng'ambi' 'Permanent Sovereignty over Natural Resources and the Sanctity of Contracts, from the Angle of Lucrum Cessans' (2015) 12(2) *Loyola University Chicago International Law Review* 153–172; Aditi Garg, 'Permanent Sovereignty over Natural Resources: An Analysis' (2021) 4(1) *International Journal of Law Management & Humanities* 624–634.

71 Andreas Bohne and Ibrahima Thiam, 'Introduction' in Andreas Bohne and Ibrahima Thiam (eds) *The Africa Mining Vision: Can Business as Usual Be Visionary? Perspectives from Four Regions* (Rosa-Luxemburg-Stiftung 2018) 4.

72 Wyk et al, 'Attainable Vision or Fairy Tale?' in Andreas Bohne and Ibrahima Thiam (eds) *The Africa Mining Vision: Can Business as Usual Be Visionary? Perspectives from Four Regions* (Rosa-Luxemburg-Stiftung 2018) 18.

Sierra Leone, Tanzania, Zambia, and Zimbabwe are at different stages of implementing it.[73] South Africa is one of the countries that never committed to implementing the AMV.[74] The resultant effect of this is that the majority of the resource-rich AU member states since the launch of the AMV with lofty objectives have maintained differing mining governance regimes which generally tend to attract foreign direct investment (FDI), support resource exports, and further enclave the resource sector rather than encourage in-country value addition and resource-based sustainable development.[75] It is in this regard that the AMV regime can be viewed as fragmented as it envisions the harmonisation of extractive sector regulation in Africa and yet does not provide implementing mechanisms on the part of the member states to achieve a harmonised extractive regulatory framework.

The reason for the non-domestication of the AMV by a majority of the AU member states is not far-fetched. Domesticating the AMV will empower citizens and civil societies to demand accountability and transparent management of resource wealth and will constrain heads of states and their cronies from rent-seeking practices, plundering resource rent, and ultimately political power.[76]

As a result of the soft law nature of the AMV, it is contended that the AMV might not achieve its chief intended purpose of regional cooperation in mining governance among African resource-rich countries to ensure that resource development in the continent facilitates socio-economic growth and development leading to industrialisation of the continent. This argument is premised on the grounds that soft law is 'something that dims in importance as the commitments of states get weaker, eventually disappearing altogether'.[77] A few examples will demonstrate the points being made here.

The AMV, for instance, stipulates the elimination of human rights violations by resource developers, elimination of the potential for conflict inherent in natural resource exploitation, and inclusion of local communities' engagement in resource exploitation decision making and benefit sharing. Nonetheless, there are no stipulated binding standards to be adopted by the member states. Each member state is at liberty to formulate any standard – that is, if the AMV is implemented nationally. This has incentivised some member states to have double standards for different categories of resources while others voluntarily require 'companies to disclose or provide information

73 Wyk et al (n 64).

74 Ibid.

75 Ben Radley, 'The End of the African Mining Enclave? Domestic Marginalization and Labour Fragmentation in the Democratic Republic of Congo' (2020) 51(3) *Development and Change* 794–814.

76 Beth Stratford, 'The Threat of Rent Extraction in a Resource-constrained Future' (2020) 169 *Ecological Economics* 1–9.

77 Andrew T Guzman and Timothy L Meyer, 'International Soft Law' (2010) 2(1) *Journal of Legal Analysis* 173.

despite the fact that history has clearly shown voluntary standards and self-imposed obligations to be insufficient'.[78] Nigeria, for instance, in communities' engagement in the development of resources on their land, has different standards for different categories of resources. For solid mineral development, developers are mandated to negotiate a community development agreement (CDA) that will stipulate the benefits that will accrue to the host communities as a condition precedent to the commencement of exploitation.[79] However, for petroleum development, industry proponents are required to incorporate a host communities' development trust that will fund community sustainable development projects to be identified based on a need assessment conducted by the industry proponents.[80]

Furthermore, it is important to note that given the soft law nature of the AMV, it must be nationally implemented by AU member states in order to become operational, and the process of implementing it nationally will necessarily require 'that the laws, policies and regulations applicable to the mineral sector and the national development planning system be aligned with the AMV'.[81] On this note, AU member states' ministers responsible for mineral resources development, at the second AU conference held in December 2011 in Addis Ababa, approved the AMV Action Plan. The Action Plan proposed nine programme clusters and an aggregate of 136 actions across a wide range of issues affecting resource exploitation and described varying types of interventions at continental, sub-regional, and national levels for the realisation of the AMV values and tenets. At the national level, this was to be realised through CMV. No member state has completely enacted a CMV save Lesotho.[82]

4.2.2 Difficulty in Domesticating the African Mining Vision 2009: Way Forward – the African Minerals Development Centre

It is worth noting that the AU appreciates the difficulty with the domestication of the AMV due to its voluntary nature, and in order to promote its implementation through national laws, the African Minerals Development

78 Andreas Bohne and Michael Reckordt, 'United in Contradiction: The Africa Mining Vision and the German Government's Raw Materials Strategy' in Andreas Bohne and Ibrahima Thiam (eds) *The Africa Mining Vision: Can Business as Usual Be Visionary? Perspectives from Four Regions* (Rosa-Luxemburg-Stiftung 2018) 43.
79 *Nigeria Mineral and Mining Act* 2007 S 116.
80 *Petroleum Industry Act* 2021 Ss 235(1) & 316. For criticisms of the differing local communities' engagement regimes in relation to solid and liquid mineral development in Nigeria, see Solomon N Obulor, *Benefit-Sharing in Relation to Energy and Natural Resource Development: A Comparative Analysis of Impact and Benefit Agreement in Nigeria and Australia* (Thesis submitted to the University of Aberdeen 2022) 250–255.
81 Wyk et al (n 64) 22.
82 Ibid 26.

Centre (AMDC) was set up in 2018.[83] The AMDC is a statutory agency of the AU Commission and is co-sponsored by the African Development Bank and the United Nations Economic Commission for Africa, entrusted with powers to coordinate the implementation of the AMV and its Action Plan.[84] Despite the setup of the AMDC, member states are not under any obligation to implement this robust policy and, as such, there is no uniform mineral law in the countries that have partially implemented it.[85]

4.3 Achieving the Objectives of the African Mining Vision 2009: A Binding Obligation on the Part of Member States to Domesticate the AMV

Premised on the weaknesses identified above, this section argues that the AMV can achieve its lofty objectives to facilitate regional cooperation in mining governance in Africa in order for 'transparent, equitable and optimal exploitation of mineral resources to underpin broad-based sustainable growth and socio-economic development' if there is an overhauling of the member states' institutions tasked with the management of resource wealth and implementation of the AMV and a binding obligation is placed on member states to implement the AMV. How to overhaul member states' institutions in order to have quality institutions capable of managing resource wealth and administering policies even in the light of a change in government is a political issue which is beyond the scope of this chapter and therefore not undertaken.

A binding obligation on member states will transform the AMV into a hard law (treaty) to be met with a sanction in the event of failure by member states to transpose it to national laws. In elevating the AMV to a treaty, the AU can rely on the *African Charter on Human and Peoples' Rights* which provides thus:[86]

1 All peoples shall freely dispose of their wealth and natural resources. *This right shall be exercised in the exclusive interest of the people.* In no case shall a people be deprived of it.
2 […]
3 The free disposal of wealth and natural resources shall be exercised without prejudice to the obligation of promoting international economic

83 United Nations Development Programme, 'African Minerals Development Centre, a New Project to Improve Sustainability on The Continent' (2019), www.un.org/africarenewal/news/african-minerals-development-centre-new-project-improve-sustainability-continent, accessed 31/05/2022; Molly Lempriere, 'A Look Inside the African Minerals Development Centre' (2019) Mining Technology, www.mining-technology.com/analysis/the-african-minerals-development-centre, accessed 30/05/2022.

84 *Statute of the African Minerals Development Centre* arts. 2(1), 3, 4, 5 & 6.

85 Wyk et al (n 64) 22.

86 *African Charter on Human and Peoples' Rights* art. 21(4)(5); italics are ours.

cooperation based on mutual respect, equitable exchange and the principles of international law.

4 *State parties to the present Charter shall individually and collectively exercise the right to free disposal of their wealth and natural resources with a view to strengthening African unity and solidarity.*

5 *State parties to the present Charter shall undertake to eliminate all forms of foreign exploitation particularly that practiced by international monopolies so as to enable their peoples to fully benefit from the advantages derived from their national resources.*

This provision ostensibly lends credence to the AMV whose primary purpose is that the exploitation of mineral resources in the African continent should facilitate continental and national sustainable socio-economic growth and sustainable development in a way that benefits every member of society.

5 Recommendation: Engaging Stakeholders in Subsoil Development Through Impact and Benefit Agreement

One of the objectives of the AMV is to promote the participation of local communities commonly referred to as host communities in the development of resources on their land.[87] It is acknowledged that there is no 'one-size-fits-all' approach to effectively ensure host communities' participation in resource mining on their land. However, one mechanism that has arguably been successful in this regard is the impact and benefit agreement (IBA).[88] It is, therefore, recommended that introducing the negotiation of IBAs in relation to resource development in Africa on the basis of the AMV will help to achieve the participation of host communities and create a tripartite partnership (host government, resource developers, and host communities) needed for resource development in Africa to fuel economic growth and sustainable development.

The importance of this recommendation is best appreciated through the lens of free, prior, and informed consent (FPIC) which is part of host communities' right to self-determination.[89] This is because each facet of FPIC provides a basis for the IBA process. For instance, *free consent*, which is the consent given by the host community without coercion or manipulation but through negotiation in good faith; *prior consent*, which is the consent of the host community that is adequately sought and obtained before resource

87 African Mining Vision 2009.

88 Vicki Beent, 'Community Benefits Agreements: A New Local Government Tool or Another Variation on the Exactions Theme?' (2010) 77(5) *The University of Chicago Law Review* 5–35.

89 Food and Agriculture Organisation of the United Nations, 'Free Prior and Informed Consent: An Indigenous Peoples' Right and Good Practice for Local Communities' (2016) FAO 11, www.fao.org/3/i6190e/i6190e.pdf, accessed 01/03/2021.

development activities commence; *informed consent*, which means that the indigenous community is properly informed of the consequences of exploitation – 'the information gap is consistently narrowed through the process'; and *consent*, defined as a collective decision reached through a decision-making process of the community.[90]

The bottom line is that provision for IBA negotiation under the AMV will profit all stakeholders (host government, resource developers, and host communities) as the aim is to establish a foundation for mutually beneficial relationships, deliver mutually rewarding sustainable benefits, and represent a new institutional framework in mitigating the well-documented negative impacts of resource development to reverse the so-called resource curse. For resource developers, IBAs negotiation helps to secure and sustain host communities' social licence to operate (SLO) which provides certainty and reduces instability and political risks.[91] For host communities, the negotiation of an IBA is, first, a sign of recognition of their *de facto* entitlement and access control to their ancestral land and resources thereon and, second, a vehicle to engage in decision making and share in resource development benefits.[92]

5.1 Contextualising Impact and Benefit Agreement: Meaning and Nature of IBA

An IBA is a private agreement which is known by different names.[93] It is viewed as a negotiated agreement between industry proponents and host communities (and sometimes the host government in a tripartite arrangement as in Australia, Canada, and Greenland) to engage host communities in decision making and benefit sharing in relation to resource development on host community land.[94]

90 John Nikolaou, *Developing a Model for Effective Community Development Agreements in the Extractive Industries* (Senior Thesis submitted to the Claremont McKenna College 2019) 10, https://scholarship.claremont.edu/cgi/viewcontent.cgi?article=3305&context=cmc_theses, accessed 01/03/2021.

91 Justine Lacey et al, 'Social Licence as Social Contract: Procedural Fairness and Forest Agreement-Making in Australia' (2016) 89 *Forestry* 489–499.

92 Courtney Fidler and Michael Hitch, 'Impact and Benefit Agreements: A Contentious Issue for Environmental and Aboriginal Justice' (2007) 35(2) *Environments Journal* 50; Eric Werker et al, 'Policies for Generating Socioeconomic Benefits from Natural Resource Extraction Projects: A Research Report for the Government of the Northwest Territories' (2017) Government of Northwest Territories, www.ntassembly.ca/sites/assembly/files/td_272-183.pdf, accessed 04/06/2022.

93 World Bank, 'Mining Community Development Agreement: Source Book' (2012) World Bank 5, https://openknowledge.worldbank.org/bitstream/handle/10986/12641/712990WP0minin00Box370065B00PUBLIC0.pdf?sequence=1&isAllowed=y, accessed 30/05/2022.

94 Such as such 'impact benefit-sharing agreements', 'community development agreement' (CDA), 'indigenous land use agreement' (ILUA), community development initiatives, partnering/partnership agreements, landowner agreements,

IBAs generally emerged as private contractual agreements and 'supra-regulatory' frameworks negotiated between resource developers and host communities, and sometimes, as already mentioned, the host government, to engage the host communities in decision making and benefit sharing in relation to the resource development within their land.[95] However, their legal nature seems imprecise due to their hybrid characteristics of contract 'supra-regulation' and administrative framework.[96] When IBA negotiation is provided for in law (soft or hard law), as proposed here, an IBA might be thought of as an administrative framework because it tends to enhance, first, the administration of onshore resource wealth for the benefit of the host communities where the resource development occurs and, second, the host communities' participation in decision making regarding the development of the resources.

As a contractual agreement, IBAs might be seen as voluntary, private, and legally binding contracts between host communities and resource developers concerning resource development on host communities' land, which serve as a conduit for the host communities to be involved in the resource development decision-making process to mitigate associated negative impacts and capture benefits, and in exchange the host communities support the resource development.[97] In this vein, IBAs impose contractual obligations and responsibilities and may be legally enforced in accordance with the country's contract law and dispute resolution mechanisms – litigation and arbitration.[98] For

community joint venture agreements, social responsibility agreements, participation agreements, exploration agreements, socio-economic monitoring agreements, community benefits agreements, and memorandum of understanding. See Adaeze L Wigwe-Chizundu, 'Are Impact and Benefit Agreements (IBAs) Worth the Hype? An Analysis of the Challenges and Benefits of the Emerging Instrument' (2018) 16(1) *OGEL* 3.

95 Ginger Gibson and Ciaran O'Faircheallaigh, *IBA Community Toolkit: Negotiation and Implementation of Impact and Benefit Agreements* (The Walter and Duncan Gordon Foundation 2010) https://namati.org/wp-content/uploads/2015/05/IBA_toolkit_March_2010_low_resolution.pdf; Jackson C Rose and Julia H Haggerty, 'Community Benefit Agreements and Funds: A Summary of Key Literature and Case Studies' (2018) Resources and Community Research Group, https://static1.squarespace.com/static/5490f581e4b040d63fb03c5f/t/5ca4b8c8104c7b0d3418db30/1554299081198/CBAreport_Final+.pdf, accessed 11/03/2022.

96 Lindsay Galbraith et al, 'Towards a New Supraregulatory Approach to Environmental Assessment in Northern Canada' (2007) 25(1) *Impact Assessment and Project Appraisal* 27.

97 Peter Siebenmorgen and Benjamin Bradshaw, 'Re-Conceiving Impact and Benefit Agreements as Instruments of Aboriginal Community Development in Northern Ontario, Canada' (2011) 9(4) *OGEL* 3.

98 Julian Gross, 'Community Benefits Agreements: Definitions, Values, and Legal Enforceability' (2007) *Journal of Affordable Housing and Community Development Law* 39; James Gathii and Ibironke T Odumosu-Ayanu, 'The Turn to Contractual Responsibility in the Global Extractive Industry' (2015) 1(1) *Business and Human Rights Journal* 79.

instance, the Greenland IBA concerning the *Tanbreez* (*Killavaat Allanguat*) project in clause 5.2 provides that the parties shall have regard to Greenland general rules and principles of law including contract and administrative law in the performance of obligations and exercise of rights under the IBA.[99]

There has been an argument about whether IBAs are *per se* contracts. This is because some available IBAs are drafted in vague terms such as 'gentlemen's agreement binding in honour' and exclude resort to litigation and arbitration, which makes breaches difficult to enforce in law, while others necessarily incorporate the elements of a valid contract.[100] For instance, clause 4.2 of the 2008 *Ahafo Social Responsibility Agreement* – a well-publicised IBA in Ghana between Newmont Ghana Gold Limited and the Ahafo Mine Local Community – and an accompanying *Ahafo Mine Local Community Local Employment Agreement* provide that no legally enforceable rights to the benefit of the parties are created in the IBA. Parties are also restricted from litigation and arbitration concerning disputes arising in implementing the IBA. However, mediation and conciliation are encouraged.[101]

Some scholars have strongly expressed the view that IBAs are akin to contracts. Gross and colleagues state categorically that IBAs are 'legally enforceable contract[s]' as they are a result of negotiation between industry proponents and host communities in which industry proponents agree to provide specific community benefits and, in exchange, the indigenous communities agree to support the project.[102] Marantz supports this view while analysing the Los Angeles Sports and Entertainment District Community Benefits Agreement (Staples Center CBA) and states that IBAs are 'a standalone, legally enforceable contract'.[103]

Other sceptic scholars argue that IBAs may not meet the fundamental elements of a contract such as the binding intention of parties to enter into legal

99 IBA concerning the *Tanbreez* (*Killavaat Allanguat*) project under the Mineral Exploitation Number 2020-54 between Tanbreez Mining Greenland A/S, Municipality of Kummune Kujalleq and the Government of Greenland 2020, https://govmin.gl/wp-content/uploads/2020/09/Tanbreez-IBA-UK-signed-version.pdf, accessed 31/03/2022.

100 Irene Sosa and Karyn Keenan, 'Impact Benefit Agreement Between Aboriginal Communities and Mining Companies: Their Uses in Canada' (2001), www.fpic.info/media/library/resources/impact-benefit-agreements-between-aboriginal-commu/Impact_benefit_agre_lQ7oRKj.pdf, accessed 06/05/2022.

101 See clause 1.2 of the Ghana Ahafo Mine Local Community Newmont Ghana Gold Ltd 2008 Social Responsibility Agreement, http://ccsi.columbia.edu/files/2016/11/Ghana-Ahafo-Mine-Local-Community-Newmont-Ghana-Gold-Ltd-2008-Social-Responsibility-Agreement.pdf, accessed 04/07/2022.

102 Julian Gross et al, *Community Benefits Agreements: Making Development Projects Accountable* (Good Jobs First and the California Partnership for Working Families 2005) 9, www.forworkingfamilies.org/sites/default/files/publications/2005CBAHandbook.pdf, accessed 01/05/2022.

103 Nicholas J Marantz, 'What Do Community Benefits Agreements Deliver? Evidence From Los Angeles' (2015) 81(4) *Journal of the American Planning Association* 252.

relations, consideration (*quid pro quo*), and voluntary and enforceable agreement (offer and acceptance).[104] With regard to the argument on consideration[105] as a fundamental element of a contract, it is important to point out that consideration is not a fundamental requirement of a contract in certain jurisdictions.[106] And in most jurisdictions where consideration is a constituent element of a contract, the English common law rules of sufficiency rather than adequacy and consideration not being past and moving from the promisee apply.[107] Applying these rules, a promise on the part of host communities not to interfere with resource development would suffice as valuable consideration, while resource developers' promises of socio-economic and financial benefits to host communities in return for the indigenous communities' promise to support the resource development will unequivocally qualify as consideration.[108] As such, the argument on consideration cannot be sustained.

The debate may appear purely academic; however, a closer examination reveals its practical relevance. This is because if an IBA cannot be enforced at law, it ceases to be a contract as one of the essentials of a contract – that is, intention to create legal relations – would be lacking.[109] However, the arguments of the sceptic scholars will not be applicable in the circumstance of our recommendation as the negotiation will be legally provided including enforcement mechanisms.[110]

Supra-regulation in relation to subsoil resource development has to do with extra-regulatory frameworks that shape (host communities' participation in) the governance of resource development on host communities' ancestral land and mainly addresses aspects of such resource development that are not or

104 Naved Sheikh, 'Community Benefits Agreements: Can Private Contracts Replace Public Responsibility' (2008) 18(1) *Cornell Journal of Law and Public Policy* 233–234.

105 *Currie v Misa* (1875) 10 Ex 153 per Lush J 162: 'a valuable consideration in the sense of the law may consist either in some right, interest, profit or benefit accruing to one party, or some forbearance, detriment, loss or responsibility given suffered or undertaken by the other'.

106 Scotland is one of the jurisdictions where consideration is not one of the requirements of a contract. See the Scottish Law Commission, 'Review of Contract Law: Discussion Paper on Formation of Contract' (2012) Scottish Law Commission Discussion Paper No 154, 14–31, www.scotlawcom.gov.uk/files/9513/3241/0632/dp154.pdf, accessed 12/07/2022; WM Gloag and Candlish R Henderson, *Law of Scotland* (14th edn, W Green, 2017) para 5.03

107 *Chappell & Co v Nestle* [1960] AC 87; *Roscorla v Thomas* (1842) 3 QB 234; *Re McArdle* [1951] Ch 669; *Eastwood v Kenyon* (1840) 11 A&E 438; *De La Bere v Pearson* [1908] 1 KB 280; *Thomas v Thomas* (1842) 2 QB 104.

108 *Thomas v Thomas* (1842) 2 QB 85; *Hamer v Sidway* (1891) 124 NY 538 27 NE 256; *Alliance Bank v Broom* (1864) 2 DR & SM 289; *Rock Advertising Ltd v MWB Business Exchange Centres Ltd* [2018] UKSC 24.

109 Gloag and Henderson (n 106) paras 5.09 & 5.10.

110 Kendra E Dupuy, 'Community Development Requirements in Mining Laws' (2014) 1(2) *The Extractive Industries and Society* 200–215.

insufficiently addressed by regulation.[111] As a 'supra-regulatory' instrument, IBAs will stipulate financial benefits to host communities and in addition to regulatory frameworks stipulate the conditions and minimum environmental standards to which resource developers must adhere while exploiting the resources or face contestation of legitimacy or revocation of their SLO by the host communities.[112] In this context, IBA helps to improve 'environmental performance, or there is at least a perception that it does' and reconciles the expectations of host communities and industry proponents.[113]

6 Conclusion

This chapter has assessed the AMV. This assessment is with a view to deciphering if it is potent in promoting cooperation among African-rich resource countries' mining governance, which in turn will facilitate resource-based sustainable growth and socio-economic development in Africa. It was found that the AMV might be the silver bullet that resource-rich African countries can leverage to ensure that resource wealth fuels socio-economic development and leads the African continent toward the path of industrialisation. However, the non-binding nature of the AMV, which leads to non-domestication by member states, has meant that the potency of the AMV in promoting cooperation in African mining governance is questionable. It, therefore, argues that, in its current form, the AMV appears to be the silver bullet that African resource-rich countries need but the AMV's ability to promote resource-based development in Africa as a region is dependent on the AMV becoming mandatory in order to have a binding effect on member states and AU member states' cooperation with the AMDC in developing requisite human capital and political institutions conducive for resource development to facilitate socio-economic sustainable development. Further, the chapter recommends the adoption of IBA from the perspective of an organic regional framework. This is to promote the tripartite partnership required for holistic socio-economic growth and sustainable development that the AMV envisions for AU member states.

111 Skylar Zilliox and Jessica M Smith, 'Supraregulatory Agreements and Unconventional Energy Development: Learning from Citizen Concerns, Enforceability and Participation in Colorado' (2016) 4(1) *Extractive Industries and Society* 69; Temple Stoellinger et al, 'Collaboration Through NEPA: Achieving a Social License to Operate On Federal Public Lands' (2018) 39(6) *Public Land & Resources Law Review* 217.

112 Galbraith et al (n 96) 29.

113 Stoellinger et al (n 111) 217–218.

9 The Potential of Digitisation in Enhancing Regulation and Sustainability in the Artisanal and Small-Scale Mining (ASM) Sector in Zambia, Nigeria, and Ghana

Selina Kudzai Zhuwarara

1 Introduction

The African continent has experienced a significant rise in Artisanal and Small-Scale Mining (ASM) activity over the last three decades, and as of June 2020, it was reported that there were over 13 million artisanal miners in Sub-Saharan Africa (SSA) alone.[1] In addition, more men, women, and even minors are seeking economic emancipation through mining activities. Therefore, the ASM sector is fast moving from the periphery of the mining economy to take centre stage in development economics in Africa. The ASM sector possesses significant potential to work as a conduit for broad-based development in African mining economies and can aid the continent's quest to foster equality and economic inclusivity within the industry.[2] However, notwithstanding significant interest and wide-ranging attempts at formalisation and enhanced economic integration undertaken over the last two decades, the embedded potential of the ASM sector has proven challenging to harness. As a result, the sector has not managed to manifest the desired transformative change it is primed to deliver in African economies.[3]

The sustained complexity in transforming the sector can be partly attributed to ineffective regulatory solutions. The difficulty in attaining effective control over the sector and the challenges in creating a viable formal economic environment for the sector are common challenges. Therefore, it is a considered view that enhancing the effectiveness of regulatory control and widening regulatory functionality beyond oversight can enable ASM host

1 Thomas Hentschel, Felix Hruschka, and Michael Priester. *Artisanal and Small-Scale Mining Challenges and Opportunities* (IIED and WBCSD, 2003).
2 African Union, *African Mining Vision, First African Union Conference of Ministers Responsible for Mineral Resources Development* (African Union, 2009).
3 Bruce G. Marshall, Marcello M. Veiga. 'Formalization of artisanal miners: Stop the train, we need to get off.' *The Extractive Industries and Society* 4(2), (2017) 300–303.

DOI: 10.4324/9781003284437-11

governments to harness the sector's full potential and improve the economic progression of the sector participants.

Regulatory capacity can benefit immensely from utilising digital proficiencies offered through advanced information communication technologies (ICT). With the demonstrated success of digital technology in various elements of the ASM value chain, such as data mapping, supply chain management, safety and health reporting, and information dissemination, there is sufficient evidence of the potential benefits of extending the scope of digital tools in regulatory frameworks.

This chapter will explore the ASM legal and policy framework in Zambia, Ghana, and Nigeria. Furthermore, it will discuss how reorganising the sector under an enhanced digital regulatory system can buttress ASM formalisation and economic integration. The chapter will also highlight the strategic nature of information in the regulatory matrix and how the benefits of improving the management of data and information can result in sustainable economic progression for the sector.

The chapter will also illustrate how a digitised regulatory system can unlock secondary rewards through the commercial use of information generated within such systems. There is significant economic value in the data produced by digital systems, as it can be further utilised to mobilise capital, services, and products for the ASM sector.

2 Regulation of the Artisanal and Small-Scale Mining Sector

2.1 Overview of the State of Artisanal Mining in Africa

Mining plays a significant role in the African economy. It is reported that the continent hosts about 30% of the world's total mineral reserves and a higher share of deposits of diamonds, vanadium, manganese, platinum, cobalt, and gold, among other minerals.[4] Consequently, the African mining industry is dominated by large-scale mining as a primary growth strategy in mineral-rich countries. The lure of higher rents and taxes from large-scale projects, coupled with the aggressive global appetite for commodities, has forced African governments to focus on large-scale mining.[5] The dominance of mining-related investments can illustrate the centrality of investment driven by large-scale mining in Africa's economy and to the overall foreign direct investment (FDI) flowing into mineral-rich countries.[6]

4 Africa Development Bank Group, https://blogs.afdb.org/fr/afdb-championing-inclusive-growth-across-africa/post/mining-industry-prospects-in-africa-10177 (20 December 2012). Accessed 19 July 2022.

5 F Bourgouin,. 'The politics of large-scale mining in Africa: Domestic policy, donors and global economics.' *The Journal of the Southern African Institute of Mining and Metallurgy 111* (2011) 525–528.

6 United Nations Conference on Trade and Development, *World Investment Report 2020: International Production Beyond the Pandemic* (United Nations 2020), 34.

The ensuing impact of concentrating on large-scale mining projects has resulted in stunted domestic capacity to extensively exploit opportunities and natural resources without reliance on FDI.[7] The dependence on FDI in the African mining industry is shown by the economic performance gap between the large-scale and ASM sectors. Consequently, the ASM sector has remained inhibited from becoming competitive because it experiences capital constraints and inadequate economic and regulatory support.[8]

ASM encompasses two principal players: artisanal miners and small-scale miners. Generally, small-scale miners operate on registered concessions and possess exclusive mining rights/licensing. On the other hand, artisanal miners may or may not have mining rights and, in most cases, do not hold exclusive rights over mining concessions. In addition, in some instances, artisanal miners work in areas that are not registered mining concessions but are areas where mineral occurrence has been informally discovered.[9]

ASM is undertaken by individuals, communities, cooperatives, or small entities. The sector is identified by the operation's size, the absence or limited level of mechanisation in mineral extraction and processing, the small number of employees, and the low annual production turnover.[10]

Most regulators in ASM host countries still encounter difficulty establishing effective oversight over numerous minuscule artisanal operations that are broadly distributed and do not follow a distinct pattern, capacity, or form. These varying characteristics usually make it challenging to successfully develop potent regulatory measures to bring the sector under effective control.

Consequently, the ASM sector has become synonymous with feeding the illicit trade of minerals, poverty, conflict, environmental degradation, severe health and safety incidents, and child labour, among other vices.[11] The continued failure to sustainably transition the sector to the formal economy and to diminish its ties with the black market has enabled these adverse effects to take root and further stifle the sector's development. Where artisanal mining is left to remain illegal or poorly regulated, such areas become a nucleus for the continued propagation of informal and illegal mining and all its ancillary adverse effects.[12]

7 Alice Lungu. *Formalisation of Artisanal and Small-Scale Mining in Sub-Saharan Africa: Challenges and Ways Forward.* Masters Dissertation, University of Manchester Faculty of Humanities School of Environment and Development (2007).

8 Gavin Hilson, Abigail Hilson, Roy Maconachie, James McQuilken, and Halima Goumandakoye. 'Artisanal and small-scale mining (ASM) in sub-Saharan Africa: Re-conceptualizing formalization and 'illegal' activity' *Geoforum* (83) (2017) 80–90.

9 IMPACT. 'Best Practices: Formalization and Due Diligence in Artisanal and Small- Scale Mining. National Resources and Infrastructure Division,' *IMPACT Transforming Natural Resource Management* (2018) 6.

10 Hentschel (n 1).

11 Extractives Hub. www.extractiveshub.org/servefile/getFile/id/4197. Accessed 10 May 2021.

12 Yuliya Zabvelina. Interview by John Jay Research. 'Illegal Mining and Organized Crime.' Illegal Mining and Organized Crime. (22 May 2020). https://johnjayresea

It is essential to acknowledge that the large-scale mining sector has thrived because of the controlled and highly regulated environment in which it exists.[13] The large-scale mining value chain's vital elements are supported and guided by comprehensive national and international rules, regulatory procedures, policy, supportive economic initiatives, valuable institutional and industrial linkages, and an organised global value chain from mine to market. As a result, the extent of standardisation and international coordination in regulating the large-scale mining sector has aided its development and growth.

The successful regulation of the ASM requires a similar level of standardisation, coordinated regulation, cooperation, and economic integration. However, as previously noted, governments and development partners have made various attempts across the continent to support the ASM sector. While positive progress is registered in some cases, these efforts have not established standardised effective control and broad-based viability of the sector.[14]

2.2 *An Overview of ASM Sector Support in Ghana, Zambia, and Nigeria*

Zambia,[15] Ghana,[16] and Nigeria[17] all legally recognise and allow artisanal mining. Table 9.1 below summarises some key highlights from the ASM regulatory structure in these countries.

Table 9.1 reflects that Zambia, Ghana, and Nigeria all have sound regulatory frameworks that extend a fair degree of structural support for the ASM sector. The support framework generally includes financial, technical, technological, and extension services. The countries also have cadastre systems providing ASM sector licensing and services. Zambia, Ghana, and Nigeria also recognise the centrality of government participation in the ASM market and have all created public entities that play an influential role in buying strategic minerals from the sector.

However, while Zambia, Ghana, and Nigeria show a relatively significant degree of government support towards the ASM sector, these countries still struggle to establish sufficient control over the sector. Common challenges include the continued proliferation of illegal mining, non-compliance, smuggling, environmental damage, poor labour conditions, health hazards, and the overall constrained economic performance of the sector. These challenges show that while the regulatory structures appear sound, underlying structural

rch.commons.gc.cuny.edu/2020/05/22/illegal-mining-and-organized-crime. Accessed April 14, 2021.

13 Bourgouin (n 5).

14 Nellie Mutemeri and Francis W. Petersen. 'Small–scale mining in South Africa: Past, present, and future.' *Natural Resources Forum 26*(4) (2002) 286–292.

15 Mines and Minerals Development, Zambia (Act 11 of 2015), Sections 29 and 42.

16 Minerals and Mining Act of Ghana (Act 703 of 2006), Sections 82–99.

17 Minerals and Mining Act of Nigeria (Act of 2007), Sections 49 and 90, https://portal.minesandsteel.gov.ng/Home/Guidelines. Accessed 8 May 2021.

Table 9.1 ASM policy highlights in Zambia, Ghana, and Nigeria

Country and ASM minerals	*Main policy highlights*
ZAMBIA Emerald, aquamarine, tourmaline, amethyst, gold, garnet, topaz, quartz, opal, manganese, copper, development minerals, cobalt, nickel, tin	**Main legislation:** Mines and Minerals Development and the General Mining Regulations of 2016. **Main policy document:** Mineral Resources Development Policy. **Main ASM administrative arm:** Director of Mining Cadastre, Ministry of Mines and Mining Development **Cadastre System:** YES **Digital/electronic licensing platform for ASM:** YES **Main licensing format:** ASM licenses are issued to individuals and miners organised in corporate form. The artisanal sector has its own distinct licensing,[18] which includes artisanal exploration licenses[19] and various formats of artisanal mining licenses – e.g. the gold panning certificate is valid for two years and is renewable. **Extension services:** YES – Legislation provides technical extension support to ASM through regional and central Ministry of Mines structures **Government financial assistance:** Intermittent. The Mining Revolving Fund is inactive[20] **Regulated/organised ASM market:** Government Buying Centres exist for key strategic minerals such as gold and precious stones Notable private and development partnered market forums include initiatives such as Africa Gem Exhibition & Conference (https://agecafrica.com), Virtu-Gem (electronic selling platform) (www.virtugem.com) **Main government buying centres:** Gold and Precious Stones ZCCM-IH (Zambia Consolidated Copper Mines –Investment Holdings) All other ASM minerals are sold to large-scale mining entities or private agents or buyers on privately negotiated terms
GHANA Gold, bauxite, manganese, diamonds, development minerals (granite aggregates), salt, precious stones	**Main legislation**: Minerals and Mining Act, The Small-Scale Gold Mining Law and associated Regulations **Main policy documents:** Minerals and Mining Policy 2016, Ghana Shared Growth and Development Agenda (GSGDA), Artisanal and Small-Scale Mining Framework 2015 **Main ASM administrative arm:** Government District Offices, Minerals Commission, and Small-Scale Mining Committees **Cadastre System:** YES

18 The Mines and Minerals Act [Chapter 213], Section 6 (g).

19 Ministry of Mines and Minerals Development Zambia, www.mmmd.gov.zm/?page_id=1097. Accessed 19 July 2022.

20 UNECA and AMDC. 'Study on ASM in Africa.' https://knowledge.uneca.org; 'Zambia.' https://knowledge.uneca.org/ASM/Zambia. Accessed 14 February 2021.

Country and ASM minerals	*Main policy highlights*
	The country also uses technology in its programmes supporting the sector, such as the Galamstop Programme, which uses drones, trackers, and software that monitors the life cycle of mining and related licenses in the small-scale mining sector and is a repository for demarcated and mapped ASM areas[21] **Digital/electronic licensing platform for ASM:** YES – through cadastre system **Main licensing format:** Community mining programmes and cooperatives for artisanal miners and small-scale licenses **Extension services:** YES **Technological assistance:** There are central processing sites supporting community mining entities. Private partners such as Commodity Monitor Limited in Ghana are pioneering clean, appropriate, and affordable technology for the ASM sector. **Government financial assistance:** Assistance is provided to cooperatives through the Minerals Commission & Minerals Development Fund **Regulated/organised ASM market** **Main government buying centres/agents:** PMMC (Precious Minerals Marketing Company) buys gold and precious stones from legal and illegal ASM miners. All other ASM minerals are sold to private agents or buyers on privately negotiated terms
NIGERIA Gold, limestone, lead, zinc, barite, gemstones, development minerals	**Main legislation:** The Minerals and Mining Act and the 2011 Minerals and Mining Regulations **Main policy documents:** 2012 Road Map for the Development of [the] Solid Minerals and Metals Sector, 2008 National Minerals and Metals Policy and the 2009 Vision 2020 National Technical Working Group on Minerals and Metals Development Report. **Main ASM administrative arm:** Nigerian Cadastre Office & Ministry of Mines and Steel Development (Artisanal and Small-scale Department) **Cadastre system:** YES **Digital/electronic licensing platform for ASM:** ASM miners can apply for licences through the Cadastre system. The Nigerian government undertakes biometric enrolment of ASM miners under its PAGMI (the Presidential Artisanal Gold Mining Development Initiative) **Main ASM licensing format:** The government prefers artisanal miners to form cooperatives to acquire small-scale mining leases.

21 Government of Ghana. 2020. https://mesti.gov.gh; https://mesti.gov.gh. Accessed 20 May 2021.

Country and ASM minerals	*Main policy highlights*
	Extension services: Extension services are offered to legal ASM miners through the Ministry of Mines and Steel Development Artisanal and Small-Scale Mining Department **Technological Assistance:** Assistance is offered under various government projects, e.g. assistance is given to gold ASM miners under the PAGMI initiative. **Government financial assistance:** Assistance is provided to registered cooperatives under the Small Grant Scheme **Regulated/organised ASM market:** Government buying centres and agents were established to buy ASM product **Main government buying centres:** There is a National Gold Purchase Programme, and various private licensed and registered buying centres and aggregators are recognised by the Ministry of Mines and Steel Development[22] All other ASM minerals are sold to private agents or buyers on privately negotiated terms

Source: Adapted from Environmental Law Institute,[23] African Minerals Development Centre,[24] Ministry of Lands and Natural Resources, Zambia,[25] and Minerals Commission, Ghana[26]

issues weaken the efforts toward empowering the sector. Table 9.2 below shows the common problems cited by ASM miners in these jurisdictions and a description of the underlying structural cause of the issues.

Following the above, the main issues affecting the ASM sector can be summarised into two overarching structural deficiencies:

a The inadequate legal construct of the sector, especially in its failure to create a solution bringing all forms of artisanal mining under effective regulation; and
b The failure to create a favourable economic environment that promotes and supports the growth of ASM enterprises in the mining industry.

22 Nigeria, Ministry of Mines and Steel Development. https://portal.minesandsteel.gov.ng. Accessed 10 May 2021.
23 Environmental Law Institute. 'Artisanal and Small-Scale Gold Mining in Nigeria: Recommendations to Address Mercury and Lead Exposure. Environmental Law Institute' (2014) 51.
24 African Minerals Development Centre. 'Report on Artisanal & Small-Scale Mining in Africa' (2017).
25 Ministry of Lands and Natural Resources Zambia. *Artisanal and Small-Scale Mining Handbook for Zambia*. (Geological Survey of Denmark and Greenland 2018).
26 Minerals Commission Ghana. Artisanal & Small-Scale Mining (ASM) Framework (2015) 31.

Table 9.2 Common ASM challenges in Zambia, Ghana, and Nigeria

Common structural deficiencies	*Description*
Financial difficulty	There is a lack of diversification of financial products and options in the financial sector for the ASM sector. Available products are not suitable or appropriately scaled to support small-scale enterprises sustainably. The cost of debt and the onerous requirements set for eligibility hinders a significant portion of individuals and small and medium-scale enterprises (SME) enterprises from accessing financing from the banking sector. Financing provided through government structures is commonly affected by limited budgets and is often impacted by poor uptake or poor repayment.[27]
Undifferentiated legal framework	Laws and processes regulating geological exploration, prospecting rights, tax, compliance, and the acquisition of concessions are cumbersome, bureaucratic, and costly, and the eligibility requirements are too stringent. The processes are not favourable to the ASM sector as they are often unattainable for most miners. The regulatory structure requires further streamlining and easing of costs and procedures relating to SMEs.[28]
Limited industrial capacity and support	The general industry that provides services, plant, and equipment for ASM entities is still limited and has not been able to integrate adequate mechanisation, competition, and product diversity. This affects the competitiveness and range of services and equipment available for the sector. The limited investment in the local manufacturing industries in Africa has consequently also stifled the advancement in diversity and cost-effectiveness of products available in the ASM sector.[29] Centralised processing centres would facilitate increased access to safe and effective technology; however, such centres are too few to cater for the sector adequately.[30]

27 Intergovernmental Forum on Mining, Minerals, Metals, and Sustainable Development (IGF). *IGF Guidance for Governments: Managing artisanal and small-scale mining*. (Winnipeg: The International Institute for Sustainable Development 2017).

28 A Hira and J. Busumtwi-Sam. *Mining Community Benefits in Ghana: A Case of Unrealized Potential*. (University of British Columbia and Canadian International Resources 2018) 31; 32.

29 Commodity Monitor Limited Ghana. 29 April 2021. https://commoditymonitor.org/2021/04/29/ghanas-president-introduced-to-mercury-free-mineral-processing-technology. Accessed May 15, 2021.

30 IGF (n 27).

Common structural deficiencies	*Description*
Inadequate regulation of ASM market	The ASM commodities market lacks adequate checks and balances through state and financial sector oversight. As a result, the market is vulnerable to collusion and manipulation to the detriment of miners. In addition, inadequate supervision and standardisation of the ASM market structures and pricing protections significantly compromise the bargaining power of the sector.[31]
Poor differentiation in tax regimes	The corporate taxes, mining tax, and levy structure requires further streamlining and re-evaluation. In most cases, while seemingly lower, the cost remains out of reach for the artisanal demographic. Further differentiation in the scope and cost of levies and taxes is required to allow for the sustained financial viability of the ASM sector. In addition, high regulatory costs place significant financial pressure on vulnerable operations.[32]
Poor policy implementation	Although good policy frameworks exist, the implementation of the policies is poor. Most key ASM institutions are inadequately resourced and funded, consistency in policy implementation is lacking, and there is often poor data/information collection and analysis from the sector. As a result, governments have not been able to implement the full extent of the support stated in their policy positions.[33]

3 The Potential Impact of Digitalisation on ASM Regulatory Efforts

3.1 The Benefits of Digitalisation in Mining Enterprises

Recent years have seen rapid adoption of digital technology in the mining industry – hence, the common reference to the significance of digital transformation in the mining industry. The term 'digital transformation' has numerous definitions; however, this chapter defines it as technology-driven processes of change derived from ubiquitous data, connectivity, and enhanced decision making.[34]

31 Duncan Murray and Benjamin Webb. 'Compliance Challenges in ASM Markets' (CCEX 2019). https://theccex.com. Accessed 10 February 2021.

32 Anuradha Joshi, Wilson Prichard, and Christopher Head. *Taxing the Informal Economy: Challenges, Possibilities, and Remaining Questions.* (Institute of Development Studies 2012).

33 Abbi Buxton. *Responding to the Challenge of Artisanal and Small-Scale Mining. How Can Knowledge Networks Help?*, (IIED 2013).

34 A Young and P Rogers. 'A Review of Digital Transformation in Mining.' *Mining, Metallurgy & Exploration* (2019) 683–699.

It is true that digitalisation has distinctly improved mining efficiency and safety through better data management. Digital technologies have allowed mines to collect, integrate, prepare, and analyse information in real time, thus improving pre-emptive decision making.[35] New technologies have introduced predictive analytics, artificial intelligence, enhanced data optimisation, analytical tools, wireless communication, cloud computing, automation, and blockchain technology. These advancements have improved productivity and safety, health, and environmental compliance, increased operational efficiency, and enhanced integrity of the minerals supply chain.[36] Syama Gold Mine in Mali is a flagship African automation project. The mine operates autonomous haulage to load and move waste and ore to the surface using Sandvik Autodig technology.[37] Such demonstrated aptitude in the operational sphere of the industry also ignites interest in how digital technology can improve the regulatory sphere of the industry, particularly the ASM sector.

3.2 The Value of Integrated Information Systems to the Development and Regulation of the ASM Sector

While digital innovation has become a common intervention for enhancing operational and information efficiency in mining enterprises, the industry's regulatory environment has been sluggish in exploring the diverse range of competencies that digital solutions can contribute to regulatory regimes. As a result, the incorporation of digital technology in the regulatory sphere has largely remained concentrated on enhancing mining title and land title databases through mining cadastre information systems or land management systems. Cadastre systems are digital/computerised systems aimed at improving the workflows that emanate from the digitalisation of regulatory and institutional systems supporting the administration of mineral and land title.[38] In some countries, the database is complemented by geographic information systems (GIS) technology which provides digital maps of designated concessions and their licensing status, such as the cadastre systems, which are operational in Nigeria, Tanzania,[39] and Mozambique.[40]

35 Alyson Cartwright and Rudy Moctezuma. 'Eclipse Mining Technologies.' Eclipse Mining. https://eclipsemining.com/2020/10/07/prepare-data-advanced-mining-technologies. Accessed 4 January 2021.

36 Young and Rogers (n 35).

37 Noel Dyson. 'Syama's automation surge.' *Mining Magazine*, 26 May 2020. www.miningmagazine.com/technology-innovation/news/1387604/syama%E2%80%99s-automation-surge. Accessed 11 May 2021.

38 Bill Feast, Jonathan Davids Toby Mills, and Paul DeVries. 'Mining Cadastre in Africa – Lessons Learnt.' 5th FIG Regional Conference. Accra, Ghana. 8–11 March 2006,

39 Tanzania Mining Cadastre GIS Map. https://portal.madini.go.tz/map. Accessed 8 May 2021.

40 Mozambique Mining Cadastre GIS Map. https://portals.landfolio.com/mozambique/en. Accessed 8 May 2021.

Cadastre systems are typically enlisted to ease the acquisition of mining rights and reinforce the security of title. A fully-fledged cadastre system is structured to streamline administrative and licensing workflows from different government departments by creating a consolidated multi-functional platform that users can use to transact and communicate with the regulator. Cadastre systems have generally proven to be more accessible, efficient, and cost-effective for investors and governments. Several mineral-rich countries have consequently adopted them in Sub-Saharan Africa.[41] Cadastre systems have been identified as a critical strategy in mining reform and attracting investment in the large-scale mining sector by improving the accessibility and integrity of the mineral rights framework.[42]

The concept of cadastre systems has also spurred motivation for the use of consolidated digital platforms that share wide-ranging socio-economic information with investors.[43] There is significant potential in harnessing and using data to improve regulation and growth in the mining industry, especially the ASM sector.

As indicated earlier, the absence of accurate socio-economic baseline information on the ASM sector has curtailed the ability of ASM host governments to fully implement policy and provide effective solutions for the regulation of the sector. The lack of accurate and comprehensive socio-economic data is also related to the unattractiveness of the ASM sector as an investment destination. Therefore, providing real-time and reliable information on the ASM sector can prove to be a key asset in the economic transformation of the ASM sector. It would make it easier for governments to implement policy effectively and for stakeholders and investors alike to see the opportunities and growth potential within the sector.

4 Valuable ICT and Digital Solutions for the ASM Sector

As discussed above, there is significant value in exploiting the benefits brought by digital competencies in the ASM regulatory framework. Digitisation can achieve flexible reach, which is required to overcome the unique challenges of governing the ASM sector. In addition, the improved management of information from the sector can become a platform for broader economic integration, thus widening prospects of transformation and sustainable growth. Below are digital tools that may well be utilised to complement regulatory

41 Enrique Ortega Girones, Alexandra Pugachevsky, and Gotthard Walser. 'Mineral Rights Cadastre, Promoting Transparent access to Mining Resources.' Extractive Industries for Development Series No. 4. (2009).

42 EITI. 'Mainstreaming transparency in Sierra Leone.' 17 July 2015. https://eiti.org/blog/mainstreaming-transparency-in-sierra-leone. Accessed 14 April 2021.

43 R C A Minnitt and F T Cawood. 'Information as an Alternative to Mineral Rights Taxation.' *The Journal of the South African Institute of Mining and Metallurgy* (1999), 347.

systems and improve the competency and capacity of regulators to manage the ASM sector more effectively.

4.1 The Potential of Biometric Authentication Systems in ASM Licensing Databases

In traditional regulatory systems, prospective miners/prospectors present themselves at designated government offices or access the cadastre system to make the relevant applications and provide the necessary supporting documentation. If the application is successful, they are allocated the appropriate mining rights and copies of the certificates/licenses evidencing their rights. The level of digitalisation of this process varies across Africa. Some countries have digitalised this process through mining cadastres, while other counties still utilise paper-based systems complemented by basic forms of digital reference filing. Therefore, the first premise in enhancing the regulation of the ASM sector is optimising the licensing processes. Licensing of artisanal miners should be accessible and cost-effective, with efficient data and information management.

Given the large number of artisanal miners, the capacity of the traditional licensing strategy should be supported by mobile licensing points located in areas where artisanal mining takes place.[44] This would improve licensing uptake. However, it inevitably increases the volume of licensees that must be processed and the volume of licensing data handled. The processing of high-volume licensing data would need to be supported by enhanced identity and data management systems that can ensure accuracy and efficiency while preserving the integrity of identity management and authentication processes. Incorporating mobile digital enrolment centres supported by biometric authentication systems can enhance identity management for artisanal miners' databases. A digital database can also allow regulators to easily cross-reference submitted information across other government databases such as police clearances, identity verification, etc.

Once enrolled on the digital database, the miners can be allocated digital identity codes linked to their unique digital profiles. The profile can become the gateway for the miner to access a host of other products and services on the government digital platform.

Biometrics is the automated identification of individuals through their behavioural or biological characteristics, such as fingerprints or iris recognition. These traits can be recorded and used as primary or secondary identification means.[45] In the case of an artisanal digital enrolment system, governments may utilise fingerprints or iris recognition for authentication

44 Minerals Commission Ghana (n 26).

45 Evangelia Micheli-Tzanakou and Konstantinos N. Plataniotis. 'Biometrics: Terms and Definitions.' *Encyclopedia of Cryptography and Security* (2011). https://doi.org/10.1007/978-1-4419-5906-5_730. Accessed 12 April 2021.

purposes. Consequently, the holder of an artisanal mining licence or certificate can further authenticate their identity through biometric confirmation at any relevant regulatory or administrative point to access products and services. The fact that such digital databases are accessible on demand improves the administrative capacity to collect, verify, and track data from individual miners.

Nigeria has enlisted biometric registration of artisanal miners under the PAGMI programme. The programme was unveiled in 2019 to facilitate the formalisation and integration of artisanal gold miners into the formal economy. It was reported that by July 2020, approximately 20,000 artisanal miners had been registered in the states of Kebbi and Osun. This reflects a rapid registration rate owing to the flexibility of the process. Nigeria further intends to utilise the accumulated identity data to assist artisanal miners in acquiring mainstream financial services.[46]

The integrity of identity data is crucial in access to financial services.[47] Therefore, digital regulatory platforms incorporating improved identity data management tools can be vital in positioning the sector for enhanced financial inclusion.

4.2 The Potential of Mobile and Wireless Telecommunications Technologies in ASM Regulatory Support

Mobile telecommunications technology has become an essential business component in the fourth industrial revolution. Sub-Saharan Africa has a mobile phone penetration rate of 45%, with a growing smartphone adoption rate of 50%, given that cheaper devices are becoming increasingly available.[48] It is reported that mobile technologies and services generated 9% of the region's gross domestic product (GDP) in 2019, amounting to more than $155 billion in economic value.[49] While various mobile devices have changed the face of modern-day commerce, the mobile phone continues to be one of the most influential and accessible devices through which commerce is increasingly being conducted. It has evolved from a social device to a powerful business medium for individuals and companies alike, especially small-scale enterprises. In the ASM sector, the mobile phone is also commonplace in the

46 Government of Nigeria. 15 July 2020. https://statehouse.gov.ng. Accessed 20 May 2021.

47 Sharmista Appaya and Minita Varghese. 'Digital ID – a critical enabler for financial inclusion,' World Bank Blogs, 20 June 2019. https://blogs.worldbank.org/psd/digital-id-critical-enabler-financial-inclusion. Accessed 21 July 2022.

48 Groupe Spécial Mobile Association. *The Mobile Economy: Sub-Saharan Africa* (GSMA 2020).

49 GSMA. *Start-Ups and Mobiles in Emerging Markets: Mobile for Development* (GSMA 2017). www.gsma.com/mobilefordevelopment/wpcontent/uploads/2018/08/Start-ups-and-Mobile-in-Emerging-Markets-Issue-1.pdf. Accessed 4 November 2020.

hands of many miners who depend on it for communication and mobile financial transactions.

The mobile phone can be a viable tool to complement and aid regulation and support commercial activities in the ASM sector. Mobile phones allow for flexibility, accessibility, and personalisation of services through various applications on handheld devices.[50] When envisaged as part of the ASM regulatory mechanisms, this device can become a medium of personalised regulatory and commercial activity between the sector participants and regulatory authorities. For example, regulators can enable the digital profiles referenced in Section 3.1 to become accessible via the miners' registered mobile phone line. A special application (app) can be developed to host these interactions, thus creating a live communication channel between the miner and the regulator without time, cost, and proximity limitations. Using mobile devices to widen regulatory reach can address the challenging characteristics that have long hampered effective regulation in this sector, such as poor communication and information dissemination.

The capacity of a versatile app to facilitate basic reporting and compliance returns has the potential to significantly improve the efficiency of communication and tracking in the ASM regulatory process. For example, simple reports and statistics can be routinely submitted via the government mobile app. Likewise, communication from the regulator in the form of verification of returns, bulletins, and notices can also be delivered to miners using mobile telecommunication channels.

The 'M-Government' (mobile government) is a concept that has long been propositioned to highlight how mobile telecommunications services can improve the effectiveness and efficiency of government in communicating with citizens, businesses, government employees, and government agencies in respect of a wide range of services and programmes.[51] The envisioned culmination of the M-Government is the creation of 'connected governance' where the government is seamlessly connected to its citizens and responsive to citizens' needs through integrated back-office infrastructure and data access.

The concept of connected governance is particularly significant to ASM sector regulation. This is because an increased level of interaction and responsiveness is required between the regulator and ASM regulatory targets, which is not ordinarily achievable with traditional regulatory mechanisms. Furthermore, as discussed earlier, it is crucial to base sector initiatives on accurate needs. Obtaining accurate needs from ASM participants demands a robust medium of communication that facilitates and supports direct and personalised interaction between the governing authority and its regulatory

50 Stuart Barnes and Euseboa Scornavacca. 'The emergence of mobile commerce' in Stuart Barnes (ed.) *E-Commerce and V-Business: Digital Enterprise in the Twenty-First Century* (Elsevier 2007) 156–178.

51 OECD and ITU. *M-Government: Mobile Technologies for Responsive Governments and Connected Societies* (2011). http://handle.itu.int/11.1002/pub/804db8ac-en. Accessed 20 January 2023.

targets. Therefore, integrating mobile telecommunication devices to enhance communication channels in the regulatory process may prove a viable and cost-effective strategy for ASM host governments to adopt.

4.2.1 Use of Mobile Phones in the ASM Sector and Development Projects in Africa

The provision of digital services supported by mobile telecommunications mediums is already operational in segments of the ASM sector value chain. These initiatives have proved to be a viable means of improving communication, safety and health reporting, and supply chain dynamics. For instance, a mobile Gold Price Message Alert project by North American company Ulula was rolled out for gold miners in Peru in 2017.[52] The mobile messaging service gave miners real-time access to the prevailing market gold prices on demand. It enabled them to negotiate fair prices for their gold. The messaging facility was offered free of charge and accessed by sending a prompt to a shortcode. This allowed the service to be available even without internet connectivity. Ulula and the International Peace Information Service (IPIS) also introduced a mobile incident reporting system in the eastern Democratic Republic of Congo. The system called 'Kufatilia' was also piloted in northwest Tanzania. The system enables individuals in ASM mining areas to anonymously report incidents or accidents through mobile messaging. The report is then uploaded to a web-based stakeholder platform that alerts civil society organisations which then move in to manage and assist.[53]

To ensure accessibility and cost-effectiveness, these initiatives predominantly use Unstructured Supplementary Service Data (USSD) protocols that enable mobile communications without an internet connection. Shortcodes have been widely adopted because of their cost-effectiveness, simple format, and ease of use, thus curtailing the digital literacy divide. The USSD format is also flexible and provides seamless roaming, making it ideal for use in the ASM regulatory space where the regulatory target is highly mobile and may operate in remote areas far from administration points.

Various other digital economic and social development services supported by mobile-based platforms are available in Africa. For example, BenBen is a mobile-based app operating in Ghana, which is a land tenure and property start-up allowing citizens to acquire land-based assets and financial transactions through USSD, Android, and web platforms. UX is a Mozambican start-up launched by Biscate to connect workers in the informal sector to employment via web and USSD protocols. It assists informal workers in

52 Andres Schipani. 'Mobile gold price text alerts aid Peru's Mercury Pollution Fight. Financial Times, 5 April 2017. www.ft.com/content/441035de-eeca-11e6-ba01-119a44939bb6. Accessed 3 November 2020.

53 IPIS. IPIS Briefing April 2020 Report. https://ipisresearch.be/weekly-briefing/ipis-briefing-april-2020. Accessed 12 January 2021.

Mozambique with limited access to the internet to have greater access to jobs and income opportunities.[54]

These projects show the viability of integrating mobile-based service protocols in supporting key socio-economic development initiatives, thereby concretising the potential of mobile telecommunications technology in supporting the ASM regulatory framework.

4.3 The Potential of Application Programming Interfaces (APIs) in Creating Conducive Economic Participation for the ASM Sector

As discussed in Section 3.2, organising the ASM sector around a digital regulatory framework can broaden governments' capacity to further integrate the sector into the broader economic framework by leveraging the valuable data the system would generate. Many governments already have such integrated systems through cadastre systems, in which case it would be beneficial to either complement the functionality of the cadastre with mobile-based telecommunications technology or to extend and modify the functionality of the cadastre system.

As previously noted, investment and enhanced economic integration of the ASM sector is commonly hampered by a lack of adequate data and information on the sector's viability as a market for capital, products, and services. Governments and prospective business partners alike find it difficult to envision artisanal or small-scale miners as sustainable business partners because of limited data and information. To address this gap, a consolidated digital regulatory platform would significantly improve the scope of information available to stakeholders. This information is critical in providing key socio-economic indicators from the sector, which constitute essential building blocks in data-led policy and economic support to the sector. Gathering and organising the sector's data through integrated software will give regulators unprecedented line of sight and access to reliable real-time data on the sector.

For governments to utilise the data emanating from the sector for improved economic integration, the different regulatory arms overseeing the sector first need to be integrated and harmonised under a single system specifically mandated to govern the ASM sector. For example, a country with an existing cadastre system already possesses the required digital platform. It would merely need to extend and enhance the system's functionality to host other regulatory agencies. However, if a country is yet to install a cadastre system, then an integrated platform would need to be developed specifically for this purpose.

The essence of an ASM integrated digital platform is to create a holistic and coordinated repository of data and information regarding the sector's regulatory status and to streamline regulatory functions efficiently. Rationalising and streamlining different regulatory authorities' processes and

54 GSMA (n 49).

reporting requirements is critical in simplifying and improving the ASM governance framework. Not only would that prevent duplicity and redundancy in regulatory operations, but it would also make compliance reporting less cumbersome. Cumbersome and complex licensing and reporting procedures have consistently been a barrier to formalisation and compliance.[55]

The need to simplify the regulatory processes and package data in simpler formats for the sector was noted during free mineral clinics, which the Nigerian Geological Services Agency facilitates for artisanal miners. The high illiteracy rates or insufficient background in geology prevented the miners from readily understanding the geological data being shared with them. This resulted in low turnout and lack of interest, and diminished the potency of the support measure.[56] Simplifying information and reporting formats is critical in ensuring that formalisation and economic integration efforts are accessible and practical to the ASM sector.

Once the vital compliance and reporting services have been simplified and digitised, the next step would be to consider the parameters and access modalities of sharing the sector's information with third-party stakeholders who are vital to the sector's development.

Various ICT protocols can aid the establishment of such multifaceted communication and transaction platforms. For instance, the ASM digital regulatory system can be structured to link with external service providers' databases through application programming interfaces (APIs). APIs form bridges between different databases, thus creating a digital ecosystem that can exchange or verify relevant information between various parties.[57] As discussed in previous sections, this increased interface on a secure and verifiable system can spur broader interest and confidence from private and public enterprises vital to the sector's growth.

An example of such a system is the Sistema de Identificación Nacional Tributario y Social (SINTyS) in Argentina which allows public bodies to integrate their databases. Although the system is primarily focused on public services integration, it is an excellent example of the capacity of a digital database to communicate with various other related databases. SINTyS has enhanced the provision of services to citizens in Argentina. The platform enables citizens' records to be linked across 13 different databases covering employment, pensions, electoral roll, social benefits, the deceased, real estate

55 Environmental Law Institute (n 23).

56 M. Ango, M. Blessing, B. Choquette, B. Erdenebat, S. Jagdish, A. Kamara, K. Tang, A Wideroth, and J. Worthington. 'Creation of a Sustainable Mining Program through Formalization of Artisanal and Small-Scale Miners.' Columbia University, 2019, 34.

57 AltexSoft. 'What is an API: Definition, Types, Specifications, Documentation.' 18 June 2019. www.altexsoft.com/blog/engineering/what-is-api-definition-types-specifications-documentation. Accessed 15 April 2021.

registries, auto registries, and poor households, along with 24 provincial civil registries – all using an individual's unique identity number.[58]

Following the same concept, ASM host governments could develop a unique multi-functional digital system that hosts public and private services for the ASM sector. Such a system would be instrumental in widening sector interface with public and private sector service providers and consequently unlock more comprehensive economic integration. Accordingly, the platform would enable key stakeholders to access and verify vital information from government agencies regarding the ASM sector and acquire direct access to an organised ASM market.

Organised and real-time information on the sector can enable bespoke services, products, or terms to be structured for the sector. This would result in the ASM sector having more expansive access to different services, skills, financial assistance, and products. Allowing the business industry access to such a system would drive greater economic integration because it would provide better insight into the state of the sector. On the other hand, a better relationship between the sector and the business industry reduces the burden on governments to provide resources and support the sector. Assisting the ASM sector to improve its capacity to transact with the broader economy sustainably is an important step in mobilising sustainable economic integration and works to strengthen the ability of the sector to mobilise resources on its own.

A simplified flowchart showing the movement of information in the proposed integrated digital ecosystem between the ASM sector, government, and external stakeholders is shown in Figure 9.1.

4.3.1 An Illustration of a Financial Transaction on Such a Government ASM Digital System

Figure 9.2 illustrates how a loan can be processed in the ASM digital regulatory platform. In this case, the figure demonstrates the transaction flow from a loan application to a bank by a registered ASM miner.

A description of the transaction sequence would be as follows. An artisanal miner applies to a bank for a loan. First, the bank accesses the Government ASM Digital System to verify the credentials of the miner, compliance standing, identity and production returns, etc. Then, through the Central Regulatory System and the Government Buying Centre, the government provides the relevant production history, identity authentification data, and a verified compliance report regarding the miner to the bank. The bank then structures appropriate lending terms based on the information provided and

58 Argentina Government. 'SINTyS – National Tax and Social Identificaiton System.' www.argentina.gob.ar/politicassociales/sintys#:~:text=Coordinamos%20el%20intercambio%20de%20informaci%C3%B3n,de%20Protecci%C3%B3n%20de%20Datos%20Personales. Accessed 8 November 2020.

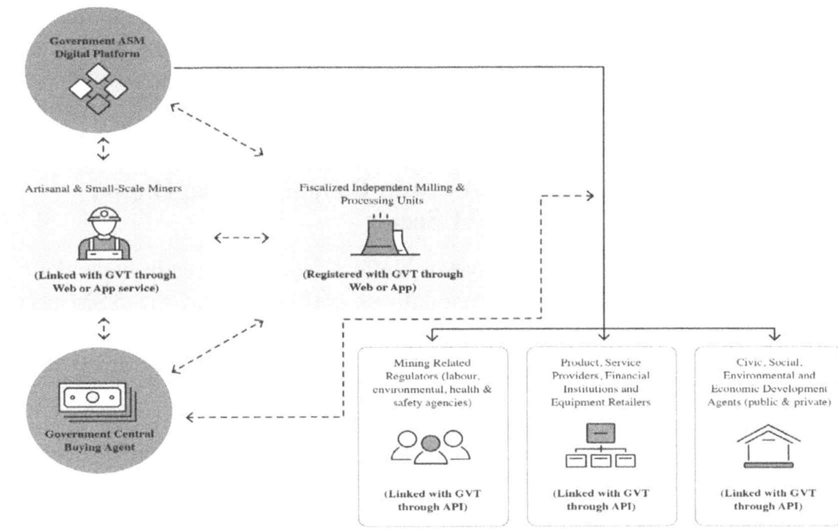

Figure 9.1 Flowchart showing the movement of information.

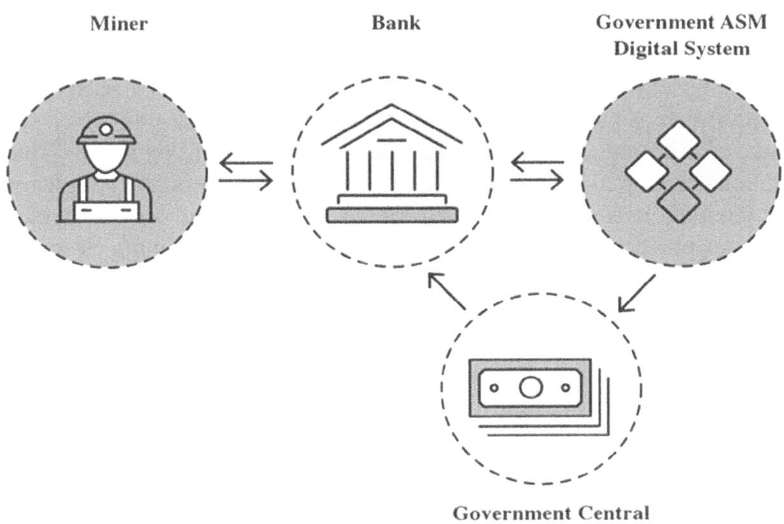

Figure 9.2 An illustration of a financial transaction

extends a loan to the miner. Finally, to facilitate repayment, the loan is lodged with the Government Buying Centre, where the miner sells his product. The appropriate repayment deductions can be made directly to the bank whenever the miner delivers his commodity or at the agreed intervals.

Taken in the context of Zambia, Ghana, and Nigeria, such systems would prove valuable for supporting policy implementation and policy effectiveness. In this case, the deployment of financial assistance is better regulated and tracked.

5 Policy Measures That Can Assist in Building Strong Digital Regulatory Systems in the ASM Sector

An enhanced digital regulatory system's effectiveness is directly dependent on the potency of the policies and regulations crafted around the ASM sector. Therefore, ASM governments must enlist supportive policy considerations to ensure the success of the digitalisation of the ASM regulatory framework.

5.1 Flexible and Accommodative Artisanal Regulatory Condition

One of the essential elements in legitimising artisanal mining is the range of the ASM licensing regime. It is vital to establish a licensing strategy that can accommodate artisanal miners as they currently present themselves in the mining industry. As highlighted previously, the entire ASM sector must be under effective control for sustainable interventions to be effective. Therefore, the proliferation of illegal artisanal mining undermines meaningful formalisation efforts in the mining industry.

Lowering eligibility requirements and offering licences for basic artisanal activities such as prospecting and panning is likely to yield positive returns on uptake of licensing. Zambia has extended the ASM licensing range to include traditional artisanal activities such as gold panning through the issuance of gold panning certificates. This wide range of licensing is commendable as it has better prospects of harnessing all forms of artisanal mining as they practically and traditionally occur in the artisanal sector.[59] Furthermore, allowing individual ASM licensing enables individuals to acquire a standalone mining right, thus allowing greater flexibility to the miner.

Most ASM host jurisdictions, such as Ghana and Nigeria, promote the creation of cooperatives and community organisations as the chief strategy for legalising artisanal activity. However, such an approach may result in low formalisation rates because a significant portion of artisanal mining is undertaken individually or in transient groupings, thus making permanently fixed groups undesirable for some miners.[60] Recognising and providing individual standalone mining rights is more likely to offer better prospects for artisanal miners to formalise their activities.

59 The Mines and Minerals Development (General) Regulations S I No. 7 of 2022, Mines and Minerals Development Act No. 11 of 2015.

60 Laura Barreto. *Analysis for stakeholders on formalization in the artisanal and small-scale gold mining sector Latin America, Africa, and Asia*. (Alliance for Responsible Mining 2011) 21.

It is also essential to proactively address digital illiteracy to ensure the success of digitised interventions. Digital illiteracy impedes the effectiveness of digital solutions. For instance, Nigeria, Ghana, and Zambia all provide a platform for the application of ASM licensing through their cadastre systems. Notwithstanding, it has been observed that the uptake of licensing through the system is minimal compared to observed artisanal activity.[61] Avoidance could signify that the audience is not well versed or finds navigating applications and services through the digital system difficult.

5.2 The Importance of State-led Structural Support to the ASM Sector

It is paramount to have state-led support in facilitating access to critical resources, skills, and services. Necessary interventions include assisted access to geological and exploration services, access to concessions, provision of mine/site planning expertise, safety, health, and environmental guidance, promotion of mechanisation, access to cost-effective management and operational technology, access to capital, and affordable and appropriate mining equipment.

Governments must facilitate and lead the mobilisation and deployment of relevant assistance to the sector.[62] Standardisation of the sector and effective support is best channelled through government structures that have the capacity to facilitate broad-based development. Private-led efforts, while important, cannot achieve broad-based transformation. Thus the centrality of the states' role in building up the capacity and standing of the ASM sector is inimitable.

5.3 Processing and Milling Agents

It is also important to acknowledge that the ASM sector has evolved into a mature production value chain with a wide range of value chain participants.[63] Therefore, it is impossible to instantaneously decommission its various elements without structuring comprehensive plans on which features will be preserved, which will be eradicated, and which can be rehabilitated. Failing to acknowledge or include some of the features within the value chain would undermine the efforts of complete integration and formalisation. Such is the case with independent processing/milling centres that process ore for ASM operations. These critical players are very influential in the ASM sector. In most cases, many of these milling centres are unregistered and operate outside the law's mandate. It is important to formalise such activities to ensure that ASM commodities find their way into national coffers, not the black market.

61 Geological Survey of Denmark and Greenland. *Federal Republic of Nigeria: Nigeria Extractive Industries Transparency Initiative (NEITI)* (GEUS 2011) Appendix C page 62.
62 Hilson and others (n 8).
63 USAID, Global Communities, and Levin Sources. *Market Analysis: Artisanal and Small-Scale Gold from Eastern DRC* (USAID 2020).

Independent processing centres have a strong relationship with ASM miners. They play multiple roles, including being financiers to artisanal miners in some cases through production-based contracts.[64] For this reason, it is vital to integrate them into formalisation and integration plans. This can potentially improve remittance of production and become a vital cog in preventing mineral leakages and improving the accuracy and transparency of ASM production statistics.

5.4 Tightening Oversight of the ASM Market

The lack of a unified and comprehensively regulated market structure in the ASM sector has undermined the sector's performance. As a result, the mineral black market has increasingly played a significant role in dominating the ASM commodities market. The black market offers convenience and unrestricted activity, but it remains primarily predatory and has significantly undermined sustainable growth.[65]

ASM commodities are prone to price manipulation, price volatility, and coercion from buyers who take advantage of the lack of adequate checks and balances over the sector. For instance, at the onset of the COVID-19 pandemic, the spot prices of gold in the artisanal mining sector plunged. Despite the high price of gold on the global market in 2020, the Artisanal Gold Council recorded a fall in gold spot prices offered to artisanal miners in the first quarter of 2020. The prices fell by between 19% and 40% in some parts of Africa, Latin America, and Asia.[66] The volatility directly impacted miners' capacity to fund operational activities and their resultant disposable income.

On the other hand, the large-scale mining sector benefitted from the rising market prices of gold during the same period. The large-scale commodities market is highly regulated, thus ensuring that governments and producers receive fair market value for their commodities. Therefore, enhancing the role of government-led ASM buying channels is essential because:

- a Government structures can facilitate broader access to ASM product and consequently result in better aggregation;
- b Increased oversight would stabilise the ASM commodities market prices against collusion and ensure the sector receives fair value; and
- c Government can facilitate broader market options in domestic and export mineral markets, thus enabling the ASM product to become more competitive in the legitimate global commodities market.

64 Selina Zhuwarara. 'The importance of increased financial inclusion of the ASM sector.' *Mining Review Africa* (2021).

65 African Natural Resources Centre and African Development Bank. *Illicit Trade in Natural Resources in Africa — A Forthcoming Report from the African Natural Resources Centre.* (ADB & ANRC October 2016) 11

66 Artisanal Gold Council. www.artisanalgold.org/gold-prices-data. Accessed 11 February 2021.

5.5 The Cost of Regulation and the Cost of Compliance

The cost of regulation and compliance is also a critical consideration to regulators or governments when dealing with the ASM sector. Therefore, the overall cost of the chosen regulatory strategy for the ASM sector must be sustainable relative to the national budget and the overall perceived benefits of the regulation.[67] Similarly, the overall cost structure of compliance must be sustainable for ASM producers to maintain.

5.5.1 The Cost of Compliance

The continued state of stunted growth of the ASM sector can also be ascribed to the prohibitive cost of compliance, which not only has created a barrier to entry but is also a barrier to growth for minor players in the sector. Mere participation in the mining sector is based on the ability to fulfil multiple requirements for the acquisition and continuous maintenance of mining rights through various compliance obligations. Consequently, the tax structure is a central topic in the ASM sector. Tax costs include all mandatory levies arising from the mining right and from conducting business in the sector. These costs include the cost for acquisition of the right (licensing fees plus renewal fees), royalties and special taxes associated with the mineral, taxes on revenues and expenses of maintaining the corporate structure, levies for access to land rights, levies from multiple compliance regulators such as safety and health regulators, labour-related taxes, mineral marketing expenses, handling levies and export taxes, and all applicable fiscal rules which build into taxable revenues.[68] The cumulative impact of such a cost structure places significant financial pressure on ASM mining entities.[69]

In this light, it is essential to continuously examine the entire spectrum of state-mandated taxes and levies and significantly streamline the regulatory burden on ASM miners.[70] Furthermore, if digitisation is to transform regulation, data and internet access costs must be rationalised and pegged at affordable prices.[71] Cumbersome regulatory cost structures will only perpetuate non-compliance and the proliferation of illegal operations.[72]

67 OECD. *Regulatory Enforcement and Inspections: OECD Best Practice Principles for Regulatory Policy.* (OECD 2014).
68 Nigeria, Statutory Instrument 47 of 2011.
69 African Minerals Development Centre (n 24).
70 Minerals Commission Ghana (n 26).
71 OECD. 'Maintaining Competitive Conditions in the Era of Digitalisation.OECD report to G-20 Finance Ministers and Central Bank Governors, July 2018.' www.oecd.org/g20/Maintaining-competitive-conditions-in-era-of-digitalisation-OECD.pdf. Accessed 31 July 2022.
72 Barnes and Scornavacca (n 50).

5.5.2 The Cost of Regulation

The administrative and fiscal costs associated with regulation are usually understated in the discussion of the formalisation and integration of the ASM sector. However, by its very nature and character, the ASM sector requires increased enforcement and monitoring compared to large-scale mining entities. Consequently, optimising the cost of regulation is necessary. The challenge for governments remains how to develop and apply enforcement strategies that achieve the best possible outcomes by achieving the highest possible levels of compliance while keeping the costs and burden as low as possible.[73] Digitalisation can assist in reducing regulatory costs if governments and businesses can keep their marginal costs low.[74] However, high marginal costs erode the economic value of digitisation. Efficient digital systems collect, interpret, store, and deploy information efficiently across regulatory structures, commercial partners and ASM miners, thus enabling significant savings on information management.[75]

5 Conclusion

The ASM experience in Zambia, Ghana, and Nigeria shows that the formalisation and economic integration of the ASM sector need to be complemented by a robust and effective regulatory system that can enhance the implementation and impact of policy and legislation.

This chapter makes a case for the digitalisation of the ASM regulatory framework. The sector has long posed significant regulatory challenges for regulators across Africa. Digitisation possesses the potential to overcome the shortcomings of traditional regulatory mechanisms. Digitalisation can improve regulatory efficiency and enable information to become a key asset in enhancing the sector's economic performance.

Digitalisation can support the ASM sector by:

a enhancing data collection, storage and use,
b improving the line of sight for policy and legislative development through real-time information,
c enabling the centralisation and coordination of the administrative and regulatory processes of the sector, thus removing redundancy and reducing costs,
d improving regulatory responsiveness, and

73 OECD (n 67).
74 Yongmin Chen. 'Improving market performance in the digital economy.' China Economic Review 62 (2020) 101482. Page 3.
75 McKinsey Digital. 'Reducing data costs without jeopardizing growth.' 31 July 2020, www.mckinsey.com/business-functions/mckinsey-digital/our-insights/reducing-data-costs-without-jeopardizing-growth. Accessed 25 July 2022.

e creating a secure and reliable repository of vital socio-economic data that commercial partners can use to tailor-make appropriate products and services for the sector.

The rapid expansion of the global digital economy necessitates a strategic shift in regulatory capacity in Africa. African economies and regulatory structures must be readied to navigate the increasing encroachment of the digital economy. It is important to increase the adoption of digital competencies in regulatory functions and public enterprises. Digitisation is especially crucial to segments of the economy and society that are ordinarily marginalised, such as the ASM sector, which can benefit immensely from optimised information networks. The ASM sector has an immutable position in the transformation of African economies, and the continent must address the constitution, management, and empowerment of the ASM sector conclusively.

10 Chinese Investments in the African Mining Sector

An Analysis of the Dispute Settlement Mechanisms

Dr Abubakar I. Umar and Dr Muhammad Bello

1 Introduction

In the past couple of decades, African countries and China have strengthened their economic relations in, among others, the areas of mineral resources development and infrastructure. China has rapidly increased its investment foothold in Africa's mining and minerals sector. In the years between 2005 and 2011, Chinese investors poured in over $50 billion into this sector.[1] The estimate in 2019 alone was around $100 billion.[2] This is not surprising at all as it reflects the growing mutual interdependence between China and Africa.[3] African countries are desperate for foreign capital to accelerate their infrastructure finance and other areas of economic development. The source of such capital, Western or Chinese, is immaterial. With widespread poverty, malnutrition, and unemployment, foreign investment into this sector has become a clear development imperative for these countries. Fortunately, they are blessed with untapped mineral resources: Africa holds 50% of the world's gold reserve, 90% of the world's platinum and cobalt, 75% of the world's coltan reserve, and 35% of the world's uranium.[4] On the other hand, China wants these and other resources in order to sustain its remarkable economic transformation. China's booming industrial sector relies heavily on African countries for its minerals and other raw materials. For instance, China's electronic and mobile phone sector relies heavily on coltan. China's major trading partners in Africa are mineral resource-rich countries such as the Democratic Republic of Congo (DRC), the Republic of South Africa (SA), Tanzania, and Nigeria.

1 'Chinese Investment in African Mining: What you need to know,' Mining Indaba (2019) www.miningindaba.com/Articles/chinese-investment-in-african-mining-what-you. Accessed 18 November 2020.
2 (n.1)
3 FOCAC, 'Beijing Declaration Toward an even Stronger China-Africa Community with a shared Future.' www.focac.org/eng/zywx_1/zywj/t1594324.htm. Accessed 24 December 2020.
4 'Chinese investment in African Mining: What you need to know,' Mining Indaba (2019) www.miningindaba.com/Articles/chinese-investment-in-african-mining-what-you. Accessed 28 February 2021.

DOI: 10.4324/9781003284437-12

In the face of this growing economic interdependence, we have seen a dramatic rise in the number of Sino-Africa bilateral investment treaties (BITs). These include the China–Nigeria Agreement for the Reciprocal Promotion and Protection of Investments 2001, China–South Africa BIT 1997, China–Zimbabwe BIT, China–Ethiopia BIT, China–Tanzania BIT, etc. In mid-2018, the United Nations Conference on Trade and Development (UNCTAD) found that 35 BITs were signed between African countries and China.[5] Many of the BITs are already in force in the territories of signatory countries.[6] These and other multilateral treaties constitute the primary governance framework for China–Africa relations.

However, as Chinese investment in the African mining sector increases, the potential for disputes is real, and therefore the dispute settlement provisions of Sino-Africa BITs require some examination. The form, substance, and structure of Sino-Africa BITs may vary from one African country to another. The standards of protection provided in those BITs may also vary. Indeed, some have questioned the adequacy of these standards.[7] As stakeholders continue to debate the reform of the investor–state dispute settlement (ISDS) regime, disputes between Chinese investors and African states are becoming increasingly likely. The Forum for China–Africa Cooperation (FOCAC) recognized this reality, which in turn led to the establishment of the China–Africa Joint Arbitration Centre (CAJAC) in 2016. In recent years, China's mining operations in Africa have raised a number of questions. In Nigeria, for instance, Chinese investors were reportedly involved in illegal mining, bribery, and other unwholesome practices.[8] Thus, governmental/regulatory intervention in Chinese investment activities may become inevitable. In addition, there has been increasing suspicion towards investment treaty arbitration hitherto recognized by many of the Sino-Africa BITs. South Africa's Protection of Investment Act 2015 (PIA) is a good example.

Therefore, the primary objective of this chapter is to examine the dispute settlement mechanisms (DSMs) under the framework for Sino-Africa mineral resource development. The existing literature tends to focus almost exclusively on the features of selected Sino-Africa BITs without specific considerations given to sectoral peculiarities.[9] Chinese activities in resource-rich African countries raise critical concerns on the efficacy and functionality of the

5 Audle Sheppard and Luis-Alexis Bret, 'Mining Arbitration in Africa' (2020) Lexology. www.lexology.com/library/detail.aspx?g=593a20a7-5f7e-49c1-9316-a7823f719025

6 Investment Policy Hub, 'China BITs,' https://investmentpolicy.unctad.org/international-investment-agreements/countries/42/china. Accessed 20 January 2021.

7 Sheppard and Bret (n 5).

8 Matthew T. Page, 'The Intersection of China's Commercial Interests and Nigeria's Conflict Landscape.' USIP Special Report, 2018. www.usip.org/sites/default/files/2018-09/sr428-the-intersection-of-chinas-commercial-interests-and-nigerias-conflict-landscape.pdf. Accessed 23 January 2023.

9 Won Kidane, 'China's Bilateral Investment Treaties with African States in Comparative Context' (2016) 49 *Cornell International Law Journal* 141, 146 and 151.

DSMs. They raise key questions worthy of consideration here. Is the framework for Sino-Africa mineral resource development adequate to address the peculiarities of Chinese transactions with African countries, especially the resource-backed loans and private investments in the mining sector? Is the CAJAC the answers to the shortcomings of investment arbitration in potential Sino-Africa mining disputes? It is important to reflect on these questions because the remarkable rise of China as a global economic power presents multiple opportunities and fundamental challenges for African countries.

Building on the above introduction, Section 2 examines the extent of Chinese investments in African countries' mineral resources development through the lens of the China Policy on Africa (CPA). Section 3 examines the key features of the Sino-Africa BITs, especially the dispute settlement provisions. In Section 4, attention will be focused on China's attitude towards mining arbitration in the context of ongoing reform of the ISDS regime and the newly established CAJAC, while Section 5 concludes the discussion with some findings and recommendations. The chapter uses the 'Angola model' to illustrate the potential for disputes in Sino-Africa mineral resource development.

2 China and the African Mining Sector

China's increasing interests in Africa's mineral resources reflects increasing Chinese domestic demands for these resources as a result of the economic expansion experienced over the past four decades. Many African countries have become major exporters of commodities to the booming Chinese markets. Some context on China's importance to Africa and the structure and health of its economy shall be our starting point.

2.1 China's Economy and the Demand for Mineral Resources

China has been one of the world's economic powers for many centuries.[10] Its history is full of economic ups and downs. In modern times, the emergence of the Communists in the first half of the twentieth century was a major turning point. The communist system proved fatal. Thus, China's relationship with Africa during this period was not impactful.[11]

10 Fu Zhufu 'The Economic History of China' (1981) 7 (1) *Modern China* 3–30.

11 The famous Tanzania–Zambia Railway built between 1970 and 1975 through a zero-interest loan of RMB 980 million is often cited as an example of China's presence in Africa. However, China was struggling with the Marxist-based central planning economic system during this period. The system was characterized by three key elements: a macro-policy environment which prioritized heavy industries thereby misaligning with the country's comparative advantage in labour resulting, partly, in a disproportionate allocation of output to investment rather than to consumption; micro-management of economic institutions through nationalization of industries and collectivization of agriculture; and a centrally planned resource allocation system that abhorred the market and its natural forces of demand and supply of products and services. See David Mason, 'China's

Following Den Xiaoping's legal and institutional reforms of the late 1970s and 1980s, China's economy began to unleash its potential. The reforms, called the 'four modernizations', focused on modernizing agriculture, industry, defence, science, and technology. With these reforms, the country transitioned from a centrally planned economic system to a more market-oriented economy. This is a hybrid economic model, which allows the state to retain substantial influence on the direction of the economy. It is a state-driven capitalism which 'negates both economic planning and laissez-faire capitalism and, instead, advocates a kind of socialism marked by strong government intervention'.[12] It involves screening of the neoclassical economic prescriptions and, where appropriate, applying them to suit Chinese interests.

Today, thanks to this system, China ranks as the second-largest economy in the world by nominal GDP and the largest by purchasing power parity (PPP).[13] China has the world's fastest-growing economy, growing at an average rate of 6% over the past 30 years. It is the most populous country in the world. In 2018, China's private sector accounted for 60% of the GDP and it was responsible for 70% of innovation, 80% of urban employment, and 90% of job creations.[14] On a per capita income basis, China ranked 73rd by GDP (PPP) per capita in 2019. Its GDP was $14,342.90 billion in 2019, representing about 12% of global GDP.[15]

China also has the fourth-largest inward foreign direct investment (FDI), and the 11th-largest outward foreign direct investment (OFDI).[16] It has the world's largest foreign exchange reserves of around $3.1 trillion.[17] In 2018, its total imports and exports were $4,623 billion, representing a 12% increase.[18]

Four Modernizations: Blueprint for Development or Prelude to Turmoil?' (1984) 11 *Asian Affairs* 47–70, 47.

12 Yang Yao, 'The political economy causes of China's successes' in Garnaut Ross, Song Ligang, and Fang Cai (eds) *China's 40 Years of Reforms and Development (1978–2018)* (Australian National University Press 2018) 75–76; Li Yang 'China's Growth Miracle: Past, Present, and Future.' https://cdn.unrisd.org/assets/events/event-aux-pages/pdf-files/chinas-growth-miracle-0808.pdf. Accessed 23 January 2023.

13 World Bank, World Development Indicators Database, https://databank.worldbank.org/reports.aspx?source=2&country=CHN#. Accessed 16 December 2020.

14 World Bank Group, *China: Systematic Diagnosis* (2018) (hereafter 'WB China Systematic Diagnosis') http://documents.worldbank.org/curated/en/147231519162198351/pdf/China-SCD-publishing-version-final-for-submission-02142018.pdf. Accessed 16 December 2020.

15 Mason (n 11).

16 World Bank Group, *China and WB 1980–2018*. http://pubdocs.worldbank.org/en/657111542261998957/F-F-en-2018.pdf. Accessed 20 December 2020.

17 IMF, 'Currency Composition of Official Foreign Exchange Reserve.' https://data.imf.org/?sk=E6A5F467-C14B-4AA8-9F6D-5A09EC4E62A4. Accessed 30 December 2020.

18 Ministry of Commerce, China (MOFCOM), 'Brief Statistics on China's Import and Export 2018.' http://english.mofcom.gov.cn/article/statistic/BriefStatistics/201903/20190302842809.shtml. Accessed 30 December 2020.

The mining sector has contributed to China's remarkably consistent economic growth. It has extensive reserves of mineral resources.[19] Further discoveries have been made in recent times.[20] In 2017, for instance, more than 170 different mineral deposits were discovered including, among others, 877 million tons of oil reserves, 555 billion m^3 of natural gas, 81 billion tons of coal, and 1,104 tons of gold.[21] According to the government, China has over 30 billion recoverable oil resources, 90 trillion m^3 geological resources, and 50 trillion m^3 of recoverable gas resources.[22] The reserves have been increasing over the years. In 2017, coal reserves increased by 4.3%, manganese by 19.1%, copper by 4.9%, bauxite by 4.9%, gold by 8.5%, fluorite by 8.9%, and crystalline graphite by 22.6%.[23]

Despite its massive resources and reserves, however, China's Ministry of Natural Resources has admitted that 'the demand has kept growing'.[24] This is not surprising at all. China is the world's largest energy consumer.[25] It is the largest consumer of primary commodities.[26] Indeed, energy consumption has been increasing over the years.[27] In 2017, for instance, coal consumption increased by 3.2%, oil consumption increased by 5.2%, and natural gas consumption increased by 14.8%.[28] A similar trend was observed for metals. For instance, in the same period, consumption of gold increased by 9.4%.[29]

China is the world's largest manufacturing economy and exporter of goods.[30] It is also the world's fastest-growing consumer market and second-largest importer of goods.[31] It is the largest trading nation in the world. It plays a dominant role in international trade and has increasingly engaged in trade and investment relations primarily through bilateral and multilateral cooperation. China became a member of the World Trade Organization in 2001. It has entered into investment and free trade agreements with several nations, including Australia, New Zealand, Pakistan, and Switzerland. It has negotiated an agreement in principle with the European Union (EU).[32]

19 Ministry of Natural Resources, Peoples Republic of China, *China Mineral Resources* (Geological Publishing House 2018) www.gov.cn/xinwen/2018-10/22/5333589/files/01d0517b9d6c430bbb927ea5e48641b4.pdf. Accessed 30 December 2020.
20 'Chinese investment in African Mining' (n 4).
21 Ibid.
22 Sheppard and Bret (n 5).
23 (n.1). See also Lorenzo Cotula et al., *China-Africa Investment Treaties: Do they work?* (International Institute for Environment and Development 2016).
24 Sheppard and Bret (n 5)
25 Mason (n 11).
26 WB China Systematic Diagnosis (n 14) above.
27 Mason (n 11).
28 Ibid.
29 Ibid.
30 Mason (n 11).
31 Ibid.
32 See EU–China Comprehensive Agreement in principle on Investment 22 January 2021, https://trade.ec.europa.eu/doclib/docs/2021/january/tradoc_159346.pdf. Accessed 20 February 2021.

For Africa, the Forum on China–Africa Cooperation (FOCAC) has become the central platform for shaping China–Africa economic relations. Although questions have been raised regarding the connection between economic growth in Africa and Chinese demands for commodities of African countries, it seems clear that African mineral resources form part of broader China–Africa economic relations.[33] Some have argued that the insatiable demand for resources is part of the Chinese development story.[34]

2.2 China Investment Policy on Africa

Modern China–Africa economic relations started in the 1950s following the Bandung Asia–Africa Conference.[35] It was essentially a south–south relation. Both state-owned enterprises and private investors participated in the outward foreign investment in Africa. In its first CPA in 2006, China's economic interests in Africa were articulated.[36] The second CPA Paper 2015 further clarifies these interests and seeks to strengthen the mutual interdependence between China and African countries.[37] Under it, China commits to

> vigorously support the efforts of African countries to improve their 'soft' and 'hard' environment for investment and development, optimizing laws and regulations for attracting and protecting foreign investment, and removing the two major bottlenecks impeding development, namely, backward infrastructure and inadequate infrastructure.[38]

Under this policy, China will 'guide, encourage and support' Chinese enterprises to jointly build economic and trade cooperation zones to attract more Chinese investments into Africa. While emphasizing 'market-based operation', the CPA seeks more trade and investment partnerships with African countries in a manner that allows for more African commodities to enter the Chinese market with a zero-tariff treatment to 97% taxable goods.[39] The government commits to supporting Chinese enterprises to expand and optimize their investment in, among others, energy, infrastructure, and industry. It

33 Deborah Brautigam et al., 'Chinese Investment in Africa: How much do we know?' PDEL Synthesis Paper No. 2 (2017). https://pedl.cepr.org/sites/default/files/PEDL_Synthesis_Papers_Piece_No._2.pdf. Accessed 20 December 2020.
34 Raphael Kaplinsky and Mike Morris, 'Chinese FDI in Sub-Saharan Africa: Engaging with Large Dragons' (2009) 21 *European Journal of Development Research* 551–569.
35 Lawal Mohammed Marafa 'Africa's Business and Development Relationship with China Seeking Moral and Capital Values of the Last Economic Frontier,' Nordiska Afrikainstitutet Discussion Paper 44 (Uppsala 2009) 6–20, 6, www.diva-portal.org/smash/get/diva2:272966/FULLTEXT02.pdf
36 China Policy on Africa 2006.
37 China Policy on Africa 2015.
38 Ibid para 3(1).
39 Ibid para 3(5).

seeks, in the spirit of mutual cooperation and sustainable development, to deepen partnerships with African countries in the areas of exploration, development and processing of mineral resources and energy.[40] Specifically, China will 'expand [a] whole-industry-chain cooperation in energy and mining sectors'.[41]

The CPA is significant for China's continuing engagements in Africa. Coordinating Chinese investments and supporting infrastructure development, which is acutely lacking in the region, define these engagements. It shows the 'directness' of China–Africa relations which prioritize economic over political matters. Political neutrality and non-intervention have been the hallmarks of this relationship. Many African leaders prefer this 'equal' treatment. In addition, the Belt and Road Initiative (BRI) is likely to further shape this relationship. It is equally likely to expose Chinese investments to political, economic, and other risks in the region.

2.3 Chinese Investments in the African Mining Sector

Data on the extent of actual Chinese investment in Africa is unclear.[42] Information from China's Ministry of Commerce and China Bureau of Statistics shows that Chinese OFDI is massive.[43] It has been argued that 'Chinese investment in Africa is not large enough to be either a calamity or a panacea'.[44] Regardless of the volume of investment, however, potential claims by and against Chinese investors cannot be discounted. This is more so because, reading the CPA, it seems that Chinese investment in Africa is largely influenced and supported by the Chinese government. For instance, in 1995, the Chinese government introduced subsidized concessional loans for Chinese investors entering Africa.[45]

In pursuance of the 2006 and 2015 policy objectives, Chinese investment in Africa has increased tremendously. In 2015, about 5% of African countries' FDI was from China.[46] Out of this, mining and construction accounted for 54%.[47] Not surprisingly, resource-rich African countries – Nigeria, DRC, South Africa, Zambia, and Sudan – were the largest recipients of Chinese FDI. Over a third of China's oil imports come from Africa.[48]

40 Ibid.
41 Ibid.
42 Brautigam et al. (n 33) 3.
43 See MOFCOM, Regular News Conference, 21 January, 2021, http://english.mofcom.gov.cn; National Bureau of Statistics of China (NBSC) www.stats.gov.cn
44 Brautigam et al. (n 33) ii.
45 Zeng Aiping and Shu Zhan 'Origins, Achievements and Prospects of the Forum on China-Africa Cooperation' (2018) 89. www.focac.org/eng/lhyj_1/yjcg/201810/P020210830627854979477.pdf
46 Brautigam et al. (n 33) 3.
47 Ibid i.
48 Wade Shepard 'What's China up to in Africa?' Forbes, 3 October 2019.

China is Tanzania's main investor. In 2015, Chinese FDI in Tanzania was $226 million.[49] It has declined to $115 million in 2019.[50] Nineteen different infrastructural projects were financed by China, the most prominent being the Bagamoyo Port.[51] The latter project has, however, been halted owing to a dispute between Tanzania and China Merchant Bank on the terms of the project, including but not limited to the following: lease period (China Merchant wants a 99-year lease while Tanzania wants a 33-year lease), tax-free status for China Merchant, special rate for water and electricity, and freedom to operate businesses without government approval.[52]

2.3.1 Nature of Chinese Investments in the African Mining Sector

From the CAP, both public and private Chinese investors participate in the African mining sector. Broadly, there are three forms of Chinese investments in Africa: infrastructure loans for resource investment, private including individual, and Chinese investment and aid. The latter is outside the scope of this chapter.

2.3.1.1 INFRASTRUCTURE FOR RESOURCE INVESTMENT

This is a peculiar Chinese investment strategy that has generated a lot of criticism.[53] It is a resource-backed loan usually given to an African government or its state-owned enterprises by the Chinese government usually through its policy banks (China Export and Import Bank, China Development Bank, and People's Bank of China) to be repaid either in kind (i.e. through mineral resources) or from future incomes to be derived from such resources.[54] In some cases, mineral resources are used as collateral for the loans. Thus, mineral resources play three important roles in this transaction: they serve as direct means of loan repayment or as a source of revenue for repayment or as collateral for the purpose of guaranteeing

49 Tanzania Invest, China-Tanzania Relations, 2021 www.tanzaniainvest.com/china#:~:text=China%20is%20Tanzania's%20N1%20investor,USD%20226%20million%20in%202015
50 Ibid.
51 Ibid.
52 John Hursh, 'A Bump in the Belt and Road: Tanzania pushes back against Chinese port project,' Center for International Maritime Security, 2 December 2019. https://cimsec.org/a-bump-in-the-belt-and-road-tanzania-pushes-back-against-chinese-port-project
53 Ana Cristina Alves, 'Chinese Economic Statecraft: A Comparative Study of China's Oil-backed Loans in Angola and Brazil' (2013) 1 *Journal of Current Chinese Affairs* 99–130.
54 David Mihalyi et al., *Resource-backed Loans: Pitfalls and Potential* (Natural Resource Governance Institute 2020) https://resourcegovernance.org/sites/default/files/documents/resource-backed-loans-pitfalls-and-potential.pdf

loan repayment.[55] Thus, this sort of investment is complex as mineral resources development is only part of a broader development finance transaction with the host government.

The case of Angola (or the 'Angola model') illustrates the intricacies of this form of transaction.[56] Under this model, the acquisition of exploitation rights usually follows China Exim Bank's loan to the Angolan government. It followed financing agreements between Angola's Ministry of Finance and China's Ministry of Trade under which oil-backed concessional loans were extended to Angola through the China Exim Bank.[57] In 2006, for instance, Angola obtained $4 billion from China through the so-called 'oil-backed loan' as China's state-owned oil firm, Sinopec, was granted exploitation rights over certain oil blocks in Angola.[58] By the terms of the loans, Chinese firms carry out infrastructural projects payable by China Exim Bank which writes down the amount against oil sales to the Chinese government.[59] Indeed, 70% of the procurement of civil engineering works for Angola's reconstruction was awarded to Chinese firms. Between 2004 and 2010, over $10 billion oil-backed credits were extended to Angola.[60] It was a convenient approach for mineral resource development for both China and Angola.[61]

In the same vein, the China Railway Group used the same model to secure mining rights over DRC's copper and cobalt.[62] Thus, through resource-backed loans, China steps in to help address African countries' infrastructure deficits. It is the largest financier of African infrastructure, financing one in five projects across the continent.[63] Nevertheless, there are instances where disputes have arisen between African governments and their Chinese counterpart or between African governments and Chinese investors. The Tanzanian government's termination of the contract for the construction of the Bagamoyo Port over alleged unfair terms is an example of Sino-Africa dispute. There were also 'disputes' over Kenya's Port of Mombasa and Djibouti's Port.

55 Ibid 3.
56 Liviu Stelian Begu et al., 'China-Angola Investment Model' (2018) 10 *Sustainability* 29–36.
57 Lucy Corkin, 'China and Angola – Strategic Partnership or Marriage of Convenience? (2011) 1 (1) *Angola Brief* 4.
58 Brookings Institute, China's Aid to Africa: Monster or Messiah (Brookings Institute 2018) www.brookings.edu/opinions/ chinas-aid-to-africa-monster-or-messiah.
59 Corkin (n 57) 2.
60 Ibid.
61 Ibid.
62 Ibid.
63 Hanna Marais and Jean-Pierre Labuschagne, 'If you want to prosper consider building roads: China's role in African infrastructure and capital projects,' *Deloitte Insight*, 22 March 2021, www2.deloitte.com/us/en/insights/industry/public-sector/china-investment-africa-infrastructure-development.html

2.3.1.2 PRIVATE CHINESE INVESTORS IN THE MINING SECTOR

Chinese private sector investment in the African mining sector has also been growing.[64] It constituted 45% of China's OFDI in 2011.[65] In 2018, the figure was $5.4 billion, an increase from $4.1 billion in 2017.[66] Studies found that in 2011 there were more than 1,500 active Chinese investment projects in Sub-Saharan Africa.[67] From two in 2002, private investments rose to 923 in 2011, representing over 55% of Chinese investment in Africa.[68] Although public/government firms engage in the mining sector more than private firms, the trend has been shifting over the years. For instance, in 2011, government firms had 25% investment in mining while private investors had 16% investment in mining, ranking third after manufacturing and trade.[69]

Similarly, government approval is required for outbound investment depending on the volume of such investment.[70] Investment above $10 million requires approvals.[71] Although incentives such as subsidies and tax incentives accompany approvals, most private investors, including individuals, tend to avoid the approval process, making the number of investors more than double the official figure.[72]

Some private Chinese mining companies engage in a joint venture with their local African counterparts.[73] Importantly, individual Chinese nationals often engage in artisanal mining, particularly in Nigeria.[74] In 2020, some Chinese nationals were arrested for allegedly engaging in illegal mining, exploiting the poor regulatory system in the country while exposing themselves to security threats.[75]

3 Dispute Settlement and Sino-Africa BITs

Features of the Sino-Africa BITs and the dispute settlement mechanisms will be examined here. However, some context on international dispute settlements system is important here.

64 Xiaofang Shen 'Private Chinese Investment in Africa: Myths and Realities.' https://openknowledge.worldbank.org/bitstream/handle/10986/12174/wps6311.pdf?sequence=1&isAllowed=y

65 Ibid 4.

66 C Tetor, 'Cash flow of Chinese direct investments to Africa 2008–2018,' Statista, 1 March 2012.

67 Ibid 5.

68 Ibid 5.

69 Ibid 6.

70 Ibid 5.

71 Ibid.

72 Ibid 10.

73 Dominic Ellis, 'Nigeria and China firms sign gold agreement,' *Mining Global*, 9 November 2020.

74 Adejumo Kabir, 'Nigeria moves against illegal Chinese miners, others,' *Premium Times*, 12 January 2020.

75 Ihouma Chiedozie, 'How Chinese prospectors exploit local miners to cash in on Nigeria's gold,' *Punch*, 17 May 2020; Olarewaju Kola, '17 Chinese arrested for illegal gold mining,' *Anadolu Agency*, 5 May 2020.

3.1 International Dispute Settlement

According to the Permanent Court of International Justice, a 'dispute' arises where there is a 'disagreement on a point of law or fact, a conflict of legal views or interests between two persons'.[76] Thus, 'international dispute' means, broadly, any dispute involving a sovereign state.[77] Keohane et al. describe it as a dispute

> between two or more parties, at least one of which, the defendant, will be a state or an individual acting on behalf of a state … [with a dispute resolution mechanism] … specifically designed to monitor and enforce the obligatory rules of the regime.[78]

A dispute resolution mechanism must be capable of engaging the jurisdiction of an adjudicative body established 'by inter-governmental agreement … or by agreement between a national government and a foreign private entity, where the court is legally situated either fully or partly outside the national juridical and governmental system of any state'.[79]

Helfer and Slaughter distinguish between 'international adjudication' and 'supranational adjudication'.[80] Keohane et al. distinguish between 'interstate' and 'transnational adjudication'.[81] In essence, an international mining dispute can be either a dispute between states (state–state/inter-state) or a dispute between a state and a national of another state. For the latter, arbitration has become the principal means of dispute resolution. Arbitration can be either institutional or ad hoc. Investment arbitration tribunals (usually under the auspices of the International Centre for the Settlement of Investment Disputes (ICSID) and the United Nations Commission for International Trade Law (UNCITRAL)) and regional economic courts and tribunals have become the driving force in international mining disputes.

76 *Mavromattis Palastine Concession (Greece v UK)* (1924) PCIJ Series A 11; *Case Concerning East Timor* (1995) *ICJ Reports* 89, 99.

77 CPR Romano, 'The Shift from the Consensual to the Compulsory Paradigm in International Adjudication: Elements for a Theory of Consent' (2007) 39 *NYU Journal of International Law & Politics* 791, 792–795.

78 RO Keohane et al., 'Legalized Dispute Resolution: Interstate and Transnational' (2000) 54 (3) *International Organisation* 457–488, 459.

79 *South China Sea Arbitration (Philippines v China)* (2016) PCA Case No 2013-19. See also N Grossman, 'The Normative Legitimacy of International Courts' (2013) 86 *Temple Law Review* 61–105.

80 LR Helfer and AM Slaughter, 'Towards a Theory of Effective Supranational Adjudication' (1997) 107 *Yale Law Journal* 273–391, 289–290. See also EA Posner and JC Yoo, 'Judicial Independence in International Tribunals' (2005) 93 (1) *California Law Review* 1–74; LR Helfer and AM Slaughter, 'Why States Create International Tribunals: A Response to Professors Posner and Yoo' (2005) 93 (3) *California Law Review* 899–956.

81 Keohane et al. (n 78) 457–458.

3.2 Features of Sino-Africa BITs

Chinese BITs are distinguished largely based on the content of their respective ISDS provisions and the extent of investor protection they provide.[82] Accordingly, there are four generations of Chinese BITs. The first-generation BITs (1982–1989), concluded with capital-exporting countries, are characterized by either the absence of ISDS provisions or provisions limiting recourse to ISDS only for the purpose of determining 'the amount of compensation for expropriation'.[83] The second-generation BITs (1989–1999) retained similar features but were concluded with capital-importing countries. The third-generation BITs (1999–2008) were concluded with both capital-importing and capital-exporting countries, but they have broader ISDS provisions and guarantees for national treatment. The fourth-generation BITs (2008 to date) symbolize China's preference for multilateralism in trade and investment matters with broader ISDS provisions for investment protection including recourse to ISDS for breach of specific treaty obligations. Thus, from its initial scepticism, China has become actively engaged in the move towards reforming the ISDS regime including the possible establishment of a multilateral investment court and an appellate body under the auspices of UNCITRAL Working Group III.[84]

Therefore, most Sino-Africa BITs contain features of the second- and third-generation BITs. It is interesting that while China started as a sceptic and turned into a proponent, some African countries started as proponents and turned into sceptics. Unfortunately, Sino-China BITs do not contain features of modern BITs (e.g. Nigeria–Morocco BIT and CETA), regarding issues of human rights, environmental and social governance, and sustainable development. The most progressive Sino-China BIT is perhaps the China–Tanzania BIT which will be examined below.

3.2.1 Substantive Provisions

For convenience, we shall examine a few Sino-Africa BITs especially those concluded with the two largest economies and resource-rich African countries: the Nigeria–China BIT and the China–SA BIT.[85]

First, the definitions of 'investment' in the China–Tanzania, China–Ethiopia, China–Nigeria, and China–SA BITs are similar.[86] The China–Nigeria

82 Kidane clams that China has 'no coherent strategy' in its BITs and that 'China may not have pursued a systematic BIT policy under preconceived BIT models in the way the United States has'. See Kidane (n 9) 146 and 151.

83 Ibid 146–147.

84 UNCITRAL WG III, 'Possible reform of investor-State dispute settlement (ISDS) Submission from the Government of China' (Thirty-eighth session Vienna, 14–18 October 2019). http://undocs.org/en/A/CN.9/WG.III/WP.177

85 China–Zimbabwe BIT. https://investmentpolicy.unctad.org/international-investment-agreements/treaty-files/3356/download

86 Art 1(1) China–Ethiopia BIT.

BIT and China–SA BIT, for instance, differ only in wording. While the former includes 'business concessions conferred by law or under contract permitted by law, including concessions to search for, cultivate, extract or exploit natural resources',[87] the latter simply mentions 'concessions conferred by law, including concessions to search for or exploit resources'.[88] Both, however, cover property rights, shares, debentures, claims to money, and intellectual property rights.

Notably absent in both are debt securities and bonds. However, the China–Tanzania BIT extends the meaning of 'investment' to 'bonds, including government issued bonds, debentures, loans and other forms of debt, and rights derived therefrom'.[89] But it includes a proviso that 'bonds, debentures and loans with an original maturity of less than 3 years shall not be deemed as investments under this Agreement'.[90] It thus contemplates long-term development projects.

In a few Chinese-initiated investment arbitrations, one of the main issues in contention was who is a 'Chinese investor' for the purpose of investment protection under a BIT.[91] It is striking that the China–Tanzania BIT's definition of an 'enterprise' includes companies and entities 'irrespective of whether it is owned or controlled by a private person or the government'.[92]

In terms of standards of treatment, there are slight variations.[93] Of course, the national treatment, most-favoured nation treatment, and fair and equitable treatment standards have been incorporated into all the BITs. However, the China–SA BIT makes an exception to these treatment standards in respect of

> domestic legislation relating wholly or mainly to taxation or programmes and economic activities specifically aimed to promote, protect and advance persons and groups of persons that have been disadvantaged as a result of past discriminatory practices in the Republic of South Africa.[94]

This seems to contemplate local legislative measures designed to address the discriminatory legacies of apartheid. Other exceptions to the treatment standards under the China–SA BIT are: 'any special arrangement to facilitate frontier trade' and the exception regarding 'special advantages to development finance institutions with foreign participation and established for the exclusive purpose of development assistance through mainly nonprofit activities'.[95] 'Frontier trade' has not been defined but, presumably, covers trading

87 Art 1(e) China–Nigeria BIT.
88 Art 1(e) China–SA BIT.
89 Art 1(1)(f) China–Tanzania BIT.
90 Ibid Art 1(1).
91 *Tza Yap Shum v Peru* ICSID Case No. ARB/07/6.
92 Art (2)(b) China–Tanzania BIT.
93 Kidane (n 9) 165–166.
94 Art 3(3) China–SA BIT.
95 Ibid art 3(4).

with member countries of the Southern African Development Community (SADC) and the Common Market for Eastern and Southern Africa (CONMESA).

All the BITs also make similar provision regarding full protection and security[96] as well as provisions prohibiting expiation except where the following conditions are met: the expropriation was for public interests, under domestic legal procedure, without discrimination, and against fair compensation.[97] However, the China–SA BIT adds that an

> investor affected by the expropriation shall have a right, under the laws and regulations of the Contracting Party making the expropriation, to promote review, by a court of law or other independent and impartial forum of that Contracting Party, of his or its case and of the valuation of his or its investment.[98]

The China–Tanzania BIT also recognizes that it is not indirect expropriation where the host state adopts legitimate regulatory measures meant to preserve or protect public health or non-discriminatory public welfare programmes.[99]

3.2.2 Dispute Settlement Mechanisms

Dispute settlement in the mining sector involves a mixture of domestic and international mechanisms depending upon the nature of the dispute. In general, arbitration is the preferred mechanism. A relatively efficient system of enforcement, arbitrators' expertise and impartiality, parties' consent, and, in the case of investment arbitration, a BIT between home and host state make arbitration a more popular dispute resolution mechanism than, say, local tribunals. For mining dispute resolution, the most popular types of investment arbitration are: arbitration under the auspices of the ICSID or ICSID Additional Facility Rules, arbitration under the UNCITRAL Arbitration Rules, and arbitration under other international arbitral institutions such as the Stockholm Chamber of Commerce.

China's attitude towards arbitration in its BITs has evolved from that of excluding investment arbitration (i.e. preferring negotiation and adjudication by local courts) to including investment arbitration.[100] As noted above, the dispute settlement provisions in Chinese BITs evolved over time.[101] The Sino-Africa BITs adopt a similar pattern. First, disputes between the states

96 Art 7 China–Tanzania BIT; Art 5 China–SA BIT; Art 5 China–Nigeria BIT; Art 5 China–Zimbabwe BIT.
97 Art 4 Nigeria–China BIT; Art 4 China–SA BIT.
98 Art 4(2) China–SA BIT.
99 Art 6(3) China–Tanzania BIT.
100 Meng Chen, 'Embracing Non-ICSID Investment Arbitration? The Chinese Perspective' (2019) 39 *Northwestern Journal of International Law & Business* 249.
101 Kidane (n 9) 152–153.

regarding the interpretation and application of the BIT shall be resolved through diplomatic channels.[102] If it cannot be resolved through this channel within six months (in the case of China–SA BIT) or 12 months (China–Nigeria BIT), then such dispute shall be submitted to an ad hoc arbitration comprising of three arbitrators, each party appointing one and the arbitrators appointing a third arbitrator as Chairman.[103] It is unclear what factors informed the variation in the months leading to the constitution of the ad hoc arbitration. There is also variation of periods of months arising from the failure to appoint the Chairman: the China–Nigeria BIT requires two months while the China–SA requires four months.[104] Thereafter, the International Court of Justice (ICJ) President may appoint the Chairman.

Importantly, a dispute between an investor and the host state shall be resolved, first, through 'negotiation'.[105] If, within six months of initiating negotiation, the dispute cannot be resolved through that method, then, in the case of China–Nigeria BIT, the dispute may be submitted to the local court of the host state or an ad hoc arbitral tribunal; in the case of China–SA BIT, it may be submitted to 'an international arbitral tribunal' provided the host state 'may require the investor to initiate administrative review procedures in accordance with its laws and regulations, and provided that the investor has not submitted the dispute to a [host state] domestic court'.[106] Reference to 'administrative review' seems to refer to the Chinese system of administrative reconsideration.[107] Thus, in both BITs preference is given to the domestic dispute resolution systems of the state parties. Where recourse to arbitration has been made, each party to the dispute would appoint an arbitrator and the two would appoint a third as the Chairman within four months. Non-constitution of the tribunal in this manner would entitle either party to invite the Secretary General of the ICSID to make the necessary appointments.[108] The tribunal would determine its procedures, but it may also be guided by the ICSID Rules.[109]

The China–Tanzania BIT, however, improves on the China–SA BIT and China–Nigeria BIT. First, it requires amicable resolution through 'negotiation' and conciliation procedures, failing which an investor may submit the dispute to a host state's local court.[110] Where the host state's law requires exhaustion of local administrative review before recourse to international arbitration, then, once that is done, the investor may submit the dispute to

102 Art 8 China–-Nigeria BIT.
103 Art 8(3) China–Nigeria BIT; Art 8(3) China–SA BIT.
104 Art 8(3)(4) Nigeria–China BIT; Art 8(3)(4) China–SA BIT.
105 Art 9(1) China–SA BIT; Art 9(1) China–Nigeria BIT.
106 Art 9(2) China–SA BIT.
107 Wang Guiguo 'Chinese Mechanisms for Resolving Investor-State Disputes' (2011) *Jindal Journal of International Affairs* 204–233.
108 Art 9(3) of China–SA BIT.
109 Art 9(4) of China–SA BIT.
110 Art 13(1) China–Tanzania BIT.

ICSID or to an ad hoc arbitral tribunal under the UNCITRAL Arbitration Rules or to any arbitration institution or ad hoc arbitral tribunal.[111] There is a three-year limitation period within which to file the claim to international arbitration.[112]

An innovative provision in the China–Tanzania BIT is the recognition that 'it is inappropriate to encourage investment by relaxing domestic health, safety or environmental measures' and that nothing in the BIT shall 'be construed to prevent a Contracting Party from adopting or maintaining environmental measures necessary to protect human, animal or plant life or health' once such measures are not arbitrary or targeting an investment.[113] This aligns with the growing trend of allowing more regulatory space for host states to strike an appropriate balance between their human rights and sustainable development commitments on the one hand and their investment treaty obligations on the other. Other Sino-Africa BITs under consideration, the China–Nigeria BIT and China–SA BIT, do not contain similar provision.

4 China's Attitudes Towards Mining Arbitrations

Virtually all Sino-Africa BITs are post-ICSID Convention BITs. China and most African countries are parties to the ICSID Convention.[114] Nevertheless, most of the BITs appear to exhibit some suspicion towards the investment treaty arbitration system. This is not surprising because, as noted above, they are widely seen as essentially south–south BITs although China's current status makes this characterization tenuous. They reflect the somewhat complex atmosphere overshadowing investment treaty arbitration in Africa. We will discuss two forms of international arbitration here: ICSID and non-ICSID arbitrations.

4.1 ICSID Arbitration

As noted earlier, China–Africa mining disputes have become increasingly likely given the volume of Chinese investment in the sector. We are not aware of any investment arbitration claim between Chinese investors and African states. However, following DRC's enactment of a new Mining Code in 2018, two Chinese companies have reportedly been seeking to negotiate with the government potentially as a prelude to an international arbitration claim.[115] Failure of this negotiation might lead to an ICSID arbitration. Of course, other ICSID mining arbitrations might shape any Chinese-initiated ICSID arbitration against any African state. In particular, the case of *Beijing*

111 Ibid Art 13(2)(3).
112 Ibid Art 13(4).
113 Ibid Art 10(1)(2).
114 ICSID Membership here.
115 'Zijin Mining sends lawyers to sort out Congo's "misunderstanding", Metal Bulletin, www.metalbulletin.com/Article/2486200/Zijin-Mining-sends-lawyers-to-sort-out-Congo-misunderstanding.html

Shougang v Mongolia[116] in which three Chinese investors sought compensation for alleged expropriation through revocation of licence is an example. The arrest of many Chinese miners for allegedly engaging in illegal mining in Nigeria might also trigger an action under the full protection and security standard.

4.2 Non-ICSID Arbitrations

For non-ICSID arbitration, the most notable, for the present purpose, is the CAJAC established following the FOCAC initiative. We will now dwell on the CAJAC.

4.2.1 CAJAC: Alternative to ICSID Arbitration?

The FOCAC Beijing Summit (Beijing Consensus) laid the foundation for the establishment of the CAJACs. It emphasized the need to 'review the traditional friendship existing between China and Africa, to observe the latest development trends of international arbitration and to envision the cooperative prospects of establishing the China-Africa Joint Dispute Resolution Mechanism'.[117] The CAJAC was founded by the Arbitration Foundation of South Africa, the Association of Arbitrators, and the Shanghai International Trade Arbitration Centre. There are presently four such centres in Beijing, Shanghai, Johannesburg, and Nairobi. They are context-specific arbitration centres established to respond to the growing significance of China–Africa economic relations.[118] In the words of the CEO, CAJAC is an 'integral part of the support structure especially crafted to foster trade and investment between China and Africa'.[119]

It is not hard to figure out China's (and African countries') preference for the new CAJACs. First, of course, the age-long legitimacy crisis bedevilling the investment arbitration regime – now undergoing reform, it might be added – creates the desire on the part of some African countries for a more Afrocentric arbitration system. For instance, SA's bitter experience with investment treaty arbitration in *Foresti v Republic of South Africa*[120] influenced its enactment of the PIA in 2015. In that case, brought under the ICSID Additional Facility (SA not being a member of ICSID), Italian

116 *Beijing Shougang v Mongolia* PCA Case No 2010-20.

117 Declaration of the Johannesburg Summit of the FOCAC, 25 December 2015, www.focac.org/eng/zywx_1/zywj/201512/t20151225_7933575.htm

118 Francis Kariuki 'The Vision Of Co-building China-Africa Joint Arbitration Centres in Different Legal Systems.' http://kmco.co.ke/wp-content/uploads/2018/08/Co-building-China-Africa-Joint-Arbitration-Centres-in-Different-Legal-Systems.pdf

119 Sadaf Habib, 'Interview with Deline Beukes, CEO of the CAJAC Johannesburg' Kluwer Arbitration Blog 26/11/18.

120 *Foresti v Republic of South Africa* (2010) IIC 445.

investors challenged the Mineral and Petroleum Resources Development Act enacted to, among others, implement the Black Economic Empowerment Policy in the mining sector, under the Italy–SA BIT and SA–Luxemburg BIT.[121] The claimants argued that the Act expropriated their investments by insisting that a certain percentage of the shares of all mining companies be owned by 'historically disadvantaged South Africans'.[122] As the matter was pending, the parties reached an agreement to discontinue. South Africa agreed to grant new mineral rights to the claimants while the latter agreed to discontinue the action and to implement an employee ownership program. As a result, SA comprehensively reviewed its investment legislation and agreements, arguing that the ISDS system subjected 'matters of vital national interest to unpredictable international arbitration that may constitute direct challenges to legitimate, constitutional and democratic policy-making'.[123]

SA's position on investment arbitration has now been provided under the PIA. Indeed, it has also terminated a number of its BITs with other countries (e.g. BITs with Austria, Belgium and Luxembourg, Denmark, France, Germany, the Netherlands, Spain, Switzerland, and the United Kingdom).[124] In the ongoing UNCITRAL WG III ISDS reform dialogue, SA proposed 'an alternative to ISDS in the form of a more modern and structured dispute settlement process'.[125]

It has also been pushing for a regional position in the SADC against ISDS. Accordingly, the SADC Finance and Investment Protocol 2016 now encourages state–state arbitration. However, the SADC tribunal suffers from a legitimacy crisis of its own following the non-compliance by Zimbabwe with the tribunal's decision in *Mike Campbell v Republic of Zimbabwe.*[126]

Like SA, Tanzania has enacted the Natural Wealth and Resources Contract (Review and Renegotiation of Unconscionable Terms) Act 2017 and the Natural Wealth and Resources (Permanent Sovereignty) Act 2017 designed to reassert sovereignty over its mineral resources and to avoid subjecting the state to the jurisdictions of foreign courts. These laws triggered some claims against Tanzania.[127] In one of the cases, it was claimed that the government allegedly banned the claimant's mineral exports on account of unpaid taxes of about $40 billion.[128]

121 Ibid.

122 Ibid.

123 Statement by Minister of Trade and Industry, Rob Davies, 'Debate on the Protection of Investment Bill 2015 in Parliament' on 17/11/15, https://www.gov.za/speeches/minister-rob-davies-debate-protection-investment-bill-2015-17-nov-2015-0000#

124 International Investment Agreements Navigator, South Africa, https://investmentpolicy.unctad.org/international-investment-agreements/countries/195/south-africa

125 UNCITRAL WG III, Possible Reforms of the ISDS: Submission from the Government of SA (38th Session, 14–18 October 2019, Vienna) para 18, http://undocs.org/en/A/CN.9/WG.III/WP.176

126 [2008] SADCT 2.

127 *Winshear Gold Corp v Tanzania* (ICSID Case No. ARB/20/25).

128 'Acacia Mining rejects $190 billion Tanzanian Tax bill,' BBC News, 25 July 2017. See also *Montero Mining v Tanzania* (ICSID Case No ARB/21/6)

Second, the legitimacy crisis extends to other issues of concern to some African countries: exorbitant costs of investment arbitration, interpretative inconsistencies, pro-investor-bias and lack of adequate representation of Africans among the elitist group of expert arbitrators in the world and other partiality concerns, etc. The ongoing UNCITRAL reform effort is unlikely to address these concerns.

Most arbitration proceedings against African countries are conducted outside Africa.[129] Indeed, African countries' investments contracts are largely governed by English law, drafted mostly by English, American, or French Lawyers, and, therefore, it has been argued, the seat of arbitration mostly would be where these experts 'are accustomed to and comfortable with but with unrewarding outcomes for their African clients'.[130] Over 90% of lawyers representing African respondents in international arbitration are non-Africans because Africans lack the requisite expertise.[131]

With respect to the cost, Nigeria also has a running battle with an oil mining investor.[132] In *Process & Industrial Development Ltd v Nigeria*,[133] the claimant claimed damages for the cancellation of a gas development project by the Nigerian government. An award of $6.5 billion with interest at the rate of 7% was awarded against Nigeria.[134] The amount was more than the country's annual budget for the year 2018. This confirms the concerns expressed by the UNCITRAL WG III on the reform of ISDS in its 34th session that exorbitant cost could divert public funds away from more compelling development imperatives.[135] Nigeria, however, has been seeking annulment of the award on account of corruption in the award of the original contract.[136]

There is no doubt that African countries are among the frequent respondents before ICSID arbitral tribunals and other 'non-African arbitral tribunals'. In the past couple of years, more htna 100 cases were brought against

129 Kariuki (n 118) 3–4..

130 Ibid.

131 Tunde Ogunseitan, 'Africa' in C Amirfar et al. (eds) *The Current State and Future of International Arbitration: Regional Perspectives* (International Bar Association 2015) 20–22.

132 *Process & Industrial Development Ltd v Federal Republic of Nigeria* (17 July 2015) https://jusmundi.com/en/document/decision/pdf/en-process-and-industrial-developments-ltd-v-the-ministry-of-petroleum-resources-of-the-federal-republic-of-nigeria-part-final-award-friday-17th-july-2015

133 *Process & Industrial Development Ltd v Nigeria* (31 January 2017) https://jusmundi.com/fr/document/decision/pdf/en-process-and-industrial-developments-ltd-v-the-ministry-of-petroleum-resources-of-the-federal-republic-of-nigeria-final-award-tuesday-31st-january-2017

134 Ibid para 112.

135 UNCITRAL WG III Possible reform of ISDS: Cost and duration, Thirty-sixth session Vienna, 29 October–2 November 2018, paras 60–65, https://undocs.org/en/A/CN.9/WG.III/WP.153

136 *Federal Republic of Nigeria v Process and Industrial Developments Ltd* [2020] EWHC 2390 (Comm).

African governments.[137] Thus, the resentment may be understandable, although some scholars argue that active engagement with non-African stakeholders to reform the system is better than 'running away' from it.[138] Despite this, empirical evidence suggests that African states actually fared well in ICSID arbitration.[139] Nevertheless, it seems fairly settled that the system was originally designed for developing countries to be 'respondents' in order to incentivize the inflow of FDI. In 2012, for instance, out of the 82 ICSID proceedings against African states, 65% of the claimants were Europeans and 16% were Americans.[140]

As noted earlier, efforts have been, and are being, made to reform the ISDS system to address some of the legitimacy concerns. Reforms of the substantive aspects have been left to the states' treaty-making processes, while some of the procedural reforms are being addressed by the ICSID and UNCITRAL. In 2014, for instance, the UN Convention on Transparency in Treaty-based Investor-State Arbitration (Mauritius Treaty) was adopted to, among others, address some of the concerns on transparency. Also, in July 2017, member states of the UNCITRAL Working Group III started work on the reform of ISDS. Work is ongoing at the time of writing.

Like some African countries, China's initial attitude towards investment treaty arbitration has been to avoid it as much as possible. Culturally, Chinese disputants prefer mediation and conciliation. Today, China is supporting the reform of the ISDS regime through a multilateral approach. This is not surprising given the extent of its OFDI.

Therefore, using the CAJAC as a dispute resolution mechanism in disputes between Chinese and African countries' nationals, entities, and authorities seems a plausible way to avoid the problematic system in favour of a context-specific, tailor-made, contemporary 'south–south'-based arbitral framework.

However, the CAJAC itself is problematic not least because, first, of its infancy and absence of comprehensive rules of procedure. Second, the complex legal pluralism in Africa, even within countries, complicates the march towards the 'Sino-Africanization' of arbitration. More than 50 countries with distinct legal traditions (common law, civil law, Roman-Dutch law, Islamic law, and customary laws) can hardly agree on uniform rules of procedure. It is instructive that the China International Economic and Trade Arbitration Commission, in 2017, developed International Investment Arbitration Rules which integrates international investment rules with Chinese dispute resolution characteristics in order to support investment disputes involving Chinese

137 ICSID, The ICSID Caseload – Statistics 2019.

138 MF Qumba, 'Assessing African Regional Investment Instruments and Investor-State Dispute Settlement' (2021) 70 *ICLQ* 197–232.

139 Karel Daele, 'African's track record in ICSID proceedings,' Kluwer Arbitration Blog, 30 May 2012, http://arbitrationblog.kluwerarbitration.com/2012/05/30/africas-track-record-in-icsid-proceedings

140 Ibid.

investors.[141] It is not clear whether the CAJACs will be guided by these rules. However, it is presumed that they might shape the development of CAJAC Rules of Procedure going forward.

Third, the plain reality of an inadequate number of arbitration experts in Africa cannot be wished away. The professional competence of the few available is equally a matter of concern. Fourth, the proliferation of arbitration centres, like the CAJAC, across Africa only adds to the problem as countries compete to become arbitration hubs despite the acute deficit in arbitral infrastructure on the continent. The CAJAC is only an addition to a long list of arbitration centres. Fifth, political instability and enforcement obstacles on account of, among others, corruption and inefficient and ineffective legal institution can derail the CAJAC itself.

Therefore, the CAJAC may not address the fears of some African countries regarding the shortcomings of the ICSID arbitration. Indeed, it may create additional problems. Unfortunately, China is not known for transparency. As the ISDS system undergoes reforms to respond to some of the criticisms, the CAJAC may not embrace the critical African concerns relating to human rights, social, and environmental governance. Indeed, UN human rights records show that African countries have been supporting China on human rights issues.[142] The CAJAC may suffer from lack of transparency. Interestingly, in its submission to the UNCITRAL WG III on the reform of the ISDS system, China indicated its willingness to support the proposal for the establishment of a Multilateral Investment Court (MIC) and an appellate body.[143] This is not surprising given the extent of Chinese OFDI especially in the context of the BRI.[144]

5 Conclusion

This chapter examined the Chinese investment in African mining sector in the context of the Sino-Africa dispute resolution mechanisms. It is hard to ignore

141 China International Economic and Trade Arbitration Commission, International Investment Arbitration Rules, 2017.

142 For instance, African countries have remained silent over China's recent human rights records. They even supported China in votes at the UNHRC. Olewe claims that being overly partisan towards Beijing's human rights records will incur no retaliatory measure from the West because Western countries cannot compete with China in terms of loans and infrastructure to African countries. He argues that African leaders' decision 'to turn to China for infrastructure funding especially in the last 20 years has transformed the continent's landscape with expansive roads, bridges, railways, ports and internet infrastructure that has ensured that the continent is not a pariah in a digital age'. See Dickens Olewe, 'Why African countries back China on human rights,' BBC News, 2 May 2021. www.bbc.com/news/world-africa-56717986

143 UNCITRAL WG III (n 84).

144 Yuwen Li and Cheng Bian 'China's Stance on Investor-State Dispute Settlement: Evolution, Challenges, and Reform Options' (2020) 67 *Netherlands International Law Review* 503–551, 505.

the real challenges facing African countries regarding the protection of investments in their respective territories. The volatility of the mining sector compounds the situation. The challenge is that each of the dispute resolution mechanisms (local courts, CAJACs, ICSID, and other forms of arbitration) is problematic in and of itself. Undoubtedly, Sino-Africa mining arbitration is good for mineral resources development and increased investment. However, its utility is likely to depend upon its capacity to respond to contemporary development challenges facing African countries. Thus, African countries cannot afford to ignore human rights, social, and environmental governance issues because they need Chinese investments. The approach adopted in some new-generation BITs (e.g. Nigeria–Morocco BIT 2016, US Model BIT 2012, and Canada–EU Trade Agreement's investment chapter) might help in this regard. On the procedural front, China–Africa mining dispute settlement mechanisms should avoid the shortcomings of investment treaty arbitration and should be flexible, open to constant reforms and innovations. China should also recognize the diversity of African countries and their areas of divergences. In other words, it would be a mistake for China, or any other power for that matter (e.g. US, Russia, EU, India, or Brazil) to assume that African countries have a common problem, a common investment strategy, and a common concern on ISDS. African countries' complex legal pluralism should be a pointer to the potential failure of any uniform approach to mining dispute resolution on the continent.

Cases

Beijing Shougang v Mongolia PCA Case No. 2010-20
Case Concerning East Timor (1995) *ICJ Reports* 89
Federal Republic of Nigeria v Process and Industrial Developments Ltd [2020] EWHC2390 (Comm)
Foresti v Republic of South Africa (2010) IIC 445
Mavromattis Palastine Concession (Greece v UK) (1924) PCIJ Series A 11
Mike Campbell v Republic of Zimbabwe [2008] SADCT 2
Montero Mining v Tanzania (ICSID Case No ARB/21/6)
Process & Industrial Development Ltd v Federal Republic of Nigeria (17 July 2015) https://jusmundi.com/en/document/decision/pdf/en-process-and-industrial-developments-ltd-v-the-ministry-of-petroleum-resources-of-the-federal-republic-of-nigeria-part-final-award-friday-17th-july-2015
Process & Industrial Development Ltd v Nigeria (31 January 2017) https://jusmundi.com/fr/document/decision/pdf/en-process-and-industrial-developments-ltd-v-the-ministry-of-petroleum-resources-of-the-federal-republic-of-nigeria-final-award-tuesday-31st-january-2017
South China Sea Arbitration (Philippines v China) (2016) PCA Case No. 2013-19
Tza Yap Shum v Peru ICSID Case No. ARB/07/6
Winshear Gold Corp v Tanzania (ICSID Case No. ARB/20/25)

Bibliography

'Zijin Mining sends lawyers to sort out Congo's "misunderstanding",' *Metal Bulletin*, www.metalbulletin.com/Article/2486200/Zijin-Mining-sends-lawyer s-to-sort-out-Congo-misunderstanding.html

Alves AC. 'Chinese Economic Statecraft: A Comparative Study of China's Oil-backed Loans in Angola and Brazil' (2013) 1*Journal of Current Chinese Affairs* 99–130.

Begu LS *et al.* 'China-Angola Investment Model' (2018) 10*Sustainability* 29–36.

Brautigam D *et al.* 'Chinese Investment in Africa: How much do we know?' PDEL Synthesis Paper No. 2 (2017) https://pedl.cepr.org/sites/default/files/PEDL_Synth esis_Papers_Piece_No._2.pdf.

Chen M, 'Embracing Non-ICSID Investment Arbitration? The Chinese Perspective' (2019) 39*Northwestern Journal of International Law & Business* 249.

Chiedozie I. 'How Chinese prospectors exploit local miners to cash in on Nigeria's gold,' *Punch*, 17 May 2020.

Corkin L. 'China and Angola – Strategic Partnership or Marriage of Convenience?' (2011) 1(1) *Angola Brief* 4.

Cotula L *et al. China-Africa Investment Treaties: Do They Work?* (International Institute for Environment and Development 2016).

Daele K. 'African's track record in ICSID proceedings,' Kluwer Arbitration Blog, 30 May 2012. http://arbitrationblog.kluwerarbitration.com/2012/05/30/africas-track-re cord-in-icsid-proceedings

Ellis D. 'Nigeria and China firms sign Gold agreement,' *Mining Global*, 9 November 2020.

FOCAC. 'Beijing declaration toward an even stronger China-Africa community with a shared future.' www.focac.org/eng/zywx_1/zywj/t1594324.htm

Grossman N. 'The Normative Legitimacy of International Courts' (2013) 86*Temple Law Review* 61–105.

Habib S. 'Interview with Deline Beukes, CEO of the CAJAC Johannesburg,' Kluwer Arbitration Blog, 26 November 2018.

Helfer LR and Slaughter AM. 'Towards a Theory of Effective Supranational Adjudication' (1997) 107*Yale Law Journal* 273–391.

Helfer LR and Slaughter AM. 'Why States Create International Tribunals: A Response to Professors Posner and Yoo' (2005) 93 (3) *California Law Review* 899–956.

Hursh J. 'A Bump in the Belt and Road: Tanzania pushes back against Chinese port project.' (Center for International Maritime Security 2019). https://cimsec.org/a-bump -in-the-belt-and-road-tanzania-pushes-back-against-chinese-port-project

IMF. 'Currency Composition of Official Foreign Exchange Reserve.' https://data.imf. org/?sk=E6A5F467-C14B-4AA8-9F6D-5A09EC4E62A4. Accessed 30 December 2020.

Kabir A. 'Nigeria moves against illegal Chinese miners, others,' *Premium Times*, 12 January 2020.

Kaplinsky R and Morris M. 'Chinese FDI in Sub-Saharan Africa: Engaging with Large Dragons' (2009) 21*European Journal of Development Research* 551–569.

Kariuki F. 'The Vision Of Co-building China-Africa Joint Arbitration Centres in Different Legal Systems.' http://kmco.co.ke/wp-content/uploads/2018/08/Co-building-China-Africa-Joint-Arbitration-Centres-in-Different-Legal-Systems.pdf

Keohane RO *et al.* 'Legalized Dispute Resolution: Interstate and Transnational' (2000) 54 (3) *International Organisation* 457–488.

Kola O, '17 Chinese arrested for illegal gold mining,' Anadolu Agency, 5 May 2020.

Li Y. 'China's Growth Miracle: Past, Present, and Future.' www.unrisd.org/80256B3C005BD6AB%2F(httpAuxPages)%2F2893F14F41998392C1257BC600385B21%2F$file%2FChina's%20growth%20miracle%200808.pdf

Marafa LM. 'Africa's Business and Development Relationship with China Seeking Moral and Capital Values of the Last Economic Frontier,' Nordiska Afrikainstitutet Discussion Paper 44 (Uppsala 2009) 6–20. www.diva-portal.org/smash/get/diva2:272966/FULLTEXT02.pdf

Marais H and Labuschagne J-P. 'If you want to prosper consider building roads: China's role in African infrastructure and capital projects,' Deloitte Insight, 22 March 2021. www2.deloitte.com/us/en/insights/industry/public-sector/china-investment-africa-infrastructure-development.html

Mason D. 'China's Four Modernizations: Blueprint for Development or Prelude to Turmoil?' (1984) 11*Asian Affairs* 47–70.

Mihalyi D *et al. Resource-backed Loans: Pitfalls and Potential* (Natural Resource Governance Institute 2020). https://resourcegovernance.org/sites/default/files/documents/resource-backed-loans-pitfalls-and-potential.pdf

Mining Indaba. 'Chinese investment in African Mining: What you need to know,' 16 October 2019. www.miningindaba.com/Articles/chinese-investment-in-african-mining-what-you

Ministry of Commerce, China (MOFCOM). 'Brief Statistics on China's Import and Export 2018.' http://english.mofcom.gov.cn/article/statistic/BriefStatistics/201903/20190302842809.shtml. Accessed 30 December 2020.

Ministry of Natural Resources, People's Republic of China. *China Mineral Resources* (Geological Publishing House 2018). www.gov.cn/xinwen/2018-10/22/5333589/files/01d0517b9d6c430bbb927ea5e48641b4.pdf. Accessed 30 December 2020.

Ogunseitan T. 'Africa' in C Amirfar *et al.* (eds) *The Current State and Future of International Arbitration: Regional Perspectives* (International Bar Association 2015) 20–22.

Olewe D. 'Why African countries back China on human rights,' BBC News, 2 May 2021. www.bbc.com/news/world-africa-56717986

Page MT. 'The Intersection of China's Commercial Interests and Nigeria's Conflict Landscape,' USIP Special Report, 2018. www.usip.org/sites/default/files/2018-09/sr428-the-intersection-of-chinas-commercial-interests-and-nigerias-conflict-landscape.pdf

Posner EA and Yoo JC. 'Judicial Independence in International Tribunals' (2005) 93 (1) *California Law Review* 1–74.

Qumba MF. 'Assessing African Regional Investment Instruments and Investor-State Dispute Settlement' (2021) 70*ICLQ* 197–232.

Romano CPR. 'The Shift from the Consensual to the Compulsory Paradigm in International Adjudication: Elements for a Theory of Consent' (2007) 39*NYU Journal of International Law & Politics* 791.

Shepard W. 'What's China up to in Africa?' Forbes, 3 October 2019.

Sheppard A and Bret L-A, 'Mining Arbitration in Africa,' Lexology, 20 April 2020. www.lexology.com/library/detail.aspx?g=593a20a7-5f7e-49c1-9316-a7823f719025

Tetor C. 'Cash flow of Chinese direct investments to Africa 2008–2018,' Statista, 1 March 2012.

UNCITRAL WG III. 'Possible reform of investor-State dispute settlement (ISDS) Submission from the Government of China' (Thirty-eighth session Vienna, 14–18 October 2019). http://undocs.org/en/A/CN.9/WG.III/WP.177

UNCITRAL WG III. 'Possible reform of ISDS: Cost and duration, Thirty-sixth session Vienna, 29 October–2 November 2018.' https://undocs.org/en/A/CN.9/WG.III/WP.153

Wang G. 'Chinese Mechanisms for Resolving Investor-State Disputes' (2011) *Jindal Journal of International Affairs* 204–233.

Won K. 'China's Bilateral Investment Treaties with African States in Comparative Context' (2016) 49*Cornell International Law Journal* 141.

World Bank Group. *China and WB 1980–2018*http://pubdocs.worldbank.org/en/657111542261998957/F-F-en-2018.pdf. Accessed 20 December 2020.

World Bank Group. *China: Systematic Diagnosis* (2018). http://documents.worldbank.org/curated/en/147231519162198351/pdf/China-SCD-publishing-version-final-for-submission-02142018.pdf. Accessed 16 December 2020.

World Bank. World Development Indicators Database, https://databank.worldbank.org/reports.aspx?source=2&country=CHN#. Accessed 16 December 2020.

Xiaofang S. 'Private Chinese Investment in Africa: Myths and Realities.' https://openknowledge.worldbank.org/bitstream/handle/10986/12174/wps6311.pdf?sequence=1&isAllowed=y

Yang Y. 'The political economy causes of China's successes' in Garnaut Ross, Song Ligang, and Fang Cai (eds) *China's 40 Years of Reforms and Development (1978–2018)* (Australian National University Press 2018) 75–76.

Yuwen L and Cheng B. 'China's Stance on Investor-State Dispute Settlement: Evolution, Challenges, and Reform Options' (2020) 67*Netherlands International Law Review* 503–551.

Zhufu F. 'The Economic History of China' (1981) 7 (1) *Modern China* 3–30.

Part III

Unpacking the Practical Issues in the Mining Sector and Conclusions

11 Climate Change Reporting among Mining Firms in Africa

Practice, Theory and Policy

King Carl Tornam Duho[1]

1 Introduction

Globally, there are concerns about environmental issues which continue to have the attention of leaders in politics, the public sector, business, and academia. A menu of such issues can be grouped under what has come to be known as the planetary boundaries. The planetary boundaries are nine environmental boundaries that need to be maintained to ensure sustainable living on earth. These earth-system processes are climate change, biodiversity loss, biogeochemical cycles, ocean acidification, land use, freshwater, ozone depletion, atmosphere aerosols, and chemical pollution.[2] Out of the seven of the boundaries quantified, the first three listed have been found to have crossed the planetary boundaries.[3] The world is at a tipping point – the thresholds for the environmental boundaries such that inaction leading to a small increment can lead to a possible catastrophe.[4] The activities of mining[5] firms have the potential to push the plenary

1 King Carl Tornam Duho, ACMA CGMA is currently a Youth Fellow with the International Monetary Fund (IMF) and an economic consultant. He is the Technical Advisor for Dataking Research Lab, Dataking Consulting, and an Honorary Fellow with the VIT Centre for TVET Policy and Education. He has many years of experience working within academia, business, and public policy. He has consulted for UNICEF, GIZ, the Royal Netherlands Embassy, AERC, Oxfam, Plan International, and IMANI. King interned with the Geneva Office of UNCTAD on international trade. He has written extensively on the financial services sector, the cocoa sector, the extractives industry, the stock market, climate change, and other wider economic issues. He has authored more than 40 research works and his research appeared in more than 15 peer-reviewed journals including *Journal of Economic Studies, International Journal of Managerial Finance*, and *Journal of Financial Crime*. https://orcid.org/0000-0002-8736-3220

2 Rockström, J., Steffen, W., Noone, K., Persson, Å., Chapin, S. I., Lambin, E., ... & Falkenmark, M. (2009) 'Planetary Boundaries: Exploring the Safe Operating Space for Humanity.' *Ecology & Society*, 14(2).

3 Ibid.

4 Ibid.

5 Some of the minerals and the African countries in which they can be found include diamonds (Botswana, Congo (Kinshasa), South Africa, Angola), gold (South Africa, Ghana, Tanzania, Mali), uranium (Namibia, Niger, South Africa), bauxite (for aluminium) (Guinea, Ghana), steel (South Africa, Egypt, Libya,

DOI: 10.4324/9781003284437-14

boundaries beyond their limit, and there are continuous calls for action to be taken.[6] Climate action is one of the most notable areas that continue to draw attention, yet there is much that is expected to be explored.

Climate change concerns have risen to be a topical issue in development and public policy.[7] Thus, it continues to receive attention from practitioners, policymakers, and researchers globally.[8] Various terms are used to describe the phenomenon including global heating, climate emergency, climate crisis, and climate risk, among others. Climate change is a natural or anthropogenic-induced change in weather patterns or related changes in ice sheets, oceans, and land surfaces over a decade or more.[9] Global warming, which is sometimes used synonymously, is a human-induced climate change driven by the emission of greenhouse gases (GHGs). It is argued that since the mid-twentieth century, humans have continued to have a significant impact on the climate, leading to global changes. The negative effects of climate change or GHGs are not limited by geographic location, industries, or economic conditions of countries. Thus, all are being advised to contribute their quota towards successful climate action. There is explicit evidence of climate risks leading to declining water supplies, erosion in coastal areas, droughts,

Algeria), aluminium (South Africa, Mozambique, Egypt), copper (mined/refined) (Zambia, South Africa, Congo (Kinshasa), Egypt), platinum/palladium (South Africa), and coal (South Africa)

6 Odell, S. D., Bebbington, A., & Frey, K. E. 2018. 'Mining and climate change: A review and framework for analysis.' *The Extractive Industries and Society*, 5(1): 201–214.

7 NASA, 'How do we know climate change is real', https://climate.nasa.gov/evidence (accessed 23 January 2023).

8 Some of the notable scholars include Keywan Riahi (International Institute for Applied Systems Analysis), Anthony A. Leiserowitz (Yale University), Pierre Friedlingstein (University of Exeter), Detlef Peter Van Vuuren (Utrecht University), James E. Hansen (Columbia University), Petr Havlík (International Institute for Applied Systems Analysis), Edward Wile Maibach (George Mason University), Josep G. Canadell (Commonwealth Scientific and Industrial Research Organisation), Sonia Isabelle Seneviratne (ETH Zurich), Mario Herrero (Commonwealth Scientific and Industrial Research Organisation), and David B. Lobell (Stanford University).

9 Houghton, E. (1996). 'Climate change 1995: The science of climate change.' *Contribution of working group I to the second assessment report of the Intergovernmental Panel on Climate Change* (Vol. 2). Cambridge University Press; Davis, W. D. (2008). 'What Does Green Mean?' *Georgia Law Review*, 43, 901; Parry, M., Parry, M. L., Canziani, O., Palutikof, J., Van der Linden, P., & Hanson, C. (Eds) (2007). 'Climate change 2007 – impacts, adaptation and vulnerability.' *Working group II contribution to the fourth assessment report of the Intergovernmental Panel on Climate Change* (Vol. 4). Cambridge University Press; IPCC (2007). 'Climate change 2007: Impacts, adaptation and vulnerability.' In: Parry, M. L., Canziani, O. F., Palutikof, J. P., van der Linden, P. J., & Hanson, C. E. (Eds) (2014). *Contribution of working group II to the fourth assessment report of the Intergovernmental Panel on Climate Change.* Cambridge University Press, p. 976.

increased heat, insect outbreaks, increased wildfires, reduced agricultural yields, and health impacts, among others.[10]

Mining is currently responsible for 4% to 7% of GHG emissions globally.[11] Air pollutants have a dire effect on ecosystems, air quality, and agriculture, as well as human and animal health. More specifically, a reduction in emissions will improve health conditions for employees and the local community and can enhance relations with stakeholders. There are even direct cost implications of the emissions which are significant in the case of regions with emissions caps. Africa alone holds over 30% of the total mineral reserve globally, a sign of the significant role of its mining industry.[12] Interestingly, relatively weak institutional quality, governance, and enforcement of laws lead to various unsustainable business practices. The extent to which different stakeholders hold businesses accountable for their actions on their communities or the environment is not great in Africa. GHG emissions and climate change concerns are some of the issues that need a response from the business community and all other stakeholders. Yet there is little knowledge about how Africa's mining firms are disclosing their contribution towards reducing GHG emissions.

The extant literature on climate change disclosure practices in Africa's mining industry is scant. This study fills the dearth in literature by providing a detailed analysis of the issues from a practical, theoretical, and policy-oriented perspective. This study explores three research questions on climate change disclosure practices:

1 What is the role of Africa's mining industry in addressing the climate change problem?
2 What is the current state of climate change disclosure among mining firms in Africa?
3 What are the determinants of climate change disclosure among mining firms in Africa?

The study builds on the institutional theory and a dataset of 21 firms over the period 2006 to 2019. The results show that corporate governance factors (board size and board independence) drive positive climate change disclosure among firms. Other factors such as firm size, the United Nations Global Compact (UNGC), human development, and corruption perception also drive the practice of climate change disclosure. Special emphasis has been laid on South Africa and Morocco which are making

10 Gates, B. (2021). *How to Avoid a Climate Disaster: The Solutions We Have and the Breakthroughs We Need.* Toronto: Knopf.
11 Henderson, K. & Maksimainen, J. 'Here's how the mining industry can respond to climate change' (McKinsey, 27 August 2020), www.mckinsey.com/business-functions/sustainability/our-insights/sustainability-blog/here-is-how-the-mining-industry-can-respond-to-climate-change (accessed 26 January 2021).
12 DLA Piper (2012). 'Mining in Africa: A legal overview', DLA Piper Law Firm.

strides, as well as Ghana. The results are relevant for understanding the current state of climate change action or GHG emissions reportage among mining firms with useful insights for practitioners, policymakers, and the academic community.

The next section discusses some theoretical and empirical findings on climate change disclosures. Section 3 provides an overview of the mining industry and the concerns about climate change. It also provides a case discussion of three African countries: Ghana, South Africa, and Morocco. Section 4 provides a detailed discussion of the methodology used for the study. Section 5 provides the results of the study, which are discussed in line with the extant literature. The penultimate section provides the conclusions while Section 7 discusses the implications of the study for practice, policy, and future research.

2 Theoretical Underpinning and Earlier Studies on Climate Change Disclosures

This section provides a brief review of the theoretical underpinning of the study and critically reviews the extant literature on climate change reporting or disclosure.

2.1 Theoretical Underpinning of the Study

Various theories have been used in the literature to explore the issue of climate change, including agency theory, signalling theory, proprietary cost theory, stakeholder theory, legitimacy theory, voluntary disclosure theory, economic theories of disclosure, and ecological modernisation theory, among others. A review study explored more than 28 different management theories applied to climate change literature, and there are many more that exist.[13] The theoretical discussions of climate change issues remain under-explored and the mining industry leaves so much to be researched.

Institutional theory explains how climate change disclosure is embedded in different political, economic, legal, and financial systems.[14] Earlier studies explored how firm-level factors, country-level factors, and global forces drive

13 Daddi, T., Todaro, N. M., De Giacomo, M. R., & Frey, M. (2018), 'A systematic review of the use of organization and management theories in climate change studies.' *Business Strategy and the Environment*, 27(4), 456–474.

14 DiMaggio, P. J., & Powell, W. W. (1983). 'The iron cage revisited: Institutional isomorphism and collective rationality in organizational fields.' *American Sociological Review*, 48(2), 147–160.; Scott, W. R. (1987). 'The adolescence of institutional theory.' *Administrative Science Quarterly*, 32(4), 493–511; Asare, E. T., Duho, K. C. T., Agyenim-Boateng, C., Onumah, J. M. and Simpson, S. N. Y. (2021), 'Anti-Corruption Disclosure as a Necessary Evil: Impact on Profitability and Stability of Extractive Firms in Africa.' *Journal of Financial Crime*, https://doi.org/10.1108/JFC-09-2020-0173

climate change disclosure practices.[15] Institutional theory argues that there are three isomorphic pressures that firms face which drive their practices.[16] The mimetic isomorphism argues that firms tend to mimic or copy from the best performers, while coercive isomorphism argues that firms are forced by regulations to apply best practices. Firms may also be exposed to normative isomorphism which is the situation that professionalism within the industry or the business ecosystem leads to a certain practice. Climate change disclosure practices, in one way or another, could be explained by any of these pressures as there is a growing number of laws, standards, and professional bodies promoting best practices, and market participants are becoming climate-conscious, leading to a catch-up effect or imitation of adoption by low performers also.

2.2 Empirical Studies on Climate Change Disclosure

Existing studies have explored climate change disclosure practices in different contexts. The broad issues covered are the nexus with firm factors, industry factors, country factors, and global factors. The studies tend to cut across industries and regions with few studies specific to the mining industry.

The first strand of studies is those that explore the determinants of climate change disclosure practices. Comyns explores the determinants of GHG reporting among oil and gas companies across the world.[17] The study found that the European Union's emissions trading scheme and the Global Reporting Initiative (GRI) guidelines enhance the quality and depth of reports. The study found that European firms tend to adopt proactive climate change strategies but do not necessarily have superior GHG reporting practices. Interestingly, media visibility and firm size do not enhance GHG reporting. Halkos and Skouloudis posit that an increasing number of firms are accounting or reporting for climate change, but most firms give superficial treatment to climate change disclosure.[18] The study argues that, in Greece, only a few firms regard climate change disclosure as a tool to empower stakeholder decision making. They found that operation in environmentally

15 Duho, K. C. T., Agyenim-Boateng, C., Asare, E. T. and Onumah, J. M. (2021). 'Convergence and determinants of anti-corruption disclosure among extractive firms in Africa.' *Journal of Financial Crime*, https://doi.org/10.1108/JFC-06-2020-0109; Comyns, B. (2016). 'Determinants of GHG reporting: An analysis of global oil and gas companies.' *Journal of Business Ethics*, 136(2), 349–369.

16 Rodrigues, L. L., & Craig, R. (2007). 'Assessing international accounting harmonization using Hegelian dialectic, isomorphism and Foucault.' *Critical Perspectives on Accounting*, 18(6), 739–757.

17 Comyns, B. (2016). 'Determinants of GHG reporting: An analysis of global oil and gas companies.' *Journal of Business Ethics*, 136(2), 349–369.

18 Halkos, G. & Skouloudis, A. (2016). 'Exploring the current status and key determinants of corporate disclosure on climate change: Evidence from the Greek business sector.' *Environmental Science & Policy*, 56(C), 22–31.

sensitive sectors, international presence, and subscription to voluntary initiatives developed externally positively influence climate change disclosure. However, factors such as firm size, sector, profitability, and the type of ownership do not significantly influence climate change disclosure.

Kılıç and Kuzey found an increase in climate change disclosure practice, but many firms have not made significant progress in incorporating climate impact in their reports.[19] In line with legitimacy and political cost theory, the study found firm size, profitability, age, and stock exchange listing to be drivers of the disclosure. The results show that the number of branches, ownership diffusion, cross-listing, and foreign ownership does not have a significant effect on climate change disclosure practices. Another study on the determinants of disclosure on GHG emissions across the world explore factors such as size, financial leverage, market-to-book ratio, return on equity, return on asset, inclusion in Dow Jones Sustainability Index, sector of operation, and acceptance of the Kyoto Protocol.[20] The study found a positive significant impact of size and market capitalisation on GHG disclosure, but an inversely significant impact from return on asset.

The second strand of studies is where corporate governance mechanisms are explored in climate change disclosure literature. Jaggi *et al.* found that Italian firms in highly polluting industries tend to disclose their carbon information.[21] The study posits that board independence, institutional shareholders, and environmental committees significantly influence voluntary carbon disclosure. In a study on the Canadian context, results show that the effectiveness of the board of directors drives climate change disclosure practices.[22]

Additionally, in the context of developing and emerging economies in the Asia Pacific, the determinants of climate change disclosure such as firm size, country of domicile, environment certification, industrial membership, board size, board independence, CEO duality, and gender were investigated.[23] The results show that only firm size and board size tend to have an insignificant

19 Kılıç, M. & Kuzey, C. (2019). 'Determinants of climate change disclosures in the Turkish banking industry.' *The International Journal of Bank Marketing*, 37(3), 901–926.

20 Prado-Lorenzo, J.-M., Rodríguez-Domínguez, L., Gallego-Álvarez, I. and García-Sánchez, I.-M. (2009). 'Factors influencing the disclosure of greenhouse gas emissions in companies world-wide.' *Management Decision*, 47(7), 1133–1157.

21 Jaggi, B., Allini, A., Macchioni, R., & Zagaria, C. (2018). 'The factors motivating voluntary disclosure of carbon information: Evidence based on Italian listed companies.' *Organization & Environment*, 31(2), 178–202.

22 Ben-Amar, W. & McIlkenny, P. (2015). 'Board effectiveness and the voluntary disclosure of climate change information.' *Business Strategy and the Environment*, 24(8), 704–719.

23 Amran, A., Periasamy, V., & Zulkafli, A. H. (2014). 'Determinants of climate change disclosure by developed and emerging countries in Asia Pacific.' *Sustainable Development*, 22(3), 188–204.

effect. In the Euro area, Giannarakis *et al.* found that independent authentication of environmental data, environmental performance and government ownership determine climate change disclosure.[24] Other factors are the firm size, board size, and, in some cases, return on investment capital.

The review shows that climate change disclosure research in the mining industry is sparse. A review study shows that the dearth in literature is more evident in developing countries.[25]

3 Overview of the Mining Industry and Climate Change Concerns

There are interesting issues to consider regarding climate change in the mining industry of Africa. This section provides general insights about climate change reporting and the regulations and policies on climate change reporting in the mining industry. Also, critical challenges and prospects are discussed in detail.

3.1 The Mining Industry of Africa – Implications for the Climate

The mining industry of Africa contributes immensely towards the provision of world metal and mineral demands. The continent endowed with valuable resources including gold, diamond, iron ore, aluminium, bauxite, zinc, manganese, lead (Pb), steel, copper, cobalt, chromite, graphite, phosphate rock, platinum, and uranium. It is now becoming clear that it is not enough for Africa to hold 30% of the world's mineral reserve, but various governance, economic, social, and environmental issues need to be considered vis-à-vis other emerging issues in the industry. There are specific industry challenges such as geopolitical risk and economic protectionism, shareholder activism, legal limits on natural resource use, regulation, and public scrutiny.[26] This draws attention to the management of critical minerals[27] which are relevant to the economic well-being of Africa and the world at large. These minerals are those needed to produce electric cars, solar panels, mobile phones, flat-

24 Giannarakis, G., Zafeiriou, E., Arabatzis, G., & Partalidou, X. (2018). 'Determinants of corporate climate change disclosure for European firms.' *Corporate Social Responsibility and Environmental Management*, 25(3), 281–294.

25 Odell, S. D., Bebbington, A., & Frey, K. E. (2018). 'Mining and climate change: A review and framework for analysis.' The Extractive Industries and Society, 5 (1), 201–214.

26 Maennling, N. & Toledano, P. 'Seven trends shaping the future of the mining and metals industry' (World Economic Forum, 20 March 2019), www.weforum.org/agenda/2019/03/seven-trends-shaping-the-future-of-the-mining-and-metals-sector (accessed 26 January 2020).

27 Some of the notable critical minerals include rare-earth elements (REE), gallium (Ga), indium (In), tungsten (W), platinum-group elements (PGE) including platinum (Pt) and palladium (Pd), cobalt (Co), niobium (Nb), magnesium (Mg), molybdenum (Mo), antimony (Sb), lithium (Li), vanadium (V), nickel (Ni), tantalum (Ta), tellurium (Te), chromium (Cr), and manganese (Mn).

screen monitors, wind turbines, and many other high-tech applications, but their supply may be at risk because of geopolitical issues, geological scarcity, trade policy, or other factors.

The current enhancements in technology will have a significant role to play in shaping the future outlook of the industry. One essential area is related to the use of big data in the mining industry. The current emphasis placed on big data, digitisation, automation, the internet of things, blockchain, and the accompanying analyses of the same in enhancing efficiency, effectiveness, optimal valorisation of existing resources or access to new ones, cannot be overemphasised. However, there are issues such as data transparency that will continue to be of concern. It is very sad to note that several mineral-rich African countries are not signatories to the Extractive Industries Transparency Initiative (EITI). This includes South Africa, Morocco, Namibia, Botswana, Libya, Zimbabwe, and Egypt. The future mining sector of Africa will require disclosure beyond the current limited number of disclosure items in the EITI framework. Governments will have to lead in guiding such disclosure efforts, but if that fails, civil society and the community must react until action is taken.

Economic models have tended to ignore negative social and environmental externalities, but in recent times much attention has been paid to them.[28] The future mining sector will be a leader in efforts to transition to a low-carbon economy. For energy transition to this low-carbon economy, demand for many minerals needs to be high to make a strong business case for this environmental protective stance. Unlike systems based on fossil fuels, low-emission energy and transportation systems tend to be mineral intensive and will need to use critical minerals. The World Bank found that using climate-friendly technologies usually requires more metals than using alternative high-carbon technologies.[29] The demand for the different minerals and metals will depend largely on which of the technologies – wind, solar, battery – have won in the marketplace. On the other hand, the rise in demand may have negative repercussions on business models, processes, and social and environmental externalities. The energy transition will be accompanied by high demand for minerals such as steel, copper, and silver, and that may lead to a change in the natural landscape.[30] There are equally negative environmental impacts that will emanate from the mining of minerals used in battery technologies, namely nickel (found in South Africa, Botswana, and Zimbabwe), cobalt

28 Chuhan-Pole, P., Dabalen, A. L., & Land, B. C. (2017). *Mining in Africa: Are Local Communities Better Off?* Washington, DC: World Bank Group.

29 Arrobas, D. L. P., Hund, K. L., Mccormick, M. S., Ningthoujam, J., & Drexhage, J. R. (2017). *The Growing Role of Minerals and Metals for a Low Carbon Future* (English). Washington, DC: World Bank Group.

30 Watari, T., McLellan, B. C., Giurco, D., Dominish, E., Yamasue, E., & Nansai, K. (2019). Total material requirement for the global energy transition to 2050: A focus on transport and electricity. *Resources, Conservation and Recycling*, 148, 91–103.

(found in Zambia, Morocco, and the Democratic Republic of the Congo (DRC)), and lithium (found in Zimbabwe and the DRC) in the course of the energy transition.

Resource development should not be feared as a competitor to a more sustainable and greener future. Yet, for African countries and businesses, the negative impacts[31] of the mining industry on climate change emissions may increase if firms fail to work towards reducing their carbon emissions. A much more value-chain focus is needed to ensure that firms work towards net-zero emission across their value chain. Climate risks can have direct (performance- and operations-based) and indirect (rising energy costs and securing of supplies) impacts on the mining industry.

3.2 Why Climate Change Reporting?

The quest to understand the relevance of climate change reporting needs to start with understanding the benefits of non-financial reports. Corporate reporting covers all official documented communications from companies that aim to give a comprehensive picture to external stakeholders. In this study, reporting and disclosure are used interchangeably. In the past, much emphasis was placed on financial indicators at the expense of non-financial factors. Thus, accounting and finance regulations tend to emphasise financial figures in line with the conventional shareholder wealth maximisation principle.[32] Yet, in recent times, it has become evident that some non-financial factors are material for decision making. The shift from shareholder capitalism to stakeholder capitalism is also driving global institutions towards calling for responsible business practices. Thus, there has been a shift from the disclosure of core financial issues based on accounting standards (e.g. International Financial Reporting Standards (IFRS), United States Generally Accepted Accounting Principles (US GAAP), etc.) to a broader corporate reporting landscape. The business community is now discussing sustainability issues, as well as concerns of environmental, social, and governance (ESG) performance.[33] This is complemented by the change from the focus on a single bottom line to a focus on the triple bottom line – the financial, social, and environmental impact.

Reporting or disclosure of corporate actions has been a means by which firms enhance transparency and accountability. The investor community

31 Some general impacts include land-use change, contamination of local streams and wetlands, deforestation, erosion, contamination and alteration of soil profiles, increased habitat fragmentation, change in migratory routes of animals, and an increase in noise level, dust, and emissions.

32 Duho, K.C.T. (2020). 'Examining Anti-Corruption Disclosures, Profitability and Financial Stability Among Extractive Firms in Africa', MPhil Thesis, University of Ghana.

33 Barker, R., Eccles, R. G., & Serafeim, G. 'The Future of ESG Is … Accounting?' (Harvard Business Review, 3 December 2020), https://hbr.org/2020/12/the-future-of-esg-is-accounting (26 January 2021).

depends on corporate reporting to make decisions about their investments. The stakeholders that need to be informed about their impact on business activities continue to increase and the reporting items continue to increase. Reports provide an account of all the activities of the firms to serve as a point of reference in the future. Valuations of stocks can depend on the sort of information included in the reports as this can determine investor confidence. A negative report about the environmental impact of mining firms on the mining communities can affect the share prices and market capitalisation of firms. Investors are now diversifying their portfolios and making decisions that integrate climate change risk.[34] The investor community now demands a constant and consistent disclosure of climate impact. Thus, disclosure has become a tool for managers to highlight the state of affairs in their businesses.

Disclosure of climate change issues is one of the indicators of how commitment to climate change can be assessed. However, there are also concerns that some firms may merely report without practically reducing their emissions, a concept known as greenwashing. Greenwashing is the practice whereby a false impression is created by firms disclosing misleading information in reports about their environmental soundness. This continues to be an issue that needs to accompany the calls for climate change disclosure, without which the contribution of business to the 2015 UN Paris Agreement will be a failure. The Paris Agreement has led to a pledge by countries to control GHG emissions in the quest to limit global warming to 2 degrees Celsius from pre-industrial levels by the year 2100 while targeting to keep warning either at or below 1.5 degrees Celsius.[35]

3.3 Climate Change Reporting Frameworks

There are different standard-setting entities guiding sustainability reporting, which have climate change disclosures as components. Reporting has been within the foci of accounting, but non-financial reporting issues tend to be multidisciplinary. There are notable sustainability frameworks developed by institutions with climate change as a component. These include the Global Reporting Initiative (GRI), the International Integrated Reporting Council (IIRC), Climate Disclosure Standards Board (CDSB), Sustainability Accounting Standards Board (SASB), and the Task Force on Climate-related Financial Disclosures (TCFD). Other generic frameworks that may be useful are the Principles of Responsible Investment (PRI), the United Nations Global Compact (UNGC), and the International Organization for Standardization (ISO) standard, the ISO 14090:2019 on *Adaptation to Climate Change.* Although

34 Krueger, P., Sautner, Z., & Starks, L. T. (2020). The importance of climate risks for institutional investors. *The Review of Financial Studies*, 33(3), 1067–1111.

35 UNFCCC, 'Paris Agreement' (United Nations Framework Convention on Climate Change, 2015), https://unfccc.int/sites/default/files/english_paris_agreement.pdf.

all the frameworks are beneficial, the uncoordinated approach of the current reporting practice leaves much to explore. The lack of harmonisation results in an increase in the reporting burden of businesses, inconsistency of data, and the inability to properly track progress in corporate emissions and reduction efforts. There are multiple formats, templates, different indicators, and requirements which make analysis or comparison an arduous task for investors, researchers, and other stakeholders.

There are some interactive and collaborative works between the institutions leading in the climate change disclosure discourse. The GRI has its GRI 305: *Emissions* which provides guidelines on climate change disclosure by firms. In the case of the IIRC, there is evidence to suggest that climate change affects all the six capitals (i.e. intellectual, natural, financial, manufactured, human, and social and relational) included under the integrated reporting framework. The revised International Integrated Reporting Framework (January 2021) which promotes further decision usefulness replaces the previous version that was developed in 2013.[36] It is argued that integrated reporting should lead to material climate change disclosures.[37] Integrated reports are forward-thinking and can shape the competitive advantage of firms as much as climate change reporting does.

The works of the CDSB towards harmonisation of climate reporting in the world is notable. The integral purpose of the institution is to promote the inculcation of climate reporting into mainstream corporate reporting. The CDSB is a forum that allows for the collaboration of how existing practices and standards can be linked to financial and climate-related information. A framework that guides the process was developed in 2010, with the most recent amendment in 2018.[38] The SASB offers an industry-based reporting framework for a total of 77 industries.[39] The standard for the mining and metals industry highlights the energy-intensive nature of the industry and the fact that mining generates significant direct GHG and carbon dioxide.[40] This

36 IIRC, 'International <IR> Framework' (International Integrated Reporting Council, January 2021), https://integratedreporting.org/wp-content/uploads/2021/01/InternationalIntegratedReportingFramework.pdf (accessed 14 February 2021).

37 Gould, S. 'Integrated Reporting and Climate Change: A Perfect Marriage' (International Federation of Accountants, 27 May 2016), www.ifac.org/knowledge-gateway/preparing-future-ready-professionals/discussion/integrated-reporting-and-climate-change-perfect-marriage (accessed 2 July 2020).

38 CDSB, 'CDSB Framework for Reporting Environmental and Climate Change Information' (Climate Disclosure Standards Board, 2019), www.cdsb.net/sites/default/files/cdsb_framework_2019_v2.2.pdf (accessed 2 July 2020).

39 The generic sectors under which the industries fall include Consumer Goods, Extractives & Minerals Processing, Financials, Food & Beverage, Health Care, Infrastructure, Renewable Resources & Alternative Energy, Resource Transformation, Services, Technology & Communications and Transportation.

40 SASB, 'Metals and Mining: Sustainability Accounting Standard' (Sustainability Accounting Standards Board, 2018), www.sasb.org/wp-content/uploads/2018/11/Metals_Mining_Standard_2018.pdf.

is mainly in activities such as fuel usage in mining, processing of ore, and smelting.

The work of the TCFD aims to further consolidate efforts towards the adoption of climate disclosure. The framework of the Task Force has four pillars, namely governance, strategy, risk management, and metrics and targets. All four relate specifically to climate change, and the metrics and targets are used to assess or manage relevant climate-related risks and opportunities. The indicators can be integrated into any corporate report or treated as a standalone. The IFRS Foundation has now started engagements to emphasise the development of international sustainability standards. By so doing, the standard-setting body may have to lead in developing a harmonised standard to achieve the consistency and uniformity that are currently lacking. The IFRS Foundation started this profound action by issuing the *IFRS Consultation Paper on Sustainability Reporting* which will be considered over time, leading to a possible single standard on sustainability.[41] Until that is done, however, reporting on sustainability issues may continue to face the challenges it faces today.

3.4 Climate Change Regulations and Policies

The reporting frameworks on climate change are voluntary, although there is a trend towards a much more mandatory application in the foreseeable future. However, binding laws, regulations, and policies tend to be biting enough to request a much swifter response. A multi-stakeholder response to climate change is needed, and governments have to be committed to achieving their nationally determined contributions (NDCs) target. Data from the Grantham Research Institute on Climate Change shows that there are over 2,096 climate laws and policies globally, 425 from Africa and 384 from Sub-Saharan Africa.[42] Moreover, since 2015, a total of 652 laws and policies have been developed globally, 117 from Africa and 107 from Sub-Saharan Africa alone.[43] It is clear that no country is acting alone, and the African continent is not being left out in the quest to mitigate climate risk using laws and policies.

The general trend is that countries may choose to rely on a legislated act of parliament on climate change, depend on executive order, or depend on strategic policy documents. Globally, 1246 (59.45%) of the laws and policies depend on executive order while the rest depend on the legislature.[44] In

41 IFRS, 'Sustainability Reporting' (IFRS Foundation, 2021), www.ifrs.org/projects/work-plan/sustainability-reporting (accessed 14 February 2021).

42 Climate Change Laws of the World database (Grantham Research Institute on Climate Change and the Environment and Sabin Center for Climate Change Law), www.climate-laws.org/legislation_and_policies (accessed 15 February 2021).

43 Ibid.

44 Ibid.

Africa, executive accented acts and policies are 316 (74.35%) but the remainder is legislated by parliament, while in Sub-Saharan Africa, 292 (76.04%) of the laws and policies depend on executive accent while the remainder is legislated art of parliament.[45] Most of the laws and policies in Africa tend to be executive documents in the forms of strategic documents, policies, and decrees and not acts of parliament.

The high number of executive or policy-type documents could be attributed to the limited institutional capacity the systemic governance difficulties which delay the development and passage of laws in Africa. Climate change laws usually require multi-stakeholder coordination and participation, which tend to be low in the continent. It is worth noting that the legislature in less democratic countries tends to be 'minimal legislatures' or merely a 'rubber stamp', which only provide latent legitimacy.[46] The concentration of power with the executive in some African countries tends to favour executive action, which may explain the preference for executive documents on climate change.[47] The political party manifestos tend to be a relevant tool in developing national policies, but little in recent political manifestos has addressed critical climate change issues. These documents feed information into national policies and need to be closely considered.

Despite the executive bias to climate change regulation and policy, the critical area of focus should be on implementing the laws and executing the policies. Currently, climate change laws and policies are usually limited and fragmented. Going forward, a more consolidated approach needs to be pursued to enhance coordinated action on climate risk. At the continental level, the Agenda 2063 of Africa recognised climate change as a major barrier to development.[48]

3.5 Selected Country Case Studies – Ghana, Morocco, and South Africa

Case studies of the status of climate change, corporate disclosures, and the mining sectors of three African countries (Ghana, Morocco, and South Africa) are provided below. These countries are selected because of their significant mineral reserves and the leading roles they play in Africa. South Africa and Morocco are notable countries making strides regarding climate change in Africa.

45 Ibid.

46 Allmark, L. (2012). 'More than rubber-stamps: the consequences produced by legislatures in non-democratic states beyond latent legitimation.' *The Journal of Legislative Studies*, 18(2), 184–202.

47 Nachmany, M., Abeysinghe, A., & Barakat, S. (2017). 'Climate change legislation in the least developing countries.' *Trends in Climate Change Legislation*.

48 WMO, 'State of the Climate in Africa (MNO-No. 1253)' (World Meteorological Organization, 2019), https://library.wmo.int/doc_num.php?explnum_id=10421 (accessed 2 July 2020).

3.5.1 Ghana

Apart from cocoa and the recently found crude oil, Ghana is known for having a natural resource reserve in gold, diamonds, manganese, and bauxite.[49] Other unexplored deposits of minerals include copper, iron ore, chrome, nickel, limestone, quartz, and mica.[50] Ghana discovered commercial quantities of the critical mineral lithium in 2018, in some parts of the Volta Region.[51] The mining sector is the foremost foreign exchange earner in Ghana, contributing 39% to total foreign exchange in 2018 and 43% in 2019.[52] There are large-scale mining activities in Ghana, as well as Artisanal and Small-Scale Mining (ASM). The large local mining firms and the multinational entities[53] tend to have comprehensive corporate reporting practices driven by their parent companies, which are usually based outside Ghana. Although the law allows for 100% foreign ownership in the mining industries, there have been various policy reforms and a Local Content and Local Participation Bill 2013 that aim at driving local participation.[54]

Mining activities have been environmentally unfriendly and lead to noise pollution, pollution of water bodies with chemicals such as arsenic, mercury, and cadmium, and the contamination of agricultural soils.[55] Illegal mining (also known as galamsey) in Ghana has been identified as a national problem and has received executive and legislative attention recently.[56] There have been concerns that although there are laws to restrict foreign operatives in the galamsey trade, there are Chinese operators who operate with Ghanaians as their proxies. Ghana's environmental menace is a sad one as the destructive activities threaten the economic returns of the country in the cocoa sector.[57]

49 GCM (2020). '2019 Mining Industry Statistics and Data', The Ghana Chamber of Mines. http://ghanachamberofmines.org/wp-content/uploads/2020/07/2019-Mining-Industry-Statistics-and-Data-for-Ghana.pdf (accessed 5 January 2021).

50 Danquah, D.Y. (2020). 'Ghana: Mining Comparative Guide', Mondaq.

51 Woods, D.B.G. (2018). 'Ghana Hopes Lithium Load Will Lure Automakers', Bloomberg Law.

52 GCM (2020). '2019 Mining Industry Statistics and Data', The Ghana Chamber of Mines. http://ghanachamberofmines.org/wp-content/uploads/2020/07/2019-Mining-Industry-Statistics-and-Data-for-Ghana.pdf (accessed 5 January 2021).

53 One of the large mining companies in Ghana is Newmont Goldcorp (American); others are Canadian and Australian. South African companies such as Goldfields and AngloGold Ashanti also operate in Ghana.

54 ITA (2020). 'Mining Industry Equipment', International Trade Administration. www.trade.gov/knowledge-product/ghana-mining-industry-equipment (accessed 5 January 2021).

55 Emmanuel, A. Y., Jerry, C. S., & Dzigbodi, D. A. (2018). 'Review of environmental and health impacts of mining in Ghana.' *Journal of Health and Pollution*, 8(17), 43–52.

56 Hilson, G. (2017). 'Shootings and burning excavators: Some rapid reflections on the Government of Ghana's handling of the informal Galamsey mining "menace".' *Resources Policy*, 54, 109–116.

57 Duho, K.C.T., Amponsah, I. A., & Opoku-Mensah, D. A. (2019). 'Exploring the Revenue Management and Producer Pricing Mechanism Within Ghana's Cocoa Sector', IMANI Centre for Policy and Education.

A particular area of interest has climate change, and stakeholders have been charged with engaging in sustainable mining practices. At a macro-level, Ghana has shown commitment to combating climate change.[58] Despite the gains archiveved, climate change disclosure in the mining sector is still limited. In most cases, the reportage is done by large or multinational entities while the ASM entities tend to have limited corporate reporting. Reporting on climate change requires some level of preparedness and provision of financial and human resources. With the increasing concerns for climate information among stakeholders, Ghanaian mining firms, whether large or small, will have to show some level of commitment to disclosing their GHG emissions and their efforts to reduce their carbon emissions. The financial system in Ghana is not yet sufficiently robust to allow for carbon pricing, and there are not yet any regulations on carbon pricing. In recent times, there have been some fiscal policies that have some environmental notion, but not yet enough to address the issue of climate change – for instance, the luxury vehicle tax of 2018 (which was withdrawn in 2019) and the pollution tax of 2021.[59]

3.5.2 Morocco

Morocco has about 70% of the global phosphate reserve and is among the top producers globally.[60] Phosphate can be used to produce fertilisers, animal feed, detergents, metal treatments, human food products, cosmetics, and electronics, among others.[61] The other minerals found in Morocco are barite, clays, cobalt, copper, fluorspar, iron ore, lead, salt, silver, talc, and zinc. The mining sector contributes about 9% of gross domestic product (GDP) and forms about 30% of total exports.[62] Morocco was the third African nation (after Ethiopia and Gabon) and the first Arab country to submit a climate action commitment (or Intended Nationally Determined Contribution –

58 EPA (2020). 'Ghana National Climate Change Policy', Environmental Protection Agency www.epa.gov.gh/epa/sites/default/files/downloads/publications/National%20Climate%20Change%20Policy.pdf.

59 MOF (2018). 'Mid-Year Fiscal Policy Review of the 2018 Budget Statement and Economic Policy', Ministry of Finance, https://mofep.gov.gh/sites/default/files/budget-statements/2018-Mid-Year-Budget-Statement-and-Economic-Policy_0.pdf; MOF (2021). 'The Budget Statement and Economic Policy of the Government of Ghana for the 2021 Financial Year', Ministry of Finance, 2021, www.mofep.gov.gh/sites/default/files/budget-statements/2021-Budget-Statement_v3.pdf.

60 DLA Piper (2017). 'Mining in Morocco: a legal snapshot', DLA Piper Law Firm.

61 Schroder, J. J., Cordell, D., Smit, A. L., & Rosemarin, A. (2019). 'Sustainable Use of Phosphorus', EU Tender ENV.B.1/ETU/2009/0025 (No. 357), Plant Research International. https://ec.europa.eu/environment/natres/pdf/sustainable_use_phosphorus.pdf (accessed 2 July 2020).

62 Brour, F. (2020). 'The Mining Industry in Morocco: A Policy Paradox that Leaves Poor People Behind', The Bard CEP ECO Reader. www.bard.edu/cep/blog/?p=11995 (accessed 2 July 2019).

INDC). Notably, the country was ranked second in the Climate Change Performance Index in 2018 and has been a notable leader in terms of efforts to combat climate change.[63] The index considered a mega-solar panel[64] which the country has built and focused on four categories: GHGs, renewable energy, energy use, and climate policy.

Morocco has also enhanced its corporate disclosure and reporting regime by increasing disclosure requirements in annual accounts and reports. Corporate entities in the country adopt and utilise reporting frameworks such as the GRI, SDG, and IIRC. Generally, the national effort to combat climate change significantly impacts the commitment of individual businesses, in line with the argument of institutional theory. Corporate disclosures are driven by global forces, country policies and programmes, the corporate boards, management strategies, and operational activities.

3.5.3 South Africa

South Africa has been a leader in Africa when it comes to corporate reporting and disclosure practices. The country has a corporate governance framework that has integrated best practices and has been noted as one of the best among global peers. The country produces more than 55 different minerals[65] across over 1,113 mines.[66] South Africa was the largest exporter of gold for many years, until 2019 when it lost the position to Ghana.[67] Historically, the mining sector has contributed about 10% of the GDP of South Africa (although COVID-19 led to a reduction of about 4%).[68]

South African mining companies have the financial wherewithal and have been able to operate in other African countries, while some are listed on different stock exchanges both in Africa and outside the continent. For instance, Goldfields is primarily listed on the Johannesburg Stock Exchange (JSE),

63 UNEP (2018). 'Morocco ranked second in Climate Change Performance Index 2018', United Nation Environment Programme. www.unep.org/news-and-stories/blogpost/morocco-ranked-second-climate-change-performance-index-2018 (accessed 5 January 2021).

64 This is the largest solar panel in the world. which was built near Ouarzazate, a city known as the gateway to the Sahara Desert. The panel has been providing electricity to approximately 650,000 people since its launch in 2016.

65 This includes gold, diamonds, platinum, coal, chrome, vanadium, and titanium among others.

66 UN SDGs (2010). 'South African Country Report for the Eighteenth Session of the United Nations Commission on Sustainable Development (CSD-18)', United Nations Sustainable Development Goals https://sustainabledevelopment.un.org/content/documents/dsd/dsd_aofw_ni/ni_pdfs/NationalReports/south_africa/SouthAfricanCSD18CountryReport.pdf (accessed 30 March 2019).

67 Whitehouse, D. (2019). 'Ghana now Africa's largest gold producer but reforms await', *The Africa Report*.

68 Minerals Council (2020). 'Facts and Figures – Pocketbook 2020', Minerals Council – South Africa. www.mineralscouncil.org.za/industry-news/publications/facts-and-figures/send/17-facts-and-figures/1366-facts-and-figures-2020-pocketbook.

with secondary listings on the New York Stock Exchange (NYSE) and the Swiss Exchange (SWX). Similarly, AngloGold Ashanti is listed primarily on the JSE but secondarily listed on the New York, Johannesburg, Accra, London, and Australian stock exchanges, as well as on the Paris and Brussels bourses. Many South African companies have committed to the various sustainable business reporting frameworks such as the GRI, UNGC, and integrated reports. Also, some companies provide a sustainable development report (SDR) which reports on corporate activities in line with the Sustainable Development Goals (SDGs).

One area of notable progress made by the country is the enactment of the Carbon Tax Act (No. 15 of 2019). The law aims to 'provide for the imposition of a tax on the carbon dioxide (CO_2) equivalent of greenhouse gas emissions, and to provide for matters connected therewith'. The implementation of the law and the policies of the government to address the issue of climate change in the mining industry is still an issue that needs critical attention.

3.6 Prospects and Critical Challenges

Climate risk has become a critical issue that needs the attention of standard setters, regulators, governments, companies, investors, the accounting profession, and academic and civil society organisations. In the mining industry, it is relevant to note the efforts being made to promote environmental disclosure and climate change mitigation and adaptation. The mining industry has a relevant role to play in the entire transition process to a low-carbon economy. The demand for minerals and metals to build the low-carbon transition technologies will provide business prospects for the industry. Investors are requesting evidence of corporate resilience to climate risk and more prospects still exist regarding better climate action.

The challenges are daunting and need to be critically addressed. The first is regarding the status of climate change disclosure. There is the need for coordination of all the many standards, leading to a harmonised framework. Moreover, there is the need to move from mere reporting to action to prevent greenwashing. Issues concerning how Africa can raise the needed climate finance or carbon pricing system remains a bleak area.[69]

There are weaknesses in the climate laws, regulations, and policies in Africa which need to be considered. The problems with the implementation of the current laws and policies cannot be overemphasised. Attention must be also paid to the extent to which each of the countries is working on meeting their NDCs. The national contribution is key, and firms need to show their

69 Duho, K.C.T. & Bonou, S.C.C. (2021). 'Climate Finance and Carbon Pricing in the Context of Africa's Continental Free Trade Area.' Dataking Working Paper Series No. WP2021-02-01. https://osf.io/h3jdq.

contributions to the national agenda. Much more partnership is needed and a more multi-stakeholder action continues to be the option at our disposal.

4 Methodology

This section provides a description of the methodology that the study employs. The study used a quantitative method with some relevant explanations provided by the review of various studies and discussion of data from relevant databases.

4.1 Data

The study utilises panel data of 21 mining companies operating in Africa, specifically in Morocco and South Africa. The data ranges over the period 2006–2019, with a special focus on corporate reports such as the financial statements, annual reports, integrated reports, sustainability reports, sustainable development reports, environmental social and governance report, GRI content index, and data on the websites of the mining companies. Macro-level data is sourced from various open sources, such as Transparency International, the World Bank, and Human Development Report.

4.2 Climate Change Disclosure Index

This study uses GRI 305: *Emissions*, which provides seven clear indicators that relate to corporate efforts on compacting climate change. The standard addresses issues regarding emissions into the air, which includes the discharge of substances into the atmosphere. The emissions include GHG,[70] ozone-depleting substances (ODS), nitrogen oxides (NO_x), and sulphur oxides (SO_x), among others. GHG emissions contribute to climate change significantly and are governed by the United Nations (UN) Framework Convention on Climate Change of 1992 and the UN Kyoto Protocol of 1997 which entered into force in 2005.[71]

The standard also made use of the GHG Protocol[72] and has two key standards, the GHG Protocol Corporate Accounting and Reporting Standard ('GHG Protocol Corporate Standard') and the GHG Protocol Corporate Value Chain (Scope 3) Accounting and Reporting Standard ('GHG Protocol Corporate Value Chain Standard'). This is also related to ISO 14090:2019 – Adaptation to Climate Change. The three layers adapted from the ISO into GRI 305 are direct GHG emissions (Scope 1), energy indirect GHG

70 The GHGs that the standard covers are carbon dioxide (CO_2), methane (CH_4), nitrous oxide (N_2O), hydrofluorocarbons (HFCs), perfluorocarbons (PFCs), sulphur hexafluoride (SF_6), and nitrogen trifluoride (NF_3).

71 GRI (2016). 'GRI 305: Emissions', Global Reporting Initiative.

72 This is developed by the World Resources Institute (WRI) and the World Business Council on Sustainable Development (WBCSD).

emissions (Scope 2), and other indirect GHG emissions (Scope 3). Additionally, issues of ODS are regulated internationally by the UN Environment Programme (UNEP) Montreal Protocol on Substances that Deplete the Ozone Layer ('Montreal Protocol').

The study uses a climate change disclosure index (CCDI).[73] The model is presented mathematically as follows:

$$CCDI_{i,t} = \frac{\sum_{j=1}^{7} score_j}{Maxscore} \tag{11.1}$$

where *i* represents the mining firms, *t* represents the year, and *j* represents the individual item in the index. The *maxscore* represents the maximum score a firm can obtain for a year. The index for each firm, *i*, at the year, *t*, should be the ratio of the sum of the individual scores to the maximum score. The score ranges from 0 to 1, and the higher the index, the higher the extent of climate change disclosure. The description of the components of the climate change indicators is presented in Table 11.1. A binary codification is used so that disclosure is recorded as 1 and non-disclosure is recorded as 0.

4.3 Explanatory Variables

The explanatory variables included in this study are board size, board independence, firm size, membership of UNGC, human development, corruption perception, and economic growth. In line with previous studies, we use the

Table 11.1. Description of dependent variable and components in line with GRI 305

GRI indicator	*Description*
CCDI	Climate Change Disclosure Index (refer to equation above)
GRI 305-1	Direct (Scope 1) GHG emissions
GRI 305-2	Energy indirect (Scope 2) GHG emissions
GRI 305-3	Other indirect (Scope 3) GHG emissions
GRI 305-4	GHG emissions intensity
GRI 305-5	Reduction of GHG emissions
GRI 305-6	Emissions of ozone-depleting substances (ODS)
GRI 305-7	Nitrogen oxides (NO_x), sulphur oxides (SO_x), and other significant air emissions

Source: Author

73 Duho, K. C. T., Agyenim-Boateng, C., Asare, E. T., & Onumah, J. M. (2021). 'Convergence and determinants of anti-corruption disclosure among extractive firms in Africa.' *Journal of Financial Crime*, https://doi.org/10.1108/JFC-06-2020-0109; Comyns, B. (2016). 'Determinants of GHG reporting: An analysis of global oil and gas companies.' *Journal of Business Ethics*, 136(2), 349–369.

natural logarithm of the total number of board members as a proxy for board size and the ratio of independent directors to total board size as the proxy for board independence.[74] It is proposed that a larger board size will influence climate change disclosure as a large size will mean the board is more diverse, and independent directors may be added to the board which can provide leadership for the emerging reporting issues such as climate change. Also, the independence of the board will drive ethical and more sustainable strategic choices, which will increase climate change disclosure. Thus, we propose positive impacts on climate change disclosure.

The size of the firm is measured as the natural logarithm of total assets.[75] The proposition is that climate disclosures have cost implications, and so larger firms have a greater tendency to disclose them than their smaller counterparts. UNGC membership is regarded as a driver of sustainable corporate behaviours as the compact provides guidance on sustainable business practices for signatories. Thus, in line with earlier studies, this study proposes a positive effect of membership on climate change disclosures.[76] The human development of countries plays a significant role in the disclosure practices of firms and so the study proposes a positive effect on climate change disclosures.[77]

Corruption perception places a drag on the sustainable business practices of firms. Therefore, the study proposes that the corruption perception index will have a positive effect on disclosure and that low levels of corruption will increase climate change disclosure.[78] Results show that firms domiciled in developed economies tend to disclose more about climate change.[79] However, economic growth may have either a positive or negative effect on climate change disclosure, depending on national efforts to ensure that economic development is sustainable.

4.4 Econometrics Model and Explanatory Variables

In line with earlier studies, a regression model has been specified to explore the determinants of climate change disclosure among the mining firms using

74 Amran, A., Periasamy, V., & Zulkafli, A. H. (2014). 'Determinants of climate change disclosure by developed and emerging countries in Asia Pacific.' *Sustainable Development*, 22(3), 188–204; Giannarakis, G., Zafeiriou, E., Arabatzis, G., & Partalidou, X. (2018). 'Determinants of corporate climate change disclosure for European firms.' *Corporate Social Responsibility and Environmental Management*, 25(3), 281–294.

75 Duho, K. C. T., Agyenim-Boateng, C., Asare, E. T., & Onumah, J. M. (2021). 'Convergence and determinants of anti-corruption disclosure among extractive firms in Africa.' *Journal of Financial Crime*, https://doi.org/10.1108/JFC-06-2020-0109

76 Ibid.

77 Ibid.

78 Ibid.

79 Amran, A., Periasamy, V., & Zulkafli, A. H. (2014). 'Determinants of climate change disclosure by developed and emerging countries in Asia Pacific.' *Sustainable Development*, 22(3), 188–204.

panel-corrected standard error regression.[80] The model is mathematically specified as follows:

$$CCDI_{i,t} = \alpha_{i,t} + \beta_1 BSIZE_{i,t} + \beta_2 BIND_{i,t} + \beta_3 SIZE_{i,t} + \beta_4 UNGC_{i,t} + \beta_5 HDI_{i,t} + \beta_6 CPI_{i,t} + \beta_7 GDPG_{i,t} + \varepsilon_{i,t} \quad (11.2)$$

where *CCDI* represents climate change disclosure index, *BSIZE* represents board size, *BIND* represents board independence, *SIZE* represents the firm size, *UNGC* represents the membership of United Nations Global Compact, *HDI* represents the human development index, and *GDPG* represents gross domestic product growth. The data varies by time, *t*, and firm, *i*, while ε represents the error term. The description of the explanatory variables is presented in Table 11.2.

Table 11.2 Description of explanatory variables

Variables	*Description*	*Source*
BSIZE (*log*)	Board size, measured as the natural logarithm of the total number of directors	Corporate reports
BIND (%)	Board independence, measured as the number of independent board of directors to total board size	Corporate reports
SIZE (*log*)	Size, measured as the natural logarithm of total assets	Corporate reports
UNGC	United Nation Global Compact membership, coded as 1, if the firm is a signatory to the UNGC; otherwise, 0	Corporate reports
HDI	Human development measured by the Human Development Index	https://hdr.undp.org/en/content/humandevelopment-index-hdi
CPI	Corruption perception, measured by the Corruption Perception Index	www.transparency.org
GDPG	Economic growth measured by gross domestic product growth	https://data.worldbank.org/indicator/NY.GDP.MKTP.KD.ZG

Source: Author

80 Beck, N. & Katz, J. N. (1995). 'What to do (and not to do) with time-series cross-section data.' *American Political Science Review*, 89(3), 634–647; Duho, K. C. T., Agyenim-Boateng, C., Asare, E. T., & Onumah, J. M. (2021). 'Convergence and determinants of anti-corruption disclosure among extractive firms in Africa.' *Journal of Financial Crime* (forthcoming), https://doi.org/10.1108/JFC-06-2020-0109.

5 Results and Discussions

This section presents the results of the analysis on climate change disclosure and the effect that some corporate factors, macroeconomic factors, and global forces place on it.

5.1 Descriptive Statistics of Climate Change Disclosure Indicators

The results in Table 11.3 indicate that, overall, the extent of disclosure on climate change is 60.8% among the mining firms surveyed. This result varies depending on time and the disclosure of the seven indicators identified under GRI 305: *Emissions*. The finding shows that GRI 305-1 which is on direct (Scope 1) GHG emissions has a score of 76.1%. Similarly, GRI 305-2 which is on energy indirect (Scope 2) GHG emissions has a score of 75.1%. In general, these two indicators are more frequently reported than the five other indicators. The GRI 305-3 indicator is on other indirect (Scope 3) GHG emissions which has a score of 69.2%. The least reported item is the GRI 305-4 which is on GHG emissions intensity and has a score of 34.3%. The other indicators, GRI 305-5 on the reduction of GHG emissions, GRI 305-6 on emissions of ozone-depleting substances (ODS), and GRI 305-7 nitrogen oxides (NO_x), sulphur oxides (SO_x), and other significant air emissions, have scores of 58.7%, 52.7%, and 59.2% respectively.

5.2 Descriptive Statistics of Explanatory Variables

The descriptive statistics in Table 11.4 indicate that the average board size of the sampled mining firms is 10.59 with a range of 4 to 18 members. The independent board of directors is an average of 6, ranging from a minimum of 1 to a maximum of 12 members. In log terms, the board size is averagely 2.3 while board independence is 56.9%. The total assets of the firms are an average of US$275 billion, which in natural logarithm terms in 24.75. This

Table 11.3 Descriptive statistics of climate change disclosure

Variables	*Obs*	*Mean*	*Std. Dev.*	*Min*	*Max*
CCDI	201	0.608	0.375	0	1
GRI 305-1	201	0.761	0.427	0	1
GRI 305-2	201	0.751	0.433	0	1
GRI 305-3	201	0.692	0.463	0	1
GRI 305-4	201	0.343	0.476	0	1
GRI 305-5	201	0.587	0.494	0	1
GRI 305-6	201	0.527	0.500	0	1
GRI 305-7	201	0.592	0.493	0	1

Source: Author, based on computations

Table 11.4 Descriptive statistics of explanatory variables

Variables	*Obs*	*Mean*	*Std. Dev.*	*Min*	*Max*
BSIZE	201	10.592	2.699	4.000	18.000
BIND	201	6.065	2.446	1.000	12.000
BSIZE (*log*)	201	2.326	0.265	1.386	2.890
BIND (%)	201	0.569	0.183	0.125	0.889
SIZE (*log*)	201	24.754	2.454	16.845	28.053
TA (US$)	201	2.75e+11	3.57e+11	2.07e+07	1.53e+12
UNGC	201	0.488	0.501	0.000	1.000
HDI	201	0.681	0.025	0.603	0.709
CPI	201	43.512	2.793	33.000	51.000
GDPG	201	1.882	1.577	-1.538	5.923

Source: Author, based on computations

shows that the companies involved are high-investment ventures, and the importance that the investor community, governments, and the general public places on them are high. On average, 48.8% of the corporate reports are for firms that are signatories to the UNGC. The average human development index is 0.681, the average corruption perception index is 43.51, while the average GDP growth is 1.88%.

5.3 Correlation and Multicollinearity

The test of multicollinearity checks for the robustness of the regression model. A correlation coefficient of less than 0.7 shows that the problem of multicollinearity may not be an issue.[81] A variance inflation factor (VIF) which is less than 10 shows that the data does not suffer from multicollinearity.[82] The results in Table 11.5 show that the highest correlation coefficient is 0.682 and the highest VIF is 2.27, which are all below the rules of thumb. Thus, the study proceeds to include all the variables in the regression models.

5.4 Determinants of Climate Change Disclosure

The results of the regression analysis are presented in Table 11.6, showing that the explanatory variables explain more than 50% of the variability in the climate change disclosure index. The chi-square (χ^2) tests show statistically significant results at 1% which shows that the model is fit for purpose.

The findings show that board size has a positive and significant effect on the climate change disclosure index. This is evident in Model 1 which is the

81 Kennedy, P. (2008). *A Guide to Econometrics*. Oxford: Blackwell Publishing.
82 Wooldridge, J. M. (2016). *Introductory Econometrics: A Modern Approach*, Boston, MA: Cengage.

Table 11.5 Pairwise correlation and variance inflation factor

Variables	*VIF*	*(1)*	*(2)*	*(3)*	*(4)*	*(5)*	*(6)*	*(7)*
(1) BSIZE (log)	1.23	1.000						
(2) BIND (%)	1.23	0.062	1.000					
(3) SIZE (log)	1.32	0.333***		0.187***	1.000			
(4) UNGC	1.57	0.295***		0.331***	0.238***	1.000		
(5) HDI	2.27	0.149**	0.119*	−0.007	0.432***	1.000		
(6) CPI	1.29	0.073		−0.238***	−0.262***	0.071	-0.013	1.000
(7) GDPG	2.00	−0.134*	−0.075	0.024	-0.277***	-0.682***	-0.166**	1.000

Source: Author, based on computations

***, **, and * represent 1%, 5%, and 10% significant levels respectively.

basic model and also in the majority of the other models that are meant to explore some sensitivities. These results suggest that large board sizes among mining firms drive the climate change disclosure practices of the firms. This is similar to some studies[83] but contrasts with others.[84] Also, board independence is found to have a positive and statistically significant impact on the climate change disclosure index of mining firms. This suggests that the independence of the board drives the climate change disclosure among the mining firms.[85]

The other firm-specific determinants in the model are firm size and UNGC membership. The results show a positive and statistically significant nexus between firm size and climate change disclosure index among the firms. This is evident in Model 1 which is the basic model and in the majority of the interaction models. This shows that larger firms or multinational enterprises, in the case of the type of dataset used in this case, disclose more on the climate change status than their smaller counterparts. This could be aligned

83 Giannarakis, G., Zafeiriou, E., Arabatzis, G., & Partalidou, X. (2018). 'Determinants of corporate climate change disclosure for European firms.' *Corporate Social Responsibility and Environmental Management*, 25(3), 281–294.

84 'Amran, A., Periasamy, V., & Zulkafli, A. H. (2014). 'Determinants of climate change disclosure by developed and emerging countries in Asia Pacific.' *Sustainable Development*, 22(3), 188–204.

85 Amran, A., Periasamy, V., & Zulkafli, A. H. (2014). 'Determinants of climate change disclosure by developed and emerging countries in Asia Pacific.' *Sustainable Development*, 229(3), 188–204; Daily, C. M., Dalton, D. R., & Cannella Jr, A. A. (2003). 'Corporate governance: Decades of dialogue and data.' *Academy Of Management Review*, 28(3), 371–382.

with the cost implications of the requirements towards climate change reporting. Although not all firms may obtain independent reviews on their climate change indicators or not all indicators may be reviewed by independent bodies, the cost implications of this practice cannot be overemphasised. Earlier studies[86] found such a positive impact while others did not.[87]

UNGC membership is one of the critical issues that are considered in disclosure studies when it comes to the use of the GRI indicators.[88] The main point made is that the compact calls on C-suite business executives to be committed to sustainable business practices which need to reflect in issues such as corporate reporting. The findings of this study show that UNGC membership has a positive and significant impact on the climate change disclosure index. This is in concordance with earlier studies, placing significance on multi-stakeholder efforts in achieving the needed result of climate action.

The results demystify some of the macro-level drivers of climate change disclosures among the mining firms, namely human development, corruption perception, and economic growth. It shows that the human development index has a positive and significant effect on the climate change disclosure index. This result is revealing because it shows the relevance of human capital and its development in driving sustainable disclosure practices. The results show that countries with better health and education – yardsticks of better human capital development – have mining firms that disclose more on climate change. It is necessary to indicate that the human development of citizens will drive the kind, type, or depth of the financial and non-financial information they demand from corporate organisations.[89] Corruption continues to undermine the progress of African countries, especially the resource-dependent ones. The findings reveal that the corruption perception index has a positive and significant effect on the climate change disclosure index. This shows that countries with a better perception of corruption by stakeholders tend to report more on climate change.[90] It shows that the call for corporate action against climate change needs to go hand in hand with the call against corruption. The results surprisingly do not show

86 Prado-Lorenzo, J.-M., Rodríguez-Domínguez, L., Gallego-Álvarez, I. and García-Sánchez, I.-M. (2009). 'Factors influencing the disclosure of greenhouse gas emissions in companies world-wide.' *Management Decision*, 47(7), 1133–1157.

87 Comyns, B. (2016). 'Determinants of GHG reporting: An analysis of global oil and gas companies.' *Journal of Business Ethics*, 136(2), 349–369.

88 Barkemeyer, R., Preuss, L., & Lee, L. (2015). 'Corporate reporting on corruption: An international comparison.' *Accounting Forum*, 39(4), 349–365; Duho, K. C. T., Agyenim-Boateng, C., Asare, E. T., & Onumah, J. M. (2021). 'Convergence and determinants of anti-corruption disclosure among extractive firms in Africa.' *Journal of Financial Crime*, https://doi.org/10.1108/JFC-06-2020-0109.

89 Duho, K. C. T., Agyenim-Boateng, C., Asare, E. T., & Onumah, J. M. (2021). 'Convergence and determinants of anti-corruption disclosure among extractive firms in Africa.' *Journal of Financial Crime*, https://doi.org/10.1108/JFC-06-2020-0109.

90 Ibid.

Table 11.6 Determinants of Climate Change Disclosure Among Mining Firms

	(1)	*(2)*	*(3)*	*(4)*	*(5)*	*(6)*	*(7)*	*(8)*
Variables	*CCDI*	*CCDI*	*CCDI*	*CCDI*	*CCDI*	*CCDI*	*CCDI*	*CCDI*
BSIZE	**0.157****	**0.190*****	**0.714**	**0.162****	**0.156****	**0.155****	**0.157****	**0.157****
	(0.068)	(0.064)	(0.668)	(0.067)	(0.066)	(0.067)	(0.068)	(0.068)
BIND	**0.339*****	**0.457*****	**0.341*****	**2.819***	**0.241****	**0.337*****	**0.353*****	**0.340*****
	(0.108)	(0.101)	(0.107)	(1.445)	(0.107)	(0.109)	(0.109)	(0.108)
SIZE	**0.040*****	**0.657*****	**0.094**	**0.099*****	**0.079*****	**0.089**	**-0.017**	**0.041*****
	(0.009)	(0.145)	(0.065)	(0.037)	(0.011)	(0.227)	(0.162)	(0.013)
UNGC	**0.101****	**0.097****	**0.104****	**0.084***	**2.189*****	**0.102****	**0.100****	**0.100****
	(0.045)	(0.043)	(0.045)	(0.047)	(0.362)	(0.045)	(0.045)	(0.045)
HDI	**6.421*****	**6.367*****	**6.448*****	**6.265*****	**7.034*****	**8.218**	**6.295*****	**6.429*****
	(1.193)	(1.127)	(1.193)	(1.176)	(1.142)	(8.495)	(1.224)	(1.195)
CPI	**0.026*****	**0.026*****	**0.027*****	**0.023*****	**0.033*****	**0.027*****	**-0.006**	**0.026*****
	(0.007)	(0.007)	(0.007)	(0.007)	(0.007)	(0.008)	(0.093)	(0.007)
GDPG	**-0.013**	**-0.011**	**-0.013**	**-0.013**	**-0.008**	**-0.012**	**-0.013**	**0.004**
	(0.018)	(0.016)	(0.018)	(0.017)	(0.016)	(0.018)	(0.018)	(0.129)
SIZE x SIZE		**-0.013*****						
		(0.003)						
SIZE x BSIZE			**-0.023**					
			(0.027)					
SIZE x BIND				**-0.100***				
				(0.058)				

	(1)	*(2)*	*(3)*	*(4)*	*(5)*	*(6)*	*(7)*	*(8)*
Variables	*CCDI*	*CCDI*	*CCDI*	*CCDI*	*CCDI*	*CCDI*	*CCDI*	*CCDI*
SIZE x UNGC					**-0.084*****			
					(0.015)			
SIZE x HDI						**-0.072**		
						(0.332)		
SIZE x CPI							**0.001**	
							(0.004)	
SIZE x GDPG								**-0.001**
								(0.005)
_cons	**-6.464*****	**-13.793*****	**-7.812*****	**-7.697*****	**-8.095*****	**-7.726**	**-4.966**	**-6.483*****
	(1.013)	(1.974)	(1.929)	(1.306)	(1.036)	(5.956)	(4.404)	(1.028)
Obs.	201	201	201	201	201	201	201	201
Firms	21	21	21	21	21	21	21	21
R-squared	0.513	0.560	0.515	0.521	0.576	0.513	0.513	0.513
Chi square ($\boxtimes^2$)	229.22***	285.81***	231.94***	232.63***	280.17***	235.29***	229.79***	234.78***

Standard errors are in parenthesis, "***". "**", "*" represents 1%, 5%, and 10% significant levels respectively.
Source: Author, based on computations.

any significant nexus between economic growth and climate change disclosures, although there were generally negative effects.

As part of conducting some sensitivity analysis, the study interacted size with the other explanatory variables. The result in Model 2 shows that size squared has a negative effect on the climate change disclosure index. This shows that while size could drive disclosure, in the long run, it can inhibit climate change disclosure. Thus, in general, size has an inverted U-shape nexus with climate change disclosure. Besides, the interaction of size with both board independence and UNGC membership yields a negative effect on the climate change disclosure index. This suggests that size may not necessarily be all that important in explaining the disclosure on climate change. Corporate governance mechanisms or UNGC membership can fail to drive disclosure among even large firms. Thus, it is needful for all firms, irrespective of their size, to be committed to addressing the global canker which needs a joint action to solve it. The interaction of firm size with board size, human development index, corruption perception index and GDP growth did not yield any statistically significant results.

6 Conclusions

There are various concerns requiring attention to be given to the nine planetary boundaries that would affect the earth if crossed. Climate change is one of these environmental boundaries which, if not addressed, may lead to serious consequences for humankind. The COVID-19 pandemic led to a decline in greenhouse gas emissions as a result of the slowdown in economic activities. This suggests that emissions may still increase if action is not taken, after the pandemic. Businesses are among the stakeholders that need to address the climate change issues, and corporate reporting is one of the means to show that commitment. For mining firms, their contribution to climate change is enormous, and so business leaders need to consider sustainable business practices to seek to mitigate it. The noteworthy examples of South Africa and Morocco as African leaders in taking action to address climate change have been discussed. Notwithstanding, there are many environmental challenges that mining activities pose as depicted in those two countries and Ghana.

This study throws light on the dearth in the literature by exploring the determinants of climate change disclosure among mining firms in Africa. The study uses a dataset of 21 mining firms (from South Africa and Morocco) covering the period 2006 to 2019, as well as the panel-corrected standard error regression.

The study found that corporate governance variables, firm size, UNGC membership, human development, and corruption perception are determinants of climate change disclosure. First, the study shows that board size and board independence positively impact the climate change disclosures of mining firms in Africa. This places emphasis on the relevance of the corporate board which is charged with the responsibility to give direction and strategic leadership. Those firms with more board members from different backgrounds and with different interests tend to increase climate change disclosure. Additionally, the independence of the board is relevant for driving climate change disclosures. Non-executive directors bring vast knowledge and expertise, which provides independent guidance for businesses towards sustainable business practice transparently and objectively.

The analyses also show the significant contribution of firm size on climate change disclosures. The study found evidence to suggest that larger mining firms are more able to disclose their climate change efforts than their smaller counterparts. Large firms have the resources, competencies, and environment to promote disclosures on climate change. The study found an inverted U-shape relationship between firm size and climate change disclosure. This means that while we may emphasise size, size may not be an end in itself, but other issues need to be considered. This assertion is reinforced by the interaction of size with board size or UNGC membership, which resulted in a negative impact on climate change disclosure. Overall, the climate change disclosure gospel needs to be communicated to organisations of all size, whether micro, small, and medium-sized enterprises, large corporations, or multinational enterprises.

The results also give revealing insights about the role of UNGC membership, human development, and corruption perception in driving climate change disclosure. There is a positive significant impact of UNGC membership on climate change disclosure. This shows that firms that are signatories of the UNGC disclose more on climate change. Since 30% of the ten principles of the UNGC are dedicated to the environment, it is not surprising to find the compact as driving climate change disclosure. The study also shows the relevance of human development in driving climate change disclosure. Human development is a function of proper investment in the health and education of citizens, which have an impact on the sophistication of the business climate. Thus, countries with better human development tend to have better disclosures on climate change. Additionally, corruption perception is a drag on the climate change disclosure practices of firms such that countries with a low perception of corruption disclose more on climate change. The results show that economic growth does not drive climate change disclosures.

7 Implications for Policy and Practice

The results of this study are relevant for practice, policy, and future research. Practice-wise, the results show generally that more than 50% of the surveyed reports disclosed climate change efforts. This is commendable and an indication that practitioners in mining firms in Africa are making some efforts to combat climate change.[91] Sadly, this study focuses on firms in the GRI database which are skewed toward reporting in line with sustainability. Thus, corporate boards and C-suite executives, in general, should still make more

91 It is worth noting that the two countries under study, South Africa and Morocco, are leaders in climate change among their African peers.

effort to increase their disclosures on climate change. Moreover, it is not enough to disclose without actually reducing net emissions, so equal attention should be paid to the reduction of net emissions. The board of directors should be at the forefront of climate change mitigation and adaptation efforts in the mining firms since that is an approach that will emphasise the issue. Notwithstanding, a multi-stakeholder approach is needed and the involvement of employees in the lower echelons of leadership is still relevant. Following corporate governance best practices enhance climate change disclosures. Also, the fight against climate change needs to be accompanied by the fight against corruption. This shows that companies should work towards implementing the ten principles of the UNGC which touch on key issues of sustainable business.

Policy-wise, the findings reveal the needed action of intergovernmental organisations, multilateral development organisations, and regional and national policymakers. The study shows that more climate action is needed by national policymakers in enhancing human development and corruption issues as they have a toll on disclosure. It is still worrying that countries like South Africa and Morocco are not signatories to the EITI which provides some guidance on the mining and petroleum industry. In the context of growing relevance on multi-stakeholder efforts, we recommend that national leaders join hands with global forces like the EITI in the fight for sustainable business practice, especially in the mining industry. Organisations like the United Nations Framework Convention on Climate Change (UNFCC), the Africa Union, the CDSB, the SASB, and the GRI have relevant insights to obtain from the study. South Africa has made strides in enacting their carbon pricing law, and Morocco has built a mega-solar panel, which are exemplary for other African countries. The results are also relevant for civil society organisations and policy think tanks in the public and private sector.

The study provides some insightful directions that future researchers could include in their research agenda. In the mining industry, future studies could apply the institutional logic theory to explain the qualitative issues regarding climate change reporting practices. Other studies can also explore the convergence of climate change disclosure among mining firms and even firms in other industries. Exploratory studies should be conducted on the general state of climate finance and carbon pricing among African countries. Researchers should tie this with concerns about carbon taxation, the inclusion of climate change in risk and investment models, and the actual reduction achieved by firms towards achieving net-zero emissions by 2050. Issues of corporate governance are diverse and should be included on the menu of future researchers.

12 A Critical Analysis of Deep-Sea Mining

An African Enviro-Legal Perspective

Leezola Zongwe[1] *and*
Obvious Terrence Mutambara[2]

1 Introduction

Virtually every sector of the economy the world over depends on the extractives sector. The extractives sector is the heartbeat of, *inter alia*, manufacturing, construction, and travel. There is no doubt that modern life is sustained by minerals and hydrocarbons. However, as the world population grows, the quest for minerals has also grown exponentially. The exhaustion of ore bodies on the surface and the increasingly poor grades of the ore have pushed the extractive industry to look for an alternative source of minerals.[3] These alternative sources are needed to avert the possibility of a reduction or complete exhaustion of the supply of minerals.[4] Moreover, the rapid growth of the manufacturing industry is giving rise to the demand for critical metals such as lithium.[5] Rare earth metals

1 Leezola Rea-Zaan Zongwe is a Property Law Lecturer at Stadio University; Doctoral Researcher and Editor, and a PhD candidate at the NRF/DST SARChI Research Chair: Mineral Law in Africa; Department of Private Law, University of Cape Town; and a Board Member: Mentorship Lead (Southern African Academic Forum) at Afronomicslaw. She holds an LL.M (Mineral and Petroleum Extraction and Use), University of Cape Town; LL.B Honours, University of Namibia; Cert. (Assessment Development in Higher Education), Stadio University; Cert. (Mineral Economics); Cert. (Sustainable Economics), University of British Columbia; Cert. (Risk Management), Deloitte; Cert. (Central Banking), Bank of Namibia; Cert. (Political Economy and Development), Leiden University; and a Certificate (International Criminal Law), Case Western Reserve University. Her passions intersect at development economics and the law of the extractive industries.

2 Obvious Terrence Mutambara is an Associate at Zuze Law Chambers. He is also a researcher at Africa China Law Desk (ACLA). He holds an LL.B (UFH) and an LL.M in Mineral and Petroleum Extraction and Use (University of Cape Town). His area of interest is oil and gas, mining, security regulations, medical negligence, and criminal law.

3 Rahul Sharma, 'Environmental Issues of Deep-Sea Mining' (2015) 11 *Procedia Earth and Planetary Science* 204–211.

4 International Copper Study Group, 'The World Copper Factbook' (2017). https://icsg.org/publications-list, accessed on 31 August 2021.

5 Sven Petersen et al., 'News from the Seabed – Geological Characteristics and Resource Potential of Deep-Sea Mineral Resources' (2016) 70 *Marine Policy*.

DOI: 10.4324/9781003284437-15

and minerals are indispensable for manufacturing smart and advanced green technologies. The largest alternative reserve for minerals is found in the deep sea.[6] The deep sea forms about 80% of the earth's oceans, which form approximately 71% of planet earth.[7]

For the last decade, deep-sea mining has been considered an experimental concept.[8] However, this is fast changing as evidenced by the active increase in interest in deep-sea mining. The International Union for Conversation of Nature predicts that the 'deep-sea mining era' will commence in 2025.[9] The pace has picked up due to the fast progression towards the energy transition. For example, there has been a sharp increase in automotive companies opting for smart technologies.[10] The automotive industry is migrating from fossil-fuelled vehicles to smart electric vehicles which are powered by batteries made up of minerals such as cobalt and lithium.[11] These minerals are found in abundance in the deep sea.

This chapter employs the definition posited by Cuyvers et al. for deep-sea mining.[12] Namely, deep-sea mining refers to the practice of employing specialized techniques and technology to excavate valuable metals and minerals from the deep sea. The metals and minerals found in the deep sea are often referred to as rare earth metals and minerals. Rare earth metals and minerals occur in three ways: (1) on the seafloor where sulphides are found; (2) in hydrothermal vents where cobalt and ferromanganese are found; and (3) on abysmal plains of the seafloor where polymetallic noodles are found.[13] Pertinently, these clusters have varying strategic minerals including, *inter alia*, nickel and cobalt. It has been discovered that these deep-sea deposits contain more minerals and are of a higher grade than surface-based sources.[14] For

6 Norman Toro et al., 'Seabed Mineral Resources, an Alternative for the Future of Renewable Energy: A Critical Review' (2020) 126 *Ore Geology Reviews*.
7 See National Geographic, 'Ocean' (2021) available at www.nationalgeographic.org/encyclopedia/ocean, accessed on 31 August 2021.
8 Marie Nvarre and Heloise Lammens, 'Opportunities of Deep-Sea Mining and ESG Risks' *Amundi Discussion Papers Series* (2017).
9 The International Union for Conversation of Nature, 'Deep-Sea Mining' (July 2018).
10 Elsa A Olivetti et al., 'Lithium-ion Battery Supply Chain Considerations: Analysis of Potential Bottlenecks in Critical Metals' (2017) 1 *Joule* 229.
11 Hanna Vikström et al., 'Lithium Availability and Future Production Outlooks' *Applied Energy* (2013) 110, 252.
12 Luc Cuyvers, Whitney Berry, Kristina Gjerde, Torsten Thiele, and Caroline Wilhem, 'Deep Seabed Mining: A Rising Environmental Challenge' *International Union for Conservation of Nature* (2018).
13 Adriana Dutkiewicz, Alexander Judge, and R. Dietmar Müller, 'Environmental Predictors of Deep-Sea Polymetallic Nodule Occurrence in The Global Ocean' *Geology* (2020) 3.
14 James R. Heina, Kira Mizell, Andrea Koschinsky et al., 'Deep-Ocean Mineral Deposits as a Source of Critical Metals for High-end Green-Technology Applications: Comparison with Land-Based Resources' *Ore Geology Reviews* (2013) 51, 3.

instance, polymetallic noodles discovered in the Clarion-Clipperton Zone, an abysmal plane stretching from Hawaii to Mexico, contain ten times more manganese than those found on land.[15] These deposits illustrate the magnitude of deep-sea rare earth metals and minerals and exemplify the economic value that deep-sea deposits possess.

The consequence of this interest combined with the evidence of unique economic value is that mining companies are starting to explore and extract minerals from the deep sea. Mining activities in the deep sea include rock cutting, sediment suction, module collection, and transportation of minerals from the ocean floor.[16] However, most of these minerals are found beyond the Exclusive Economic Zones (EEZ) of countries.[17] As such, they fall under the auspices of the United Nations Convention on the Law of the Sea (UNCLOS) through the formulation of the International Seabed Authority (ISA).

This chapter explores the why, who, what, where, and how of deep-sea mining. It discusses deep-sea mining as a rich alternative source of exploitable minerals. It also addresses concerns about the 'colonization' of the deep sea by mining companies and attempts to answer the pertinent legal and environmental issues surrounding deep-sea mining.

2 Why Deep-Sea Mineral Resources?

The deep sea covers approximately 70% of the earth.[18] Its geomorphologic, geological, and ecosystems are largely unknown to the surface world.[19] Although several studies have been undertaken to explore the deep sea, conflicting accounts have been published. A 2013 study found that the deep sea hosts rich mineral deposits similar to those found on the earth's surface.[20] Conversely, in 2018 research found that the deep sea contains minerals that

15 Daniel Ackerman, 'Deep-Sea Mining: How to Balance Need for Metals with Ecological Impacts' *Scientific American* (31 August 2020). www.scientificamerican.com/article/deep-sea-mining-how-to-balance-need-for-metals-with-ecological-impacts1, accessed on 31 August 2021.

16 The International Bank for Reconstruction and Development/World Bank 'Precautionary Management of Deep-Sea Minerals' (2017).

17 Michael Lodge, 'The International Seabed Authority and Deep Seabed Mining' *United Nations* (2016). www.un.org/en/chronicle/article/international-seabed-authority-and-deep-seabed-mining, accessed on 31 August 2021.

18 The Maribus, 'World Ocean Review: Living with The Oceans.' https://worldoceanreview.com/wp-content/downloads/wor1/WOR1_en.pdf, accessed on 31 August 2021.

19 Damian Carrington, 'Is Deep Sea Mining Vital for A Greener Future – Even If It Destroys Ecosystems?' (4 June 2017) *The Guardian*. www.theguardian.com/environment/2017/jun/04/is-deep-sea-mining-vital-for-greener-future-even-if-it-means-destroying-precious-ecosystems, accessed on 31 August 2021.

20 James R. Heina, Kira Mizell, Andrea Koschinsky et al., 'Deep-Ocean Mineral Deposits as A Source of Critical Metals for High-end Green-Technology Applications: Comparison with Land-Based Resources' *Ore Geology Reviews* (2013) 51.

are unique to the deep sea.[21] This new 'rush' has been exacerbated by the fact that the deep sea has been found to contain rich deposits of minerals such as polymetallic noodles, polymetallic sulphides, cobalt, and magnesium, among others.[22]

The ocean floors contain valuable minerals and metals that are critical for electronics and smart technologies such as clean energy generation, telecommunications, and transportation.[23] Therefore, as the human population grows and surface minerals are outstripped, mining companies are looking to the deep sea to cater to the demand for rare earth minerals and metals that the green revolution has created.[24] The United Kingdom's National Oceanography Centre has highlighted that to foster a low-carbon future, the world needs to tap into the deep sea for metals and minerals that support the production of green energy.[25]

Some deposits of the minerals that are found in the deep sea, such as rare earth minerals, are in short supply on the earth's surface.[26] Consequently, with the surface's short supply of critical minerals, most countries are unable to meet their industrial needs. The world's leading rare earth minerals producer is China, and the country has placed export quotas on its rare earth minerals to ensure that it meets its domestic needs.[27] Furthermore, the scarcity of these resources has led to geopolitical tensions that have resulted in several disputes being lodged at the Dispute Settlement Body of the World Trade Organization.[28] This illustrates the skewed nature of the demand and supply of rare earth metals. The demand significantly outweighs the supply. Thus, seeing that the surface supply of these metals and minerals is unable to satisfy the global industrial need, the authors concur with the United Kingdom's National Oceanography Centre that mining the deep sea will provide an alternative rich supply source of these vital resources.

21 Luise Heinrich, Klaus Boehnke, Christopher Cohrs et al., 'Deep-Sea Mining: Interdisciplinary Research on Potential Environmental, Legal, Economic, and Societal Implications' *Integrated Environmental Assessment and Management* (2018) 999.

22 Rahul Sharma, *Deep-Sea Mining: Current Status and Future Considerations.* Springer International Publishing (2017), 3.

23 The Institute for Sustainable Futures, 'Responsible Minerals Sourcing for Renewable Energy' (2019).

24 The Secretariat of the Pacific Community (SPC) 'Deep Sea Minerals and the Green Economy' (2013).

25 Peter D. Chaniotis, Laura M. Robson, Anaëlle J. Lemasson et al., 'UK Deep-Sea Conservation: Progress, Lessons Learned, and Actions for the Future' *Aquatic Conservation: Marine and Freshwater Ecosystems* (2019), 377.

26 Georgios Charalampides, Konstantinos I. Vitalis, Baklavaridis Apostoplos et al., 'Rare Earth Elements: Industrial Applications and Economic Dependency of Europe' *Procedia Economics and Finance* (2015) 24, 24.

27 Yuzhou Shen, Ruthann Moomy and Roderick G. Egger, 'China's Public Policies Toward Rare Earths, 1975–2018' *Mineral Economics* (2020) 33, 1.

28 Nabeel A. Mancheri, 'World Trade in Rare Earths, Chinese Export Restrictions, and Implications' *Resources Policy* (2015) 46, 267.

Africa has not been left from the new frontier of mining. For illustration purposes, seabed – albeit not in international waters – diamond mining activities in Namibia have been ongoing since 1961.[29] South Africa and Namibia have made significant headway in marine diamond mining off their coasts.[30] Namibia has also recorded discoveries of deposits of phosphate offshore.[31] Additionally, in 2018 Nigeria drafted policies geared toward deep-sea mining activities.[32] Mauritius and Madagascar have recorded a significant increase in exploration activities in their seabed.[33] As early as 2000, Murton had compiled a list of the various mineral available in a few countries on Africa's continental shelf,[34] and by 2012 the African Union had already adopted the 2050 Africa's Integrated Maritime Strategy (2050 Aim Strategy).[35]

3 Historical Development of Deep-Sea Mining

Deep-sea metals and minerals were first discovered during the expedition of the *HMS Challenger* between 1872 and 1876. Thompson described the dredge hauls of polymetallic nodules as peculiar 'black oval bodies'.[36] Similarly, the International Geophysical Year (IGY) has released reports describing the strange life forms and vast mineralogical resources contained in the deep sea.[37] However, it was only in 1976, after samples of these resources were analyzed, that Buchannan predicted that the polymetallic noodles found in the deep sea will stimulate economic activities.

The IGY was established to provide the world with a deeper understanding of the planet's geophysical characteristics.[38] The IGY was established to

29 Good Governance Africa (2018). https://gga.org/deep-sea-mining-africas-new-frontier, accessed on 31 August 2021.

30 C. Savage et al., 'Comparative Meta-Analysis of The Impact of Offshore Marine Mining on Macrobenthic Communities Versus Organic Pollution Studies' *Marine Ecology Progress Series* (2001) 221, 221.

31 Rosanna Carver, 'Resource sovereignty and accumulation in the blue economy: The case of seabed mining in Namibia' *Journal of Political Ecology* (2019) 26, 3.

32 Good Governance Africa (2018) available at https://gga.org/deep-sea-mining-africas-new-frontier

33 Alex Benkenstein, 'Seabed Mining: Lessons from The Namibian Experience' *SAIIA Policy Briefing* (2014) 87.

34 Nicholas N. Kimani, 'Enacting National Seabed Mining Laws in Africa: Importance of a Practitioner's Perspective' *Environment and Development Law Journal* (2015) 11/2, 93.

35 Africa's Integrated Maritime Strategy 2050 Version 1.0 (2012).

36 Rahul Sharma, 'First Nodule to First Mine-Site: Development of Deep-Sea Mineral Resources from The Indian Ocean' *Current Science* (2010) 99, 750–759.

37 Roger D. Launius, James Rodger Fleming, and David H. DeVorkin, *Globalizing Polar Science Reconsidering the International Polar and Geophysical Years* (Palgrave Macmillan, 2010).

38 Holi Riebeek, 'Paleoclimatology: A Record from the Deep' (27 September 2005) *Nasa Earth Observatory*. https://earthobservatory.nasa.gov/features/PaleoclimatologySedimentCores, accessed on 31 August 2021.

better understand the deep sea and the oceanic floor. The IGY's activities included scientists dredging the deep sea and photographing the ocean floor.[39] After the IGY's report was released, leading extractive companies, such as Rio Tinto and Consolidated Goldfields, developed a huge interest in marine minerals. This created the so-called 'rush to the deep sea'; however, despite the rush, the fall of metal prices in the 1970s halted the interest in mining the deep sea.[40]

During the ferromanganese deposits conference in 1972, the issue of mineralogy and geochemistry data of the ocean floor was heavily emphasized. This resulted in numerous data collection exercises. These extensive exploration programmes led to several organizations and companies laying claim to large areas of the seafloor in anticipation of the great mineral haul that the data indicated are present in the deep sea.

With all these ongoing activities, environmentalists became more and more concerned about the rush to the deep sea. The deep sea is vast and largely unknown; humanity's lack of knowledge of the deep sea thus presents a huge challenge for mining safety and sustainability. Even with the data that has been collected so far, none of the studies performed could estimate within a margin of error what the possible negative outcomes would be on the ocean ecosystem if the deep sea is mined.[41] This concern led to the establishment of the UNCLOS which culminated in the creation of the ISA in Jamaica in 1994. ISA regulates activities beyond countries' EEZ and national jurisdictions.[42]

The Preparatory Commission of the UNCLOS has been involved in the negotiation of the terms and obligations of pioneer pnvestors since 1988. The pioneer investors (PIs) are, *inter alia*, Russia, Japan, China, France, and Korea. The PIs are primarily interested in polymetallic modules. However, a small discovery found in nationally controlled deep waters in 2017 reignited interest in mining in the deep sea.[43] A 2018 report shows that the number of countries and organizations interested in the deep sea has grown exponentially from eight in 2010 to 25 in 2015.[44] These new players have diverse interests which include sulphides, crusts, and noodles.[45]

39 Ibid.
40 Paul A.J. Lusty and Bramley J. Murton, 'Deep-Ocean Mineral Deposits: Metal Resources and Windows into Earth Processes' *Elements* (2008) 14(5) 301, 30.
41 Jennifer M. Durdena, Laura E. Lallierc, Kevin Murphye et al., 'Environmental Impact Assessment Process for Deep-Sea Mining in "The Area"' *Marine Policy* (2018) 87, 194–202.
42 The International Telecommunications Union 'Opportunities and Legal Challenges' (2012).
43 Luc Cuyvers, Whitney Berry, Kristina Gjerde, Torsten Thiele, and Caroline Wilhem, 'Deep Seabed Mining: A Rising Environmental Challenge' *International Union for Conservation of Nature* (2018).
44 EarthWorks 'The Seabed Mining' (2018).
45 Ibid.

However, mining in international waters will only come to fruition in a few years.[46] By the end of 2020, the ISA had issued 21 exploration contracts covering an area of about 1 million km^2 of deep-sea area.[47] Data for 2021 show that in less than a year, ten more exploration contracts have been added, bringing the current total of exploration contracts to 31, issued to 22 different contractors, and covering a total surface area of more than 1.3 million km^2.[48] The estimated valuation of deep-sea metal and mineral reserves is from $8 trillion to more than $16 trillion.[49]

4 Environmental Impact of Deep-Sea Mining

It is no secret that extractive industries' activities are a significant contributor to environmental degradation.[50] Deep-sea mining involves the excavation and loosening of the seafloor.[51] As highlighted above, the deep-sea ecosystem possess rich natural resources such as polymetallic noodles. These resources are found on the international seabed at least 1,000 km from the land.[52] Rare earth metals and minerals are found at depths of about 1.5 to 6 km below the seafloor and they also occur in the extreme conditions of the deep sea.[53]

The turn of the century saw an increased interest in the preservation of the environment.[54] Extensive studies with near-accurate answers concerning the

46 The International Telecommunications Union, 'Opportunities and Legal Challenges' (2012).

47 International SeaBed Authority, 'Exploration Contracts' available at https://isa.org.jm/exploration-contracts, accessed on 21 November 2020.

48 See The International Seabed Authority's official website at https://isa.org.jm/exploration-contracts, accessed on 30 August 2021. Also, see ZeroHedge 'Deep Sea Mining Is An $8 Trillion Opportunity' (16 August 2021) *OilPrice*. https://oilprice.com/Metals/Commodities/Deep-Sea-Mining-Is-A-8-Trillion-Opportunity.html accessed on 30 August 2021.

49 ZeroHedge, ibid.

50 Daniel A. Bagah, William Angko and John Paul Tanyeh, 'Environmental Degradation and Small-Scale Mining Nexus: Emerging Trends and Challenges in Northern Ghana' *Developing Country Studies* (2016) 2, 36.

51 Olive Heffernan, 'Seabed Mining Is Coming – Bringing Mineral Riches and Fears of Epic Extinctions' *Nature*, 24 July 2019. www.nature.comOliveHefferna n/articles/d41586-019-02242-y, accessed on 30 August 2021.

52 Elizabeth M. De Santo, Elizabeth Mendenhall, and Elizabeth Nyman, 'A rush is on to mine the deep seabed, with effects on ocean life that aren't well understood' *The Conversation* (17 August 2020). https://theconversation.com/a-rush-is-on-to-mine-the-deep-seabed-with-effects-on-ocean-life-that-arent-well-understood-139833, accessed on 30 August 2021.

53 Thomas Kuhn, Alan J. Jamieson, Pedro Martinez Arbizu et al., 'Manganese Nodule Fields from the Northeast Pacific as Benthic Habitats.' In Harris, P. and Baker, E. (eds) Atlas of Seafloor Geomorphology as Benthic Habitat, 2nd edition. Elsevier

54 Peter Ehlers, 'Blue Growth and Ocean Governance – How to Balance the Use and The Protection of The Seas' *World Maritime University Journal* (2016) 15, 194.

impact that deep-sea mining will have on the environment are vital before any activity is sanctioned in the area. Embarking on mining in the deep sea without these answers can be detrimental to the ocean ecosystem because, contrary to popular belief, the sea cannot assimilate all the pollution without incurring harm.[55]

The world has turned its attention to the preservation and sustainable use of water, air, and land.[56] The presence of rich mineral resources in the deep sea and the award of contracts for exploration and mining of the deep sea by the ISA have steered sustainability efforts toward the preservation of the deep-sea environment.[57] Environmental degradation may lead to habitat loss and seabed pollution. Moreover, plans for mining the deep sea include ordinary mining activities such as establishing tailings dams which have a significant negative effect on the environment.[58] This process of deep-sea tailings disposal is an ongoing concern; land mines have in recent years also moved to deep-sea disposal of tailings because of lack of land availability. Although this chapter does not discuss this issue, the authors thought it prudent to highlight this area that would benefit from future research.

Furthermore, activities such as suction lifts that remove sediments on the seafloor of the ocean have also caused great concern because of the inability of the equipment used to differentiate between the minerals being mined and other fauna and flora which may be vital to the deep-sea ecosystem.[59] The ocean has several clusters on which different species survive; there are species of marine life that can only survive in the deep sea.[60] Moreover, the discharge from plumes and sediments during the lifting and recovery stage affects the euphotic zone in the seabed region, and the sound disruption by the equipment can also disorient the sea life and cause considerable damage to feeding and mating habits.[61] Thus far, studies show that an amalgamation of sediments and benthic water can potentially increase the phytoplankton and

55 The National Academy of Sciences, *The Pollution Is the Release of Undesirable Substances* (2007).

56 Intergovernmental Oceanographic Commission of UNESCO, 'International (UN) Decade of Ocean Science for Sustainable Development: Towards the Ocean We Need for The Future We Want' (13 November 2017).

57 Rahul Sharma and Samantha Smith, 'Deep-sea Mining and the Environment – An Introduction' inSharma, R. (ed.) Environmental Issues of Deep-Sea Mining. Springer International Publishing.

58 Deep-Ocean Stewardship Initiative, www.dosi-project.org, accessed on 30 August 2021.

59 Gerd Schriever and Hjalmar Thiel, 'Tailings and their Disposal in Deep-Sea Mining' *ISOPE Ocean Mining and Gas Hydrates Symposium* (2013) 13.

60 J.J. Childress, 'Are There Physiological and Biochemical Adaptations of Metabolism in Deep-Sea Animals?' *Trends in Ecology and Evolution* (1995) 10, 30–36.

61 Jeffrey C Drazen, Craig R Smith, and Kristina Gjerde, 'Report of the Workshop Evaluating the Nature of Midwater Mining Plumes and Their Potential Effects on Midwater Ecosystems' *Rio Journals* (2019).

organisms found in pelagic zones, and these zones are found in areas with high a concentration of polymetallic nodules.[62]

Additionally, the turbidity caused by sediments alters the marine ecosystem.[63] Several studies now show the devasting effect of deep-sea mining on large sea mammals. Large marine life – such as sharks, whales, and seals, among others – is particularly sensitive to anthropogenic activities.[64] Anthropogenic noise creates a new form of pollution. The noise and activities of exploration and mining affect deep-sea dwellers' feeding, breeding, and migration.[65] Some marine life uses sonar to identify food and prey, and reports by the Earth Island Institute show that the noise caused by exploration and mining activities has a devasting effect on sonar and eco-location communication.[66] Studies on whales have found that sound pollution by deep-sea mining equipment can lead to food sources being driven away, the destruction of habitats, and the complete ceasing of all mating activities, leading to dangerously low population numbers.[67] This poses an endangerment of marine life that can lead to extinction. Other sea dwellers rely on dark energy; hence, the light created by mining equipment adversely affects these creatures.[68] The stress on marine life caused by the light and the noise created by mining operations can lead to the death and extinction of several marine populations.[69] The consequences of such devastation for surface life can only be predicted to be severe and possibly cataclysmic.

62 Richard A. Frank, 'Environmental Aspects of Deepsea Mining' *Virginia Journal of International Law* (1975) 15:4 815, 818.

63 Michael E. Kjelland, Christa M. Woodley, Todd M. Swannack et al., 'A review of the potential effects of suspended sediment on fishes: Potential dredging-related physiological, behavioural, and transgenerational implications.' Environment Systems and Decisions (2015) 35

64 Chao Peng, Xinguo Zhao, and Guangxu Liu, 'Noise in the Sea and Its Impacts on Marine Organisms' *International Journal Environmental Research and Public Health* (2015), 2. European Environment Agency, 'Anthropogenic,' www.eea.europa.eu/archived/archived-content-water-topic/wise-help-centre/glossary-definitions/anthropogenic-processes, accessed on 7 December 2020. Pollution is the kind of pollution that emanates from human activities. Anthropogenic is mostly used in the context of environmental infractions from humans. See Richard B. Aronson, Sven Thatje, and James B. McClintock, *Anthropogenic Impacts on Marine Ecosystems in Antarctica Annals of the New York Academy of Sciences* (2011) 1223, 1.

65 Bob B.M. Wonga and Ulrika Candolinb, 'Behavioural responses to changing environments' *Behavioral Ecology* (2015) 26, 667.

66 See the Earth Island Institute, 'Fish, mollusks and other sea animals use of sound, and the impact of anthropogenic noise in the marine acoustic environment' (2002).

67 Russell Lande, 'Anthropogenic, Ecological and Genetic Factors in Extinction and Conservation' *Researchers on Population Ecology* (1998) 40, 1.

68 W. Bach, K.J. Edwards and J.M. Hayes et al., 'Energy in the Dark: Fuel for Life in the Deep Ocean and Beyond' *Eos* (2006) 87, 3.

69 Peng (n 64).

5 United Nations and the Position of International Law

With growing interest in deep-sea mining, legal issues such as ownership and regulatory authority have emerged. The earliest concern and desire to regulate was raised by Arvid Pardo, the Maltese ambassador to the United Nations in 1967. He proposed that deep-sea mining should be regulated by being designated as a common heritage for all mankind and that the exploitation of these resources should benefit all mankind.[70] He advocated for the peaceful exploitation of the deep sea and considered that this may lead to a shared prosperous future for all world citizens. At the core of Pardo's discussion was the need for the internationalization of the ocean and seabed floor beyond the limit of national jurisdiction.[71] Following Pardo's submissions, the United Nations set an ad hoc review committee to look into deep-sea mining.[72] After three years, the United Nations pronounced international waters and the deep sea a common heritage for all mankind. This position exacerbated the rush for the exploitation of the deep sea. And, as discussed in Section 4 above, deep-sea mining is poised to damage the deep-sea environment and the ocean's ecosystem irreparably and have unintended consequences for land life.

Protection of the world's oceans has its genesis in 1926 when the international community assembled in Washington, DC to discuss international regulations dealing with oil spills.[73] The international effort to protect the oceans was concluded in 1954.[74] Ever since the first global effort to preserve the marine environment, the world has become more aware of the dangers that adversely affect the marine environment.[75] The quest to protect the marine environment has led to the exponential growth of international environmental law.[76] The 1972 Convention on the Prevention of Marine Pollution by Dumping of Wastes and Other Matter (CPMDWO) is the first international convention to protect the marine environment.[77] It provided for the effective protection of all forms of marine life by regulating waste dumping.[78] The convention set the scene for the Stockholm Action plan which

70 Dr. Arvid Pardo 'Father of Law of Sea' Conference United Nations (1999).

71 J. Henry Glazer, 'The Maltese Initiatives Within the United Nations – A Blue Planet Blueprint for Transnational Space Authority' *Ecology Law Quarterly* (1974) 4, 285.

72 Helmut Tuerk, 'The Common Heritage of Mankind After 50 Years' *Indian Journal of International Law* (2017) 57, 259–283.

73 Djamchid Momtaz, 'The United Nations and the Protection of The Environment: from Stockholm to Rio de Janeiro' *Political Geography* (1996) 15, 265.

74 Tuerk (n 72).

75 Momtaz (n 73).

76 Tuerk (n 72).

77 The International Marine Organization, 'Convention on the Prevention of Marine Pollution by Dumping of Wastes and Other Matter.' www.imo.org/en/OurWork/Environment/Pages/London-Convention-Protocol.aspx, accessed on 11 December 2020.

78 Ibid.

emphasized the importance of the protection of the marine sphere.[79] Its recommendation devoted great protection to the marine environment.[80]

Pardo argued for the ecological unity of the ocean.[81] He based his arguments on the interpretations of the 1958 Geneva Continental Shelf Convention which pointed out the need for a new United Nations body to administer the area.[82] This new body was tasked with the distribution of the proceeds emanating from the exploitation of minerals from the hydrospace to developing countries.[83]

The crux of the idea was primarily to substitute the freedom of the sea principle with a principle of a common heritage for all mankind.[84] With the rise of national territories and sovereignty, a long debate ensued within the clusters of international law scholars and policymakers regarding the classification of the seas and outer space. Two schools of thought emanated. On the one hand, the argument arose that the seas did not belong to anyone but can be appropriated; thus it should be classified as *res nullius*.[85] On the other hand, it was argued that the high seas belonged to all nations because the high seas cannot be appropriated but can be used for the common good of all mankind.[86] However, as international law and the international bodies that enforce it developed, some parts of the earth were declared as spheres outside the bounds of all nations.[87] The high seas formed part of this. As such, they were classified as *res communes*.[88] This classification exacerbated the rush for deep-sea resources.

The rush is described as alarming and threatening to the ecological sustainability of the deep sea.[89] This was worrying because very little is known

79 Momtaz (n 73).

80 The United Nations Environment Programme, 'Action Plan for The Protection and Sustainable Development of The Marine and Coastal Areas of The East Asian Region' (1994).

81 L.F.E. Goldie, 'A Note on Some Diverse Meanings of "The Common Heritage of Mankind"' *Syracuse Journal of International Law and Commerce* (1983) 10, 69.

82 Harry M. Collins, 'Mineral Exploitation of the Seabed: Problems, Progress, and Alternatives' *Natural Resources Lawyer* (1979) 12, 617.

83 E. Wenk, *The Politics of the Ocean* (1972), 260.

84 Alexandre Kiss, 'The Common Heritage of Mankind: Utopia or Reality?' *International Journal* (1985) 40.

85 John Macdonell, 'Occupation and Res Nullius' *Journal of the Society of Comparative* Legislation (1899) 1, 284.

86 H. Wallace Roberts, 'International Law – The High Seas, the Continental Shelf, and Free Navigation' *North Carolina Law Review* (1957) 15, 525.

87 Isaak I. Dore, 'International Law and the Preservation of the Ocean Space and Outer Space as Zones of Peace: Progress and Problems' *Cornell International Law Journal* (1982) 15, 4.

88 Dinah L. Shelton and Alex Kiss, 'Guide to International Environmental Law' in Alexandre Kiss and Dinah Shelton, *Guide to International Environmental Law* (Martinus Nijhoff, 2007) 15.

89 Jochen Halfar and Rodney M. Fujita, 'The Danger of Deep-Sea Mining' *Science* (2007) 316.

about the deep sea. However, the sparse research available suggests that the deep-sea ecosystem is very fragile and that it takes significant time to recover from any substantial harm suffered.[90] Cognisant of the destructive nature of the rush for the deep sea, the United Nations – knowing that these activities could not go unregulated – designed a system in which all interested parties have an inalienable right to participate in the exploitation of these resources.[91] Consequently, the United Nations General Assembly proposed a framework that promotes the sharing of benefits as well as the sustainable exploitation of the deep sea.[92]

The concept of deep-sea minerals as *res communes* was finally entrenched in the UNCLOS in 1982. The principle is discussed at length in the preamble. Article 137 provides that 'no state can claim or impose sovereign rights over any part of the area'. The prohibition concerning the occupation and appropriation articulated in Article 137 has been recognized as *jus cogens.*[93] Further, Article 140 emphasizes the nature of the deep-sea resources and provides that activities in the deep seas should be for the benefit of mankind regardless of the geographical location of the state, whether coastal or landlocked.

In addition, Part XII of the UNCLOS is fully committed to the protection of the deep-sea ecosystem. The UNCLOS lays down the framework that creates an obligation for states to protect the marine environment. In addition, the United Nations Environment Programme (UNEP) addresses, *inter alia*, the environmental degradation of the seas and the oceans. The United Nations has become the leading proponent of the protection of the deep sea. This quest has been made apparent by the Rio Declaration on Environment and Development wherein world leaders are committed to protecting the environment. The Rio Conference envisages reconciling environmental protection and economic demand for exploitation of the earth's resources. In this regard, deep-sea mining activities should be carried out in a manner that complies fully with international environmental obligations. However, with the depletion of surface minerals and the surge in demand for minerals to support the green revolution, it is going to be difficult to strike a healthy balance. Consequently, deep-sea mining has become inevitable, and its regulation is now more pressing than ever.

However, developed and developing states have contrasting ideas regarding the development and exploitation of the deep sea. Developed nations favoured a licensing approach as advocated for by the Maltese Ocean Space

90 F. Althaus, A. Williams, T.A. Schlacher et al., 'Impacts of Bottom Trawling on Deep-Coral Ecosystems of Seamounts Are Long-Lasting' *Marine Ecology Progress Series* (2009) 397, 280.

91 Tuerk (n 72).

92 A Pardo, 'Law of the Sea Conference – What went wrong?' in R L Friedheim (ed.), *Managing Ocean Resources: A Primer* (Boulder, CO: West View Press, 1979) 139–140.

93 R Wolfrum, 'Common Heritage of Mankind' in R Bernhardt (ed) *Encyclopaedia of Public International Law* vol. I, 693.

Treaty (MOST). Conversely, developing countries favoured the position that the ISA should create an enterprise that will solely carry out deep-sea mining operations.[94] A compromise, known as the 'parallel system', was eventually reached. Under the parallel system, the mine applicant has to divide the mine site into two portions of equal value and has to designate the party reserved for the activities through the enterprise.[95]

Africa has accepted the potential of the blue economy with the adoption of the 2019 Africa Blue Economy Strategy. African leaders are aware of the wealth that deep-sea mining can generate and the economic value that would have for the continent.[96] However, investment in the blue economy is still lagging. The 2018 Sustainable Blue Economy Conference made ten commitments to the sustainable use of the marine environment. One of the commitments was to spur fisheries growth and reduction of oceanic pollution. The commitment to reduce pollution, however, clashed with the reality that deep-sea mining will increase pollution and destroy the fishing industry. The African Union (AU) and the United Nations Economic Commission for Africa (UNECA) are now involved in disseminating information about deep-sea mining activities.[97] Deep-sea mining is still a subject of contentious ongoing consultations and although the AU has adopted the 2050 Aim Strategy and its Blue Economy Strategy, no formal continental exploration and exploitation regulations have yet been issued.

6 The International Seabed Authority

6.1 Establishment

The preamble of the UN General Assembly Resolution 2749 (XXV) provided the bedrock for the UNCLOS. It called for the creation of international machinery that controls the exploitation of mineral resources found in international waters. UNCLOS created the ISA which is charged with regulating and overseeing deep seabed mining. ISA is tasked with the formulation of a mining code that regulates the life cycle of hydrospace mining. ISA is also responsible for the regulation of the area and granting of mining contracts. The ISA is a fully autonomous international organization tasked to organize and control activities in areas with resources within international waters. The instrument creates norms to be observed by all parties.

94 J.M. Van Dyke, 'Sharing Ocean Resources in a Time of Scarcity and Selfishness' in H.N. Scheiber (ed.), *Law of the Sea, the Common Heritage and Emerging Challenges* (Martinus Nijhoff, 2000) 5

95 Tuerk (n 72).

96 Jennifer J. Silver et al., 'Blue Economy and Competing Discourses in International Oceans Governance' *The Journal of Environment and Development* (2015) 24, 141.

97 Institute for Security Studies. 'Harnessing Africa's Blue Economy' (2019).

The UNCLOS does not specify conditions for membership of the ISA. The devotion of a state to the convention is enough to make it a member of the authority. The purposive reading of Article 157(2) presumes that the authority has dual powers: main powers (those conferred by UNCLOS) and supplementary powers that are mostly unwritten which are necessary for the authority to operate efficiently in tandem with powers conferred to it by its establishing authority.[98]

6.2 Distributing Resource Revenues by ISA

The primary conundrum that confronts the ISA is the determination of the value of the resources mined.[99] The ISA is tasked with establishing a mechanism for the equitable sharing of profits and marine resources. However, the ISA has not yet settled on a definition for 'equitable sharing'. Future research into what constitutes equitable sharing would greatly benefit this area of study. Consequently, the ISA is currently grappling with the formulation of a payment model for the financial returns from deep-sea resources between states and contractors.[100]

Another hurdle that the ISA faces is the formulation of a distribution model for revenue collected by the ISA from players involved in mining in the transnational space.[101] The UNCLOS obligates the ISA to develop a framework for the equitable sharing of the revenue generated by mining companies for the benefit of mankind.[102] This framework is yet to be determined and its design is still being debated. The main contention revolves around the principles to be employed to determine the validity and value of the claims that countries have on the resources generated.[103] Another obstacle for the ISA concerns the type of mechanism to be employed for the distribution of the funds, such as whether the distribution should be a direct payment to a project that countries will benefit from or to individual countries for them to use the funds as they wish.[104] The heated ongoing debates around these obstacles make it nearly impossible to predict when an outcome will be reached, but what one can predict is that an outcome that all states are happy with is not likely to be reached anytime soon.

98 F. Armas-Pfirter, 'The Management of Seabed Living Resources in the Area Under UNCLOS' *Revista Electronica de Estudios Internacionales* (2006) 11, 12.
99 Ibid.
100 Michael W. Lodge, Kathleen Segerson, and Dale Squires, 'Sharing and Preserving the Resources in the Deep Sea: Challenges for the International Seabed Authority' *The International Journal of Marine and Coastal Law* (2017), 32.
101 J. Otto, C. Andrews, and J. Tilton et al., *Mining Royalties: A Global Study of Their Impact on Investors, Governments, and Civil Society* (Washington, DC: World Bank, 2006).
102 Joanna Dingwall, *Commercial Mining Activities in the Deep Seabed beyond National Jurisdiction: The International Legal Framework* (2020), 141.
103 Lodge (n 100) 432.
104 Ibid.

7 Ocean Floor Mining in Africa

Africa has made some significant movements in submarine mining. Almost ten years ago, in 2012, the AU developed Africa's 2050 Integrated Maritime Strategy, and in 2015, the AU adopted Agenda 2063. These documents both address the need to develop and benefit from Africa's oceanic resources. However, no real movement was made to realize these goals until 2017, at the United Nations Ocean Conference when the ISA, the African Union, and the Government of Norway (NORAD) registered a voluntary commitment to support Africa's blue economy.[105] The project aims to organize five sub-regional workshops across the African continent; thus far, three workshops have been held to develop capacity on the continent regarding deep-sea and seabed mining.[106] The objective of the project is for the continent to sustainably mine and develop deep-sea and seabed resources found within the African continental shelf and in international waters.[107] Sustainable development and mining of the deep sea and seabed are envisaged to be vital to Africa's blue economy.[108] This commitment was formalized in 2019 in the Africa Blue Economy Strategy.

Seabed mining in Africa is not a new phenomenon. South Africa and Namibia have been engaged in mining diamonds off the ocean floor since the early 1900s.[109] Moreover, these two countries are both currently exploring the possibility of safe and sustainable offshore phosphate mining. However, these mining and exploring activities have not been met without harsh criticism. Environmental activists in Namibia have raised concerns over the possible impact phosphate mining in Namibian waters will have on, *inter alia*, the rich fishing industry of the country. Offshore phosphate mining in Namibia is the world's first.[110] Concerns around pollution are the hallmark of the sharp criticism of sea mining.

The fishing industry in Namibia is rightfully concerned that phosphate mining activities will destroy the industry.[111] As discussed above, pollution of the sea harms fish breeding and threatens the marine environment and the

105 The International Seabed Authority, 'Third Workshop for The Promotion of the Sustainable Development of Africa's Deep-Seabed Resources in Support of Africa's Blue Economy Strategy Concludes Successfully' (4 June 2021). www.isa.org.jm/news/third-workshop-promotion-sustainable-development-africas-deep-seabed-resources-support-africas, accessed on 30 August 2021.

106 Ibid.

107 Ibid.

108 Ibid.

109 Kevin Richardson, 'A Perspective of Marine Mining Within De Beers' *Journal of the Southern African Institute of Mining and Metallurgy* (2007) 107, 1.

110 Detlef P. Van Vuuren et al., 'Phosphorus Demand for the 1970–2100 Period: A Scenario Analysis of Resource Depletion' *Global Environmental Change* (2010) 20, 429.

111 Food and Agricultural Organisation of the United Nations, 'The State of World Fisheries and Aquaculture' (2020). www.fao.org/documents/card/en/c/ca9229en, accessed on 30 August 2021.

fishing industry.[112] Moreover, the significant impact that the ocean has on sustaining human life goes well beyond providing sustenance and fuelling economic activity. Scientists estimate that approximately 50% of the earth's oxygen comes from the ocean.[113] As such, playing fast and loose by mining the oceans – in the absence of concrete scientific evidence that such mining will not be harmful to the marine ecosystem – is simply something that the human race cannot risk.

As far back as 1985, Chapman and Shannon presented evidence that the southwestern coast of Namibia (where ocean floor mining for diamonds happens) experiences a seasonal low concentration of oxygen annually. Thus, deep-sea mining could lead to a reduction in oxygen levels in the ocean.[114] The reduction in oxygen along the shelf affects the microbenthic life in oceans which threatens the marine environment. The phosphate mining project is the subject of a legal battle in Namibia. The matter has been reserved for judgment. The opposing parties unanimously agree that the mining activity will irreparably harm the marine ecosystem.[115]

Considering the above, deep-sea mining in Africa is still in its infancy, and many countries are still waiting for ISA to pass regulations on deep-sea exploitation. Thus far, the ISA has only passed regulations pertaining to exploration activities. Shabahat advocates for the creation of a deep-sea mining code, stressing the need for the ISA to finalize the mining code for activities to be regulated to avoid catastrophic events such as the Deepwater Horizon oil spill.[116]

Apart from the regulations that Nigeria was working on with the ISA in 2018, there are no records of any other African country passing legislation to exclusively regulate deep-sea mining. Currently, these mining licences are awarded within the traditional (land) mining frameworks. Several risks can be identified in this approach, incuding the non-applicability of the existing regulatory frameworks to the deep-sea mining conditions.

8 Conclusion

The ongoing negotiations between parties interested in mining the deep sea and the ISA are vital to settling the pertinent enviro-legal issues identified

112 Lisa A. Levin et al., 'Defining "Serious Harm" to the Marine Environment in the Context of Deep-Seabed Mining' *Marine Policy* (2016) 74, 255.

113 National Ocean Service Science Report (2021). https://oceanservice.noaa.gov/about/nos-science-report/2021, accessed on 30 August 2021.

114 Savage (n 30) 273.

115 Felix Zvomunya, 'A Legal Fight in Namibia That Could Reshape the Future of Our Seas' *The Oxygen Project* (16 October 2020). www.theoxygenproject.com/post/a-legal-fight-in-namibia-that-could-reshape-the-future-of-our-seas, accessed on 30 August 2021.

116 Elham Shabahat, 'A Mining Code for the Deep Sea' *Hakai Magazine* (25 June 2021). www.hakaimagazine.com/news/a-mining-code-for-the-deep-sea, accessed on 30 August 2021.

throughout this chapter. However, there are certain issues – such as more extensive testing in the deep-sea environment of the effects of mining and the ways to limit the harm inflicted or to rehabilitate the area mined – that only dedicated, generously funded, and extensive scientific, economic, and legal research will address.

The deep sea has potential resources that could support civilization sustainably. However, it is a source that needs to be mined with greater care and foresight than that employed in land mining.[117] Certainly, tapping into the deep sea as a source of minerals would serve the world's move toward the energy transition.[118] However, to combat the risks identified, miners need to present detailed and scientifically backed risk management strategies that reflect near-accurate depictions and projections of their activities on the deep-sea ecosystem, as well as the consequential damage to the greater oceanic atmosphere and life on land.

Furthermore, African countries need to take intentional steps toward regulating ocean floor mining with their EEZs. A first step, in lieu of the ISA completing the Deep-Sea Mining Code's exploitation regulations, would be developing a continental framework (most likely drawing on the international conventions) for safe and sustainable mining of the ocean floor. Countries can then develop national frameworks or nationalize the existing international agreements. This chapter has shown that whether these steps are taken or not, deep-sea mining will happen – the questions that remain are: will countries (and mining companies) do the work to ensure that deep-sea mining is done safely and sustainably; and will African countries and companies be in a position to benefit from and play an active role in ensuring the safe and sustainable mining of the deep sea? The answers to both these questions depend largely on the progression of the ISA's negotiations.

References

Africa's IntegratedMaritime Strategy 2050 Version 1.0 (2012).

Benkenstein, A. 'Seabed Mining: Lessons from the Namibian Experience.' Policy Briefing 87, Governance of Africa's Resources Programme (2014).

Bibliography and Index to Literature on Manganese Key to Geophysical Records Documentation No. 6, U.S. Department of Commerce National Oceanic and

117 In this regard, environmental activist groups, such as Conservation International and the World Wildlife Fund for Nature, have proposed a moratorium on the adoption of the ISA's Mining Code on deep-sea exploitation and on deep-sea mining as a whole for a ten-year period, citing the need for more time to understand and study the risks associated with deep-sea mining. See Elham Shabahat, 'A Mining Code for the Deep Sea' *Hakai Magazine* (25 June 2021). www.hakaimagazine.com/news/a-mining-code-for-the-deep-sea, accessed on 30 August 2021.

118 The OECD Roundtable on Corporate Responsibility, 'Transition to A Low-Carbon Economy: Public Goals and Corporate Practices' (10 July 2010).

Atmospheric Administration Environmental Data Service National Geophysical and Solar-Terrestrial Data Centre 'Nodules:1874–1975' Boulder (April 1976).
Corpolongo, A. 'Deep-Sea Mining Impacts on Diverse Ocean Ecosystems.' *Sciworthy*, 1 September 2020. https://sciworthy.com/deep-sea-mining-threatens-the-foundation-of-diverse-ocean-ecosystems
Dutkiewicz, A., Judge, A., and Dietmar Müller, R. 'Environmental Predictors of Deep-Sea Polymetallic Nodule Occurrence in the Global Ocean.' *Geology* (2020) 48(3): 293–297.
Kiss, A. 'The Common Heritage of Mankind: Utopia or Reality?' *International Journal* (1985) 40(3): 423–441.
Pardo, A. 'Law of the Sea Conference – What went wrong?' In Friedheim, R.L. (ed.) *Managing Ocean Resources: A Primer* (Boulder, CO: West View Press, 1979) 137–148.
Belyakov , A. *The Role of the International Seabed Authority in the Regime of the Protection of the Living Resources Within Areas Beyond National Jurisdiction*, LLM Thesis, University of Norway (2013).
Wonga, B.B.M. and Candolin, U. 'Behavioural responses to changing environments.' *Behavioral Ecology* (2015) 26(3): 665–673.
Savage, C., Field, J.G., and Warwick, R.M. 'Comparative meta-analysis of the impact of offshore marine mining on macrobenthic communities versus organic pollution studies.' *Marine Ecology Progress Series* (2001) 221: 265–275.
Peng, C., Zhao, X., and Liu, G. 'Noise in the Sea and Its Impacts on Marine Organisms.' *International Environmental Research and Public Health* (2015) 12*(*1): 12304–12323.
Committee of U.S. Citizens in Nicaragua v Reagan, 859 F.2d 929, 940 (D.C. Cir. 1988)
Carrington, D. 'Is deep sea mining vital for a greener future – even if it destroys ecosystems?' *The Guardian*, 4 June 2017. www.theguardian.com/environment/2017/jun/04/is-deep-sea-mining-vital-for-greener-future-even-if-it-means-destroying-precious-ecosystems
Ackerman, D. 'Deep-Sea Mining: How to Balance Need for Metals with Ecological Impacts.' *Scientific American*, 31 August 2020. www.scientificamerican.com/article/deep-sea-mining-how-to-balance-need-for-metals-with-ecological-impacts1
Bagah, D.A., Angko, W. and Tanyeh, J.P. 'Environmental Degradation and Small-Scale Mining Nexus: Emerging Trends and Challenges in Northern Ghana.' *Developing Country Studies* (2016) 6(2).
Van Vuuren, D.P., Bouwman, A.F., and Beusen, A.H.W. 'Phosphorus demand for the 1970–2100 period: A scenario analysis of resource depletion.' *Global Environmental Change* (2010) 20(3): 428–439.
Development of the United Nations Convention on the Law of the Sea and the Legal and Institutional Framework for Ocean Affairs in Belize: Sustainable Marine Fisheries, Marine Aquaculture, Seafood Processing, Marine and Coastal TourismUnited Nations Conference on Trade and Development (2019).
Dr. Arvid Pardo 'Father of Law of Sea' ConferenceUnited Nations (1999).
Shelton, D.L. and Kiss, A. 'Guide to International Environmental Law.' In Kiss, A. and Shelton, D. (eds) *Guide to International Environmental Law* (Martinus Nijhoff, 2007).
Division for Ocean Affairs and the Law of the Sea: Office of Legal Affairs, *The United Nations Convention on the Law of the Sea: A Historical Perspective* (2012).
Momtaz, D. 'The United Nations and the protection of the environment: From Stockholm to Rio de Janeiro.' *Political Geography* (1996) 15(3–4): 261–271.

Wenk, E. *The Politics of the Ocean* (University of Washington Press, 1972).

Stocker, M. 'Fish, mollusks and other sea animals use of sound, and the impact of anthropogenic noise in the marine acoustic environment.' *Journal of the Acoustical Society of America* (2002) 112(2431).

EarthWorks. (2018). 'Seabed Mining.'https://earthworks.org/issues/seabed-mining

Olivetti, E.A., Ceder, G., Gaustad, G.G., Fu, X. 'Lithium-ion Battery Supply Chain Considerations: Analysis of Potential Bottlenecks in Critical Metals.' *Joule* (2017) 1 (2): 229–243.

Litchman, E. 'Resource Competition and the Ecological Success of Phytoplankton.' *Science Direct* (2007).

Shabahat, E. 'A Mining Code for the Deep Sea.' *Hakai Magazine*, 25 June 2011. www.hakaimagazine.com/news/a-mining-code-for-the-deep-sea

Baroncini, E. 'The China-Rare Earths WTO Dispute: A Precious Chance to Revise the China-Raw Materials Conclusions on the Applicability of GATT Article XX to China's WTO Accession Protocol.' *Cuadernos de Derecho Transnational* (2012) 4.

De Santo, E.M., Mendenhall, E., and Nyman, E. 'A rush is on to mine the deep seabed, with effects on ocean life that aren't well understood.' *The Conversation*, 17 August 2020. https://theconversation.com/a-rush-is-on-to-mine-the-deep-seabed-with-effects-on-ocean-life-that-arent-well-understood-139833

Franckx, E. 'The 200 Mile Limit: Between Creeping Jurisdiction and Creeping Common Heritage?' *George Washington International Law Review* (2007) 39.

Franckx, E. 'The International Seabed Authority and the Common Heritage of Mankind: The Need for States to Establish the Outer Limits of their Continental Shelf.' *International Journal of Marine and Coastal Law* (2010) 25.

European Environment Agency. 'Anthropogenic.' www.eea.europa.eu/archived/archived-content-water-topic/wise-help-centre/glossary-definitions/anthropogenic-processes

Explanatory Notes for the Geologic Map of the Circum-Pacific Region, Pacific Basin Sheet.

Althaus, F., Williams, A., Schlacher, T.A., *et al.* 'Impacts of Bottom Trawling on Deep-Coral Ecosystems of Seamounts Are Long-Lasting.' *Marine Ecology Progress Series* (2009) 397.

Armas-Pfirter, F. 'The Management of Seabed Living Resources in the Area Under UNCLOS.' *Revista Electronica de Estudios Internacionales* (2006) 11.

Zvomunya, F. 'A Legal Fight in Namibia That Could Reshape the Future of Our Seas.' The Oxygen Project, 16 October 2020. www.theoxygenproject.com/post/a-legal-fight-in-namibia-that-could-reshape-the-future-of-our-seas

Boon, F.K. 'The Rio Declaration and Its Influence on International Environmental Law.' *Singapore Journal of Legal Studies*(1992): 347–364.

Food and Agricultural Organisation of the United Nations. 'The State of World Fisheries and Aquaculture' (2020). www.fao.org/documents/card/en/c/ca9229en

Frank, R.A. 'Environmental Aspects of Deepsea Mining.' *Virginia Journal of International Law* (1975) 15: 4.

Geneva ContinentalShelf Convention (1958) *Article* 1, U.N. Doc. A/Conf. 13/L. 55, U. S.T.: as 5578; 15 U.S.T. 47

Schriever, G. and Thiel, H. 'Tailings and their Disposal in Deep-Sea Mining.' ISOPE Ocean Mining and Gas Hydrates Symposium (2013).

Charalampides, G., Vitalis, K.I., Apostoplos, B., and Ploutarch-Nikolas, B. 'Rare Earth Elements: Industrial Applications and Economic Dependency of Europe.' *Procedia Economics and Finance* (2015) 24: 126–135.

Goldberg, E.D. 'Diamonds and Plastics Are Forever?' *Marine Pollution Bulletin* (1994) 28.

Goldberg, E.D. 'The Health of The Oceans.' *Update Chemical Ecology* (1994) 10.

Good Governance Africa. 'Deep-sea mining: Africa's new frontier' (2018). https://gga.org/deep-sea-mining-africas-new-frontier

Roberts , H.W. 'International Law – The High Seas, The Continental Shelf, and Free Navigation.' *North Carolina Law Review* (1957) 35(4): 15.

Vikström, H. *et al.* 'Lithium Availability and Future Production Outlooks.' *Applied Energy* (2013) 110.

Collins, H.M. 'Mineral Exploitation of the Seabed: Problems, Progress, and Alternatives.' *Natural Resources Lawyer* (1979) 12.

Gupta, H.K. *Encyclopedia of Solid Earth Geophysics*. (Springer Science, 2011).

Tuerk, H. 'The Common Heritage of Mankind After 50 Years.' *Indian Journal of International Law* (2017) 57.

Riebeek, H. 'Paleoclimatology: A Record from the Deep.' Nasa Earth Observatory, 27 September 2005. https://earthobservatory.nasa.gov/features/Paleoclimatology_SedimentCores

Odishaw, H. 'The International Geophysical Year and World Politics.' *Journal of International Affairs* (1959) 13.

International Copper Study Group. *The World Copper Factbook* (2017). https://icsg.org/copper-factbook

United Nations Environmental Program. 'Inputs to Secretary General's Annual Report to the General Assembly on the Ocean and Law of the Sea Part 1: "The Effects of Climate Change on Oceans".' (1998) www.un.org/Depts/los/general_assembly/contributions_2017/UNEP.pdf

The International Bank for Reconstruction and Development/The World Bank. *Precautionary Management of Deep-Sea Minerals* (2017). https://openknowledge.worldbank.org/handle/10986/28298

International Economic Development Council. 'Analysis of the Electric Vehicle Industry' (2013).

Bernhardt, R., Macalister-Smith, P./Max-Planck-Institut für Ausländisches Öffentliches Recht und Völkerrecht. *Encyclopedia of Public International Law*. (Amsterdam/New York/Oxford: North-Holland Publishing Co., 1992).

The International Marine Organization. 'Convention on the Prevention of Marine Pollution by Dumping of Wastes and Other Matter' (1972). www.imo.org/en/OurWork/Environment/Pages/London-Convention-Protocol.aspx

Intergovernmental Oceanographic Commission of UNESCO. 'International (UN) Decade of Ocean Science for Sustainable Development: Towards the Ocean, We Need for The Future We Want.' 13 November 2017.

Institute for Security Studies. 'Harnessing Africa's Blue Economy' (2019).

The Institute for Sustainable Futures. 'Responsible Minerals Sourcing for Renewable Energy' (2019).

International Seabed Authority. 'Exploration Contracts.' https://isa.org.jm/exploration-contracts

International Seabed Authority. *Deep Seabed Mining Payment Regime Workshop Conference Report*, Scripps Institution of Oceanography, La Jolla, CA, 17–18 May 2016.

International Seabed Authority. Implementation of Article 82 of the United Nations Convention on the Law of the Sea: Report of an International

Workshop No. 12, Kingston, Jamaica (2013). www.isa.org.jm/sites/default/files/files/documents/ts12-web.pdf

The International Telecommunications Union. 'Opportunities and Legal Challenges' (2012).

The International Union for Conversation of Nature. 'Deep-Sea Mining' (2018).

Dore, I.I. 'International Law and the Preservation of the Ocean Space and Outer Space as Zones of Peace: Progress and Problems.' *Cornell International Law Journal* (1982) 15.

Childress, J.J. 'Are There Physiological and Biochemical Adaptations of Metabolism in Deep-Sea Animals?' *Trends in Ecology and Evolution* (1995) 10.

Glazer, J.H. 'The Maltese Initiatives Within the United Nations – A Blue Planet Blueprint for Transnational Space Authority.' *Ecology Law Quarterly* (1974) 4.

Otto, J., Andrews, C., and Tilton, J. *Mining Royalties: A Global Study of Their Impact on Investors, Governments, and Civil Society.* (Washington, DC: World Bank, 2006).

Dingwall, J. 'Commercial Mining Activities in the Deep Seabed beyond National Jurisdiction: The International Legal Framework.' In Banet, C. (ed.) *The Law of the Seabed.* (Leiden: Brill Nijhoff, 2020).

Halfar, J. and Fujita, R.M. 'The Danger of Deep-Sea Mining.' *Science* (2007) 316.

Silver, J.J. *et al.* 'Blue Economy and Competing Discourses in International Oceans Governance.' *The Journal of Environment and Development* (2015) 24.

Van Dyke, J.M. 'Sharing Ocean Resources in a Time of Scarcity and Selfishness.' In Scheiber, H.N. (ed.) *Law of the Sea, the Common Heritage and Emerging Challenges.* (The Hague: Martinus Nijhoff Publishers, 2000) 3–36.

Macdonell, J. 'Occupation and "Res Nullius".' *Journal of the Society of Comparative Legislation* (1899) 1.

Heina, J.R., Mizell, K., Koschinsky, A. *et al.* 'Deep-Ocean Mineral Deposits as A Source of Critical Metals for High-end Green-Technology Applications: Comparison with Land-Based Resources.' *Ore Geology Reviews* (2013) 51.

Drazen, J.C., Smith, C.R., Gjerde, K. *et al.* 'Report of the Workshop Evaluating the Nature of Midwater Mining Plumes and Their Potential Effects on Midwater Ecosystems.' *Research Ideas and Outcomes* (2019) 5. https://riojournal.com/article/33527

Durdena, J.M., Lallierc, L.E., Murphye, K. *et al.* 'Environmental Impact Assessment Process for Deep-Sea Mining In "The Area".' *Marine Policy* (2018) 87.

Noyes, J.E., Jésus, J.L., Scholz, W.S. *et al.* 'Deep Seabed Mining: The Work of the Reparatory Commission's Annual Meeting.' *American Society of International Law* (1988).

Derraik, J.G.B. 'The Pollution of The Marine Environment by Plastic Debris: A Review.' *Marine Pollution Bulletin* (2002) 4.

Richardson, K. 'A Perspective of Marine Mining Within De Beers.' *Journal of the Southern African Institute of Mining and Metallurgy* (2007) 107.

Kindt, J.W. 'The Environmental Aspects of Deep Seabed Mining.' *UCLA Journal of Environmental Law and Policy* (1989) 8.

Kindt, W. 'International Environmental Law and Policy: An Overview of Transboundary Pollution.' *San Diego Law Review* (1986) 23.

Mizell, K., Koschinsky, A., and Conrad, T. 'Deep-ocean mineral deposits as a source of critical metals for high-and green-technology applications: Comparison with land-based resources.' *Ore Geology Reviews* (2012) 50.

Van Nijen, K., Van Passel, S., Brown, C.G. *et al.* 'The Development of a Payment Regime for Deep Sea Mining Activities in the Area through Stakeholder Participation.' *The International Journal of Marine and Coastal Law* (2019) 34.

Goldie, L.F.E. 'A Note on Some Diverse Meanings of "The Common Heritage of Mankind".' *Syracuse Journal of International Law and Commerce* (1983) 10.
United Nations Conference on Trade and Development. 'The Legal and Institutional Framework Governing Ocean-Based Economic Sectors in Barbados' (2019).
Levin, L.A., Amon, D.J., and Lily, H. 'Challenges to the Sustainability of Deep-Seabed Mining.' *Nature Sustainability* (2020) 3: 784–794.
Levin, L.A. *et al.* 'Defining "Serious Harm" to the Marine Environment in the Context of Deep-Seabed Mining.' *Marine Policy* (2016) 74.
Cuyvers, L., Berry, W., Gjerde, K., Thiele, T., and Wilhem, C. *Deep Seabed Mining: A Rising Environmental Challenge.* (International Union for Conservation of Nature, 2018).
Koschinksy, A., Heinrich, L., Boehnke, K., Cohrs, C. *et al.* 'Deep-Sea Mining: Interdisciplinary Research on Potential Environmental, Legal, Economic, and Societal Implications.' *Integrated Environmental Assessment and Management* (2018) 14(6): 672–691.
The Maribus. 'World Ocean Review: Living with the Oceans.' https://worldoceanreview.com/wp-content/downloads/wor1/WOR1_en.pdf
Gavouneli, M. 'Protection standards for the marine environment: Updating Part XII of the Law of the Sea Convention?' *SSRN Electronic Journal* (2017). https://dx.doi.org/10.2139/ssrn.3073263
Nvarre, M. and Lammens, H. 'Opportunities of Deep-Sea Mining and ESG Risks.' *Amundi Discussion Papers Series* (2017).
Kjelland, M.E., Woodley, C.M., Swannack, T.M. *et al.* 'A review of the potential effects of suspended sediment on fishes: Potential dredging-related physiological, behavioural, and transgenerational implications.' *Environment Systems and Decisions* (2015) 35.
Lodge, M. 'The International Seabed Authority and Deep Seabed Mining.' United Nations (2016). www.un.org/en/chronicle/article/international-seabed-authority-and-deep-seabed-mining
Lodge, M.W. Statement to the Open Meeting of the Commission on the Limits of the Continental Shelf. United Nations, New York, 10 March 2017. www.isa.org.jm/files/documents/EN/SG-Stats/2017/CLCS-10Mar17.pdf
Lodge, M.W., Segerson, K., and Squires, D. 'Sharing and Preserving the Resources in the Deep Sea: Challenges for the International Seabed Authority.' *The International Journal of Marine and Coastal Law* (2017) 32.
Mancheri, N.A. 'World Trade In Rare Earths, Chinese Export Restrictions, And Implications.' *Resources Policy* (2015) 46.
Finsa. 'Namibia. Marine Phosphate Mining Saga Back in Court.' 6 July 2020. www.fishingindustrynewssa.com/2020/07/06/namibia-phosphate-mining-saga-goes-back-to-court
The National Academy of Sciences. 'Pollution is the release of undesirable substances' (2007)
National Geographic. 'Ocean.'www.nationalgeographic.org/encyclopedia/ocean
National Ocean Service Science Report (2021). https://oceanservice.noaa.gov/about/nos-science-report/2021
National Science Foundation. 'Dark Energy – Life Beneath the Seafloor: Scientists Present Recent Findings on The Subsurface Biosphere.' *Science Daily*, 19 November 2012. www.sciencedaily.com/releases/2012/11/121119163458.htm

Toro, N., Robles, P., and Jeldres, R.I. 'Seabed Mineral resources, an alternative for the future of renewable energy: A critical review.' *Ore Geology Reviews* (2020) 126.

Kimani, N.N. 'Enacting National Seabed Mining Laws in Africa: Importance of a Practitioner's Perspective.' *Environment and Development Law Journal* (2015) 11.

OECD Roundtable on Corporate Responsibility. 'Transition to a Low-Carbon Economy: Public Goals and Corporate Practices' (2010).

Heffernan, O. 'Seabed Mining Is Coming – Bringing Mineral Riches and Fears of Epic Extinctions.' *Nature*, 24 July 2019. www.nature.comOliveHeffernan/articles/d41586-019-02242-y

Chapman, P. and Shannon, L.V. 'The Benguela ecosystem. II: Chemistry and Related Processes.' *Oceanography and Marine Biology* (1985) 23.

Bond, P. 'Blue Economy Threats, Contradictions and Resistances Seen from South Africa.' *Journal of Political Ecology* (2019) 26.

Pardo's statements in 22 U.N. GAOR, U.N. Doc. A/C.1/PV.1515.

Lusty, P.A.J. and Murton, B.J. 'Deep-Ocean Mineral Deposits: Metal Resources and Windows into Earth Processes.' *Elements* (2008) 14(5): 301.

Chaniotis, P.D., Robson, L.M., Lemasson, A.J. *et al.* 'UK Deep-Sea Conservation: Progress, Lessons Learned, and Actions for the Future.' *Aquatic Conservation: Marine and Freshwater Ecosystems* (2019) 30(2): 375–393.

Ehlers, P. 'Blue Growth and Ocean Governance – How to Balance the Use and The Protection of The Seas.' *World Maritime University Journal* (2016) 15.

Princz v. Federal Republic of Germany 26 F.3d1166 (D.C. Cir. 1994).

Sharma, R. *Deep-Sea Mining: Current Status and Future Considerations.* (Springer International Publishing, 2017).

Sharma, R. 'Development of Environmental Management Plan for Deep-Sea Mining.' in Sharma, R. (ed.) *Environmental Issues of Deep-Sea Mining.* (Springer International Publishing, 2019).

Sharma, R. 'Environmental Issues of Deep-Sea Mining.' *Procedia Earth and Planetary Science* (2015) 11.

Sharma, R. 'First Nodule to First Mine-Site: Development of Deep-Sea Mineral Resources from The Indian Ocean.' *Current Science* (2010) 99.

Sharma, R. and Smith, S. 'Deep-sea Mining and the Environment – An Introduction.' in Sharma, R. (ed.) *Environmental Issues of Deep-Sea Mining.* (Springer International Publishing, 2019).

Aronson, R.B., Thatje, S., and McClintock, J.B. 'Anthropogenic impacts on marine ecosystems in Antarctica.' *Annals of the New York Academy of Sciences* (2011) 1223.

Launius, R.D., Fleming, J.R., and DeVorkin, D.H. *Globalizing Polar Science Reconsidering the International Polar and Geophysical Years.* (New York: Palgrave Macmillan, 2010).

Carver, R. 'Resource sovereignty and accumulation in the blue economy: the case of seabed mining in Namibia.' *Journal of Political Ecology* (2019) 26.

Lande, R. 'Anthropogenic, Ecological and Genetic Factors in Extinction and Conservation.' *Researchers on Population Ecology* (1998) 40.

Muckelston, S. 'Strip-Mining Reclamation Requirements in Montana – A Critique.' *Montana Law Review* (1971) 32.

Nandan, S.N., Lodge, M.W., and Rosenne, S. 'The Development of the Regime for Deep Seabed Mining.' International Seabed Authority (2002).

Petersen, S. *et al.* 'News from the Seabed–Geological Characteristics and Resource Potential of Deep-Sea Mineral Resources.' *Marine Policy* (2016) 70.

Baker, E., Beaudoin, Y., and Pendleton, L. 'Deep-Sea Minerals and the Green Economy.' The Secretariat of the Pacific Community (SPC) (2013). https://nicholasinstitute.duke.edu/sites/default/files/publications/chapter_8_deep_sea_minerals_and_the_green_economy.pdf

Kuhn, T., Jamieson, A.J., Martinez Arbizu, P. *et al.* 'Manganese Nodule Fields from The Northeast Pacific as Benthic Habitats.' In Harris, P. and Baker, E. (eds) *Atlas of Seafloor Geomorphology as Benthic Habitat*, 2nd edition. (Elsevier, 2020).

Aqorau, T. 'Obligations to Protect Marine Ecosystems Under International Conventions and Other Legal Instruments.' *Reykjavik Conference on Responsible Fisheries in the Marine Ecosystem*, 2001. https://doi.org/10.1079/9780851996332.0025

United Nations Convention on the Law of the Sea 1982.

United NationsGeneral Assembly Resolution 2749 (XXV).

United Nations Environment Programme. 'Action Plan for The Protection and Sustainable Development of The Marine and Coastal Areas of The East Asian Region' (1994).

Bach, W., Edwards, K.J., Hayes, J.M. *et al.* 'Energy in the Dark: Fuel for Life in the Deep Ocean and Beyond.' *Eos* (2006) 87(7): 73–78.

Lavelle, W., Ozturgut, E., and Baker, E.T. *et al.* 'Discharge and surface plume measurements during manganese nodule mining tests in the North Equatorial Pacific.' *Marine Environmental Research* (1982) 7.

International Seabed Authority. 'Workshop: Development of Standards and Guidelines for the Mining Code.' Pretoria, South Africa, 13–17 May 2019. www.isa.org.jm/index.php/node/19116

Wolfrum, R. 'Common Heritage of Mankind.' In Bernhardt, R., Macalister-Smith, P./Max-Planck-Institut für Ausländisches Öffentliches Recht und Völkerrecht. *Encyclopedia of Public International Law.* (Amsterdam/New York/Oxford: North-Holland Publishing Co., 1992).

World Trade Organisation. 'China – Measures Related to The Exportation of Rare Earths, Tungsten, and Molybdenum' (WT/DS433/R/Add.1).

Shen, Y., Moomy, R., and Egger, R.G. 'China's Public Policies Toward Rare Earths, 1975–2018.' *Mineral Economics* (2020) 33.

13 Blue Economy's Deep Seabed Mining

The Future of Energy?

Pablo Ferrara

1 Introduction

Simply and schematically, it is possible to assert that the economy has been deployed – that is, practised, promoted, and regulated – from onshore areas (land), through offshore areas (oceans), all the way through time into outer space.[1] It is also possible to assert, simply, that regulatory frameworks are dyadically divided in all three areas into private and public norms – depending on a broad number of direct and indirect conditions.

When referring to the exploitation of offshore areas, economic models can be divided between artisanal and hyper-industrial. In between these two extremist models evolved a new theory of harmonic, sustainable development: Blue Economy (BE). As recalled by O. Gbadegesin and S. Akintola:

> The term 'blue economy' was first used by Professor Gunter Pauli in his book 'The Blue Economy: 10 Years, 100 Innovations, 100 Million Jobs' to describe a wealth creation strategy using nature-inspired derivatives on the basis of environmental correctness. According to Pauli, the 'blue economy' broadly encompasses activities geared towards achieving economic growth and development based on ocean activities whilst also considering the social and environmental outcomes of these activities.[2]

Although not originally, Blue Economy suits the United Nations Sustainable Development Goals (SDGs) (aka the 'United Nations 2030 Agenda').[3] Of course, SDGs are a rather newly schematic and marketing-oriented way to promote the United Nations basic milestones, but they are also rather useful

1 N. Ferguson, *The Ascent of Money: A Financial History of the World* (Penguin 2018).

2 Opeyemiyi Gbadegesin and Akintola Simisola, 'Charting the Course for a Blue Economy in Nigeria: A Legal Agenda' (2021) 1(7) Journal of Environmental Law and Policy. https://doi.org/10.33002/jelp001.01, accessed 12 December 2021.

3 Sustainable Development Goals. www.un.org/sustainabledevelopment/sustainable-development-goals, accessed 12 December 2021.

DOI: 10.4324/9781003284437-16

for the industrial and business approach.[4] Why? Because they are *innovative, sustainable, clustered, and – of course – they can be modelled.*[5]

Clustering and modelling, as well as tagging and #hashtagging, became a fashionable way of deploying explanations with a scientific-like colour.[6] Also, as mentioned elsewhere, any modelling in social sciences is only that: modelling. This means social sciences modelling is nothing more than a way of theorizing. In this sense, it is possible to assert this work does also theorize, cluster, and model. What about? Especially in relation to Blue Economy's biodiversity cluster.[7]

When referring to economic clusters – biodiversity, in this case – three milestones should be particularly fundamental: (i) the product's and economy's essentials to be considered (e.g. GDP incidence, governance, political and financial risk, etc.); (ii) the product's prospective role for a particular country's economic model (e.g. Is the country simply looking to supply its own needs? Is it looking to become an export titan? Is it looking for product replacement?); and (iii) the product's industrial and trade standing points. Only a clear and precise understanding of these criteria will countenance the next stage: discussion, selection, and implementation of the most suitable cluster development model. Of course, this work aims at examining some of the possible answers, always bearing in mind that the real future is infinite and invariably unforeseeable.

Methodologically, this work will follow four stages: (i) an introduction to the basics of Blue Economy; (ii) a review of the foundations of biodiversity according to the United Nations Convention on the Law of the Sea and its related documents; (iii) a review and comparison of possible modelling through existent and prospective developments in the Indian, Pacific, and Atlantic Oceans; and (iv) presentation of findings and conclusions, including special suggestions regarding Indian Ocean prospective projects.

4 In 2014, developing nations met in Abu Dhabi to explore the BE approach within the UN Sustainable Development Knowledge Platform. Based on the discussion, a document including a 'framework for sustainable development' was presented. The participating nations stated that safeguarding socio-economic development from environmental degradation is the basis of BE and that the value of the blue capital in the economy should be assessed. Since then, BE has gained recognition at the global level confirming the role of the ocean in many of the global challenges faced by the planet, such as food security and climate change (OECD, 2016). Gianluca Sampaolo, Dominique Lepore, and Francesca Spigarelli, 'Blue Economy and the quadruple helix model: the case of Qingdao' (2021) 23 Environment, Development and Sustainability. https://doi.org/10.1007/s10668-021-01378-0, accessed 12 December 2021.

5 Sustainable Development Goals (n 3).

6 Robert E. Goodin, *The Oxford Handbooks of Political Science* (Oxford University Press 2018).

7 Pablo Ferrara, *Don't Be Afraid, It's Nature: A Theory of Thermodynamic Pragmatism* (Arraes 2020).

2 Blue Economy Basics: Theorizing, Clustering, and Modelling

Within normal and reductionist frame-working, an undertaking to analyse off-shore (i.e. maritime, as opposed to marine and oceanic) development models and operating policies raises an endless succession of queries.[8] Conventional approaches – even worse when compounded by undue haste – could simply lead to considering a model's convenience. However, haste is not the best counsellor in social studies – even less so when the debate is on the substrate, basis, and founding of modern economy trade and production – that is, commodities supply and consequent wealth (i.e. capital) enlargement.

2.1 Theorizing

Any decent public policy analysis would start by considering the real feasibility of certain clustered economic production models, then compare both the necessity and the benefits of certain production model *vis-à-vis* other models (short, medium-, or long-term), and finally provide recommendations for its implementation (or not) with any particularity. Thus, it is logically possible to infer the following: *The more technologically intensive a business activity is, the more capital-intensive it will be, increasing risks and decreasing feasibility.*

First of all, it is essential to understand that Blue Economy is tempting not only because of the inconvenience of onshore business development, but also because it is an increasingly profitable activity with an annual 8% rate.[9]

Second, like any development model, Blue Economy theorizing does also allow clustering. For the purpose of this work, it is possible to recognize or define a minimum of 20 necessary clusters: (i) administration of *coasts*; (ii) administration of *ports*; (iii) maritime space planning (*MSP*); (iv) ports' *infrastructure*; (v) ports' *urban* planning and infrastructure; (vi) logistics planning (*rails, roads, and infrastructure*); (vii) shipping and maritime transport routes planning and development (*shipping and transportation*); (viii) fisheries, fishing farms (*fisheries and seafood*); (ix) submarine laying of cables and pipelines (*submarine laying; telecommunications; oil and gas transportation*); (xi) exploration and exploitation of offshore hydrocarbons (*oil and gas*); (xii) *deep seabed mining*; (xiii) marine environmental and development planning (*environment*) (xiv) planning and development of biodiversity (*biodiversity*); (xv) *geology*; (xvi) *oceanography*; (xvii) *cartography*; (xviii) *finance*; (xix) policymaking and normative regulation (*public policy and law*); and (xx) *management*.

8 This chapter defines the model as a set of political, regulatory, and trade measures implemented as part of offshore activities.

9 J. Ellsmoor, 'Rethinking Our Oceans: Investing in The Blue Economy.' Forbes, 26 January 2019. www.forbes.com/sites/jamesellsmoor/2019/01/26/rethinking-our-oceans-investing-in-the-blue-economy/?sh=26de41423531, accessed 12 December 2021.

Biodiversity,[10] fisheries,[11] deep seabed mining,[12] maritime transportation,[13] and telecommunications[14] are only some of the most relevant economic clusters. Table 13.1 shows their incidence in the total global gross development product (GDP).

Table 13.1 Blue Economy industry world GDP incidence

	Total global GDP[15]	*Pharma industry*[16]	*Cosmetology industry*[17]	*Shipping industry*[18]	*Fisheries industry*[19]	*Telecommunications industry*[20]	*Mining industry*[21]
GDP	84.705T	1,27T	532B	8.7B	159.32B	1,657.7B	159.74B
GDP%	100%	1.5%	0.63%	0.01%	0.18%	1.9%	0.18%

Source: Compiled by author

10 Miranda C. Jones and William W.L. Cheung, 'Multi-model ensemble projections of climate change effects on global marine biodiversity' (2015) 72(3) ICES Journal of Marine Science. https://doi.org/10.1093/icesjms/fsu172, accessed 12 December 2021.
11 FAO Fishing Zones, www.fao.org/3/u8480e/U8480E4o.jpg, accessed 12 December 2021.
12 ISA, https://isa.org.jm/index.php/minerals/exploration-areas, accessed 12 December 2021.
13 Shipping Density, www.ammoniaenergy.org/topics/carbon-footprint, accessed 12 December 2021.
14 Interactive Submarine Cable Map, https://mapscaping.com/blogs/geo-candy/interactive-submarine-cable-map
15 https://data.worldbank.org/indicator/NY.GDP.MKTP.CD?start=2020&view=chart, accessed 12 December 2021.
16 www.statista.com/statistics/263102/pharmaceutical-market-worldwide-revenue-since-2001, accessed 12 December 2021.
17 Pamela N. Danzinger, '6 Trends Shaping The Future Of The $532B Beauty Business.' Forbes, 1 Septemer 2019. www.forbes.com/sites/pamdanziger/2019/09/01/6-trends-shaping-the-future-of-the-532b-beauty-business/?sh=36f7934d588d, accessed 12 December 2021.
18 Marcin Placek, 'Protected size of the container shipping market worldwide 2020–2028.' Statista, 21 April 2022. www.statista.com/statistics/1097059/global-shipping-containers-market-size, accessed 12 December 2021.
19 M. Shahbandendeh, 'Global seafood market value 2020–2025.' Statista, 16 December 2021). www.statista.com/statistics/821023/global-seafood-market-value, accessed 12 December 2021.
20 'Telecom Services Market Size, Share & Trends Analysis Report by Service Type (Mobile Data Services, Machine-To-Machine Services), by Transmission (Wireline, Wireless), by End-use, by Region, and Segment Forecasts, 2021–2028.' Grand View Research. www.grandviewresearch.com/industry-analysis/global-telecom-services-market > accessed 12 December 2021.
21 'Mining Equipment Market Size, Share & Trends Analysis Report by Equipment (Surface, Underground), by Application (Metal, Non-metal, Coal), by Region, and Segment Forecasts, 2020–2027.' Grand View Research. www.grandviewresearch.com/industry-analysis/mining-equipment-industry, accessed 12 December 2021.

According to Table 13.1, *biodiversity* industries (pharma and cosmetology) – as defined by the *Convention on Biological Diversity*[22] –represent the highest world GDP incidence of the economic clusters reviewed with 2.13% – even over telecommunications (1.9%).[23] This provides a hint of the potential of an exponential sustainable development of offshore biodiversity industry, considering that '[o]nly about 0.0001 percent of the deep-seafloor has been subject to biological investigation' (United Nations Environmental Program, 2007).[24]

2.2 Modelling

Like any other business development start-up analysis, each and every specific cluster implies a different level of technology, capital, risk, and feasibility. These levels will vary in relation to several variables, under the umbrella of momentum (economic and political timing) and the country (governance, geopolitics, culture, history, etc.).

Before modelling, it is necessary to understand standing points. In this sense, world trade statistics (WITS) provide an idea of current trade and production scenarios, allowing us to evaluate future Blue Economy models worldwide and understand their potential trade development power (PTDP):[25]

Of course, Blue Economy modelling, like any other public policy, also implies choosing a viable business corporative model for a particular country according to the particularities of its society, the analysed cluster, and the global momentum of the analysis. Therefore, it is possible to deduce preliminarily that every modelling will allow a more or less centralized, regulated, and monopolized model – not to mention the degree of development of

22 Convention on Biological Diversity. '"Biological diversity" means the variability among living organisms from all sources including, inter alia, terrestrial, marine and other aquatic ecosystems and the ecological complexes of which they are part; this includes diversity within species, between species and of ecosystems.' CBD, Article 2, www.cbd.int/convention/articles/?a=cbd-02, accessed 12 December 2021.

23 Covering 64% of the surface of the ocean, and providing nearly 95% of its volume, marine areas beyond national jurisdiction (ABNJ) are home to an important part of the world's biodiversity, support significant fisheries, and play a critical role in stabilizing global climate. Modalities for Advancing Cross-Sectoral Cooperation in Managing Marine Areas Beyond National Jurisdiction. UNEP (DEPI)/RS.12 /8.

24 P.P.E. Weaver, A. Benn, P.M. Arana, J.A. Ardron, D.M. Bailey, K. Baker, D.S.M. Billett, M.R. Clark, A.J. Davies, P. Durán Muñoz, S.D. Fuller, M. Gianni, A.J. Grehan, J. Guinotte, A. Kenny, J.A. Koslow, T. Morato, A.J. Penney, J.A.A. Perez, I.G. Priede, A.D. Rogers, R.S. Santos, and L. Watling, 'The impact of deep-sea fisheries and implementation of the UNGA Resolutions 61/105 and 64/72' (Report of an international scientific workshop, National Oceanography Centre, Southampton, 2011), http://hdl.handle.net/10013/epic.37995, accessed 12 December 2021.

25 https://wits.worldbank.org/trade-visualization.aspx?lang=es, accessed 12 December 2021.

Table 13.2 Blue Economy products scenarios

Country	*GDP*[26]	*Per capita GDP*[27]	*Risk classification*[28]	*Control of corruption (–2.5 to +2.5)*[29]	*Trade diversification (+50%/3 Products)*[30]	*Rate of technology per product*	*Rate of investment per product*	*Government system*
World	84.705T	10,925	n/a	n/a	n/a	n/a	n/a	n/a
Argentina	383,066.98M	8,441	7	–0.07	Low (3 products)	Low	Low	Federal constitutional republic
Australia	1,330,900.93M	51,812	n/a	1,81	Low (3 products)	Low	Low	Federal constitutional monarchy
Brazil	1,444,733.26M	6,796	5	–0.33	Medium (4 products)	Medium	Medium	Federal constitutional republic
China	14,722,730.70M	10,500	2	–0.32	Low (2 products)	High	High	Socialist unitarian uni-party republic
France	2,603,004.40M	38,625	n/a	1.30	Low (3 products)	High	High	Constitutional republic
Germany	3,806,060.14M	45,723	n/a	1.90	Low (3 products)	High	High	Federal constitutional republic
India	2,622,983.73M	1,900	3	–0.23	Medium (4 products)	Medium	Medium	Federal constitutional republic
Indonesia	1,058,423.84M	3,869	3	–0.42	Medium (4 products)	Medium	Medium	Constitutional republic

Country	*GDP*[26]	*Per capita GDP*[27]	*Risk classification*[28]	*Control of corruption (–2.5 to +2.5)*[29]	*Trade diversification (+50%/3 Products)*[30]	*Rate of technology per product*	*Rate of investment per product*	*Government system*
Japan	5,064,872.88M	40,113	n/a	1.48	Low (2 products)	High	High	Parliamentary monarchy
Korea	1,630,525.01M	31,489	n/a	0.76	Low (2 products)	High	High	Constitutional republic
Madagascar	13,720.63M	495	7	–1.01	Low (2 products)	Low	Low	Constitutional republic
Mexico	1,076,163.32M	8,346	3	–0.82	Low (2 products)	High	High	Federal constitutional republic
Netherlands	912,242.34M	52,304	n/a	2	Medium (4 products)	High	High	Parliamentary monarchy
Nigeria	432,293.78M	2,097.1	6	–1.09	Low (1 product)	Low	Low	Constitutional republic
Portugal	231,255.59M	22,439	n/a	0.76	Medium (4 products)	High	High	Constitutional republic
Russia	1,483,497.78M	10,126	4	–0,83	Low (1 product)	Medium	Medium	Federal parliamentary republic
Singapore	339,998.48M	59,797	0	2.16	Low (2 products)	High	High	Parliamentary republic

Country	GDP[26]	Per capita GDP[27]	Risk classification[28]	Control of corruption (−2.5 to +2.5)[29]	Trade diversification (+50%/3 Products)[30]	Rate of technology per product	Rate of investment per product	Government system
South Africa	301,923.64M	5,090	4	0.08	Medium (4 products)	Medium	Medium	Parliamentary republic
Spain	1,281,199.09M	27,057	n/a	0.65	Medium (4 products)	High	High	Parliamentary monarchy
UK	2,707,743.78M	40,284	n/a	1.77	Medium (4 products)	High	High	Parliamentary monarchy
USA	20,936,600.00M	63,543	n/a	1,22	Medium (4 products)	High	High	Federal constitutional republic

Source: Compiled by author

26 World Bank national accounts data and OECD National Accounts data files, 'GDP (current US$).' https://data.worldbank.org/indicator/NY.GDP.MKTP.CD?start=2020&view=chart, accessed 12 December 2021.

27 World Bank national accounts data and OECD National Accounts data files, 'GDP per capita (current US$).' https://data.worldbank.org/indicator/NY.GDP.PCAP.CD?view=chart, accessed 12 December 2021.

28 OECD, 'Country Risk Classifications of the Participants to the Arrangement on Officially Supported Export Credits.' www.oecd.org/trade/topics/export-credits/documents/cre-crc-current-english.pdf, accessed 12 December 2021.

29 World Bank, 'Worldwide Governance Indicators.' https://info.worldbank.org/governance/wgi/Home/Reports, accessed 12 December 2021.

30 World Bank, 'Detailed Country Analysis.' http://wits.worldbank.org/es/visualization/detailed-country-analysis-visualization.html, accessed 12 December 2021.

each cluster.[31] In other words, every country can engage in Blue Economy, but their engagement will be more or less difficult according to its current development, evolution capacities, and public policy orientation (Table 13.2).

WITS show very few insular states other than Madagascar (e.g. Seychelles), which are also low trade, low tech, and low financially diversified. However, case reviews show that it is mainly insular states (e.g. Tonga, Nauru, Cook Islands, Kiribati, Seychelles) that partner up with developed states' corporations – mostly from their former colonial periods – in order to pursue deep seabed mining Blue Economy developments. It is time to focus on biodiversity, and some study cases, to analyse their potential.

3 The Fundamentals

3.1 Business

As any other ecosystemic element, marine minerals provide a natural service as well as upholding a prospective industrial use (Table 13.3 and Table 13.4).

Table 13.3 Typography of deep sea ecosystems[32]

Supporting services	
Habitat	Diverse topography across a huge expense of physical space at varying depths, temperatures, level of light, etc. provide habitat for myriad species.
Nutrient cycling	Deep-sea storage, transport and recycling of various elements underpin organic processes, ecosystems and living organisms. An example: nutrients brought to the surface to feed producer organisms that eventually feed fisheries.
Primary production from chemicals	Bacteria and other microorganisms living beyond the reach of sunlight produce primary material in the food web from chemicals; this primary production supports diverse organisms at higher trophic levels within the deep sea and beyond.
Resilience	Though less understood, the resilience of deep-sea ecosystems contributes (via its various services) to the resilience of other ecosystem services.
Provisioning services	
Fish and other species	Deep-sea habitats are home to a number of commercially fished species; In addition, they provide crucial links in food webs for fish and species that also live in the upper water column (including large pelagics like tuna as well as large charismatic marine mammals).

31 World Bank, 'Detailed Country Analysis.' https://wits.worldbank.org/trade-visualization.aspx?lang=es; http://wits.worldbank.org/es/visualization/detailed-country-analysis-visualization.html, accessed 12 December 2021.

32 A. Jadhav, *Minerals from the Sea* (Dakshin Foundation 2016) 10.

Marine minerals	An obvious focus for this paper, ocean minerals can be found in abundance at various depths and topographies. This including minerals already exploited – namely oil and gas – as well as products expected in the near (e.g. polymetallic sulphides) or distant (methane hydrates) future.
Industrial and pharmacological compounds	As bio-prospecting continues, the deep sea may represent a vast storehouse of compounds that may be found useful in the future for their industrial or medicinal value. Already, some discoveries have resulted from the study or harvest of corals, sponges and bacteria.
Regulating services	
Gas absorption and regulation of climate	The oceans act as a great sink of organic materials; this is particularly important for the sequestration of carbon and the consumption of methane from various sources.
Waste and pollution storage and sanitation	Legions of marine organisms breakdown, store, decompose or otherwise "transform" waste materials, chemicals, sewage and the like, mostly entering the ocean from land.
Biological control	Deep-sea species may be linked to the regulation of other pests and pathogens, such as those spread through transporting ballast waters on ships.
Active carbon dicoded storage	In addition to its natural carbon sequestration ability, the deep ocean may also other the (debated and controversial) possibility of human-directed carbon capture and storage in seawater, on the seafloor or within geological structures.
Cultural services	
Human social dependence	The deep sea supports various human activities such as exploration, scientific research, cultural productions (Including literature) and even tourism (e.g. whale watching).

Source: Armstrong C et. al. "Services from the deep: Steps towards valuation of deep sea goods and services," 4–8.

From the living resources perspective, Gbadegesin and Akintola accurately recall:

> Oceans and seas are the source of a huge variety of life forms including macro and micro-organisms. Living marine resources have huge potential for developing new food, biochemical, pharmaceutical, cosmetics and bio-energy applications. According to Suttle, the growing commercial interest in marine bio-prospecting to discover new plants and animal species from which medicinal drugs and other commercially valuable compounds can be derived is expected to increase with developments in science, technology and innovation since the first drugs from marine organisms were commercialized over a decade ago. The Nigerian marine environment is rich in biodiversity with a wide range of novel species of microorganisms. This rich biodiversity if harnessed and commercialized through the application of scientific and engineering principles has the potential to become a potent revenue source.

Table 13.4 Prospective uses of marine minerals[33]

Type of deposit	*Contained minerals and materials*	*Known locations*	*Potential use*
Calcareous sands	Calcium carbonate/lime	Continental shelf offshore to Gujarat, Maharashtra, Tamil Nadu, Andhra Pradesh and Orissa	Cement, plaster, paints, building sand, other industrial materials
Lime mud	Lime	Continental shelf offshore to Maharashtra, Karnataka, Tamil Nadu, Andhra Pradesh and Orissa	Cement, plaster, paints.
Heavy mineral/ placer deposits			
Sand	Silica	Continental shelf offshore Kerala	Construction sand
Phosphorites	Phosphorus	Beyond the shelf off Gujarat, continental shelf offshore to Maharashtra, Karnataka, Tamil Nadu	Agricultural fertilizers
Micro-manganese nodules	Manganese, Iron, copper, lead, zinc, nickel and cobalt	Deep seabeds west of Lakshadweep	Metals production, Industrial uses
Ferromanganese crusts	Cobalt, manganese, Iron, cerium and other potential elements	Andaman Sea limited study to date)	Metals production, Industrial uses
Gas hydrates	Methane	Beyond the shelf off Andhra Pradesh and the Andaman Sea	Conversion to natural gas
Polymetallic nodules	Manganese, Iron, copper, lead, zinc, nickel and cobalt	International waters of the Central Indian Ocean Basin	Metals production, Industrial uses
Polymetallic sulphides	Copper, lead, zinc, gold and silver	International waters of the Indian Ocean Ridge	Metals production, Industrial uses

Source: Compiled by author

Devoting resources to the development of technology to aid bio-prospecting in Nigeria has the potentials for poverty eradication and national development through marine sustainable aquaculture and fisheries, providing sustainable alternative sources of energy, pharmaceutical discoveries that would improve human health amongst others.[34]

33 Ibid 33.
34 Gbadegesin and Simisola (n 2).

It has been mentioned that economy extended from onshore to offshore areas. In UNCLOS terms, the offshore economy may be understood to cover from a coastal state's territorial sea through its Exclusive Economic Zone (EEZ) and Continental Shelf (CS), all the way into the Area and the High Sea:

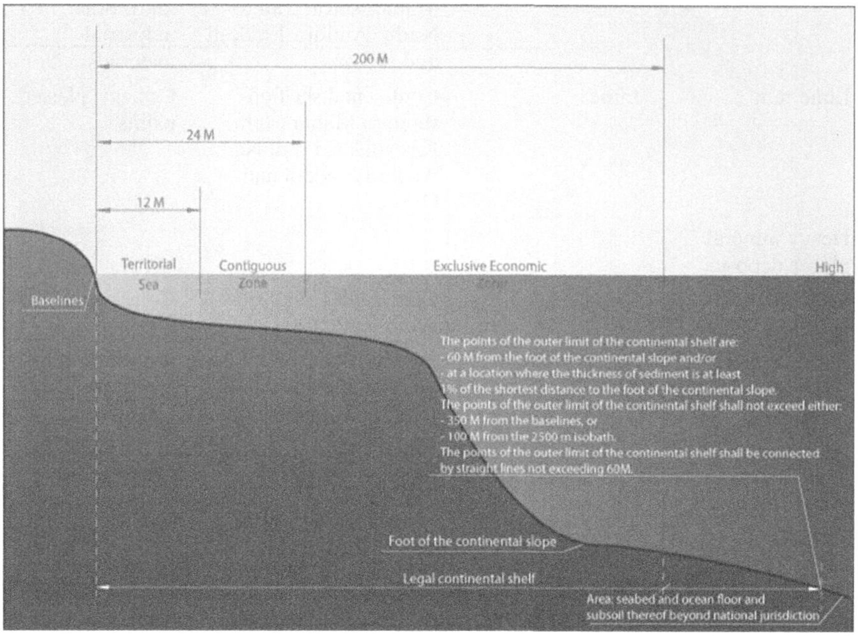

Figure 13.1 UNCLOS Marine Areas: Territorial Sea – EEZ – Continental Shelf – High Sea

It has also been asserted:

> One of the most challenging issues in ocean affairs today is managing marine genetic resources in the high seas. Discussions on biodiversity beyond national jurisdictions is complex and vast, encompassing fundamental questions such as the meaning of conservation and management of biodiversity, the meaning of marine genetic resources, transfer of technology, benefit sharing, the meaning and implication of a legally binding instrument, and many other rather complex matters.[35]

35 Myron H. Nordquist and Ronán Long, *Marine Biodiversity of Areas Beyond National Jurisdiction* (Brill Nihjoff 2021); Jeff A. Ardron, Rosemary Rayfuse, Kristina Gjerde, and Robin Warner, 'The sustainable use and conservation of biodiversity in ABNJ: What can be achieved using existing international agreements?' (2014) 49 Marine Policy. http://dx.doi.org/10.1016/j.marpol.2014.02.011, accessed 12 December 2021.

Why is it so complex? Because:

> [A]ctivities in the areas beyond national jurisdiction will have a direct impact on areas under our jurisdiction and also *vice versa*. Conservation and management of living resources as well as pollution are of particular concern, because the marine environment as an ecosystem will affect them regardless of legal boundaries established in accordance with UNCLOS 1982. Fish have no passports.[36]

In addition:

> [A]lthough our planet Earth has more waters than lands, we do not have a single organization that deal with ocean. We have UNGA, ITLOS, ISBA, DOALOS, IMO, FAO, IOC, Regional Fisheries Organizations in many regions, and many others.[37]

Core normative regarding marine biodiversity can be found in the 1982 UN Convention on the Law of the Sea (UNCLOS). It is relevant to have a minimum understanding of these UNCLOS definitions, rights, and obligations of coastal and outer (i.e. flagship) states.[38]

Thus, in a nutshell, there is: (i) a different set of deep seabed mining *corpus juris* per state, which may also vary between each state's interior waters and territorial sea; (ii) a different set of deep seabed mining *corpus juris* per ptate's EEZ that must respect the minimum standards of UNCLOS and other international deep seabed mining *corpus juris* (e.g. customary law, treaties, jurisprudence, principles); (iii) a set of uniform deep seabed mining minimum standards applying to the Area according to UNCLOS and other international *corpus juris* (e.g., customary law, treaties, jurisprudence, principles, reports, recommendations, template contracts); (iv) a set of uniform deep seabed mining-related activities minimum standards applying to the High Sea according to UNCLOS and other international *corpus juris* (e.g. customary law, treaties, jurisprudence, principles).[39]

3.1 The United Nations Convention on the Law of the Sea (UNCLOS)

It is essential to review basic UNCLOS definitions and legal implications. In the first place, Article 2 of UNCLOS affirms:

> 1. The sovereignty of a coastal State extends, beyond its land territory and internal waters and, in the case of an archipelagic State, its archipelagic waters, to an adjacent belt of sea, described as the territorial sea.

36 Nordquist and Long (n 35).
37 Ibid.
38 Ardron, Rayfuse, Gjerde, and Warner (n 35).
39 Ibid.

> 2. This sovereignty extends to the air space over the territorial sea as well as to its bed and subsoil.
> 3. The sovereignty over the territorial sea is exercised subject to this Convention and to other rules of international law.[40]

In addition, Article 21 of UNCLOS states:

> 1. The coastal State may adopt laws and regulations, in conformity with the provisions of this Convention and other rules of international law, relating to innocent passage through the territorial sea, in respect of all or any of the following:
>
> (a) the safety of navigation and the regulation of maritime traffic;
> (b) the protection of navigational aids and facilities and other facilities or installations;
> (c) the protection of cables and pipelines;
> (d) the conservation of the living resources of the sea;
> (e) the prevention of infringement of the fisheries laws and regulations of the coastal State;
> (f) the preservation of the environment of the coastal State and the prevention, reduction and control of pollution thereof;
> (g) marine scientific research and hydrographic surveys;
> (h) the prevention of infringement of the customs, fiscal, immigration or sanitary laws and regulations of the coastal State.
>
> 2. Such laws and regulations shall not apply to the design, construction, manning or equipment of foreign ships unless they are giving effect to generally accepted international rules or standards.
> 3. The coastal State shall give due publicity to all such laws and regulations.
> 4. Foreign ships exercising the right of innocent passage through the territorial sea shall comply with all such laws and regulations and all generally accepted international regulations relating to the prevention of collisions at sea.[41]

Article 55 of UNCLOS affirms:

> The exclusive economic zone is an area beyond and adjacent to the territorial sea, subject to the specific legal regime established in this Part, under which the rights and jurisdiction of the coastal State and the rights

40 UNCLOS, www.un.org/Depts/los/convention_agreements/texts/unclos/closindx.htm> accessed 12 December 2021; A. Proelss, *UNCLOS: A Commentary* (Hart-NOMOS 2016).
41 UNCLOS (n 40); Proelss (n 40).

and freedoms of other States are governed by the relevant provisions of this Convention.[42]

In addition, Article 56 states:

> 1. In the exclusive economic zone, the coastal State has:
>
> (a) sovereign rights for the purpose of exploring and exploiting, conserving and managing the natural resources, whether living or non-living, of the waters superjacent to the seabed and of the seabed and its subsoil, and with regard to other activities for the economic exploitation and exploration of the zone, such as the production of energy from the water, currents and winds;
> (b) jurisdiction as provided for in the relevant provisions of this Convention with regard to:
>
> (i) the establishment and use of artificial islands, installations and structures;
> (ii) marine scientific research;
> (iii) the protection and preservation of the marine environment;
>
> (c) other rights and duties provided for in this Convention.
>
> 2. In exercising its rights and performing its duties under this Convention in the exclusive economic zone, the coastal State shall have due regard to the rights and duties of other States and shall act in a manner compatible with the provisions of this Convention.
> 3. The rights set out in this article with respect to the seabed and subsoil shall be exercised in accordance with Part VI.'[43]

However, Article 58 of UNCLOS asserts:

> 1. In the exclusive economic zone, all States, whether coastal or land-locked, enjoy, subject to the relevant provisions of this Convention, the freedoms referred to in article 87 of navigation and overflight and of the laying of submarine cables and pipelines, and other internationally lawful uses of the sea related to these freedoms, such as those associated with the operation of ships, aircraft and submarine cables and pipelines, and compatible with the other provisions of this Convention.
> 2. Articles 88 to 115 and other pertinent rules of international law apply to the exclusive economic zone in so far as they are not incompatible with this Part.

42 UNCLOS (n 40); Proelss (n 40).
43 UNCLOS (n 40); Proelss (n 40).

> 3. In exercising their rights and performing their duties under this Convention in the exclusive economic zone, States shall have due regard to the rights and duties of the coastal State and shall comply with the laws and regulations adopted by the coastal State in accordance with the provisions of this Convention and other rules of international law in so far as they are not incompatible with this Part.[44]

On the other hand, Article 76 of UNCLOS determines:

> The continental shelf of a coastal State comprises the seabed and subsoil of the submarine areas that extend beyond its territorial sea throughout the natural prolongation of its land territory to the outer edge of the continental margin, or to a distance of 200 nautical miles from the baselines from which the breadth of the territorial sea is measured where the outer edge of the continental margin does not extend up to that distance.[45]

Article 76 also provides for the possibility and procedure of the coastal state to establish a continental shelf between 200nm and 350nm.

The jurisdiction of coastal states is determined by the provisions of UNCLOS, which preserves for the benefit of all states the freedom of navigation, overflight, and of the laying of submarine cables and pipelines, as well as other internationally lawful uses of the sea related to these freedoms. In the exclusive economic zone, the coastal state has sovereign rights for the purpose of exploring and exploiting, conserving, and managing the natural resources of the seabed and subsoil. Similarly, in connection with the conservation of living resources, and as provided in Article 61 and related articles, UNCLOS provides as follows:

> 1. The coastal State shall determine the allowable catch of the living resources in its exclusive economic zone.
> 2. The coastal State, taking into account the best scientific evidence available to it, shall ensure through proper conservation and management measures that the maintenance of the living resources in the exclusive economic zone is not endangered by over-exploitation. As appropriate, the coastal State and competent international organizations, whether subregional, regional or global, shall cooperate to this end.
> 3. Such measures shall also be designed to maintain or restore populations of harvested species at levels which can produce the maximum sustainable yield, as qualified by relevant environmental and economic factors, including the economic needs of coastal fishing communities and the special requirements of developing States, and taking into account

44 UNCLOS (n 40); Proelss (n 40).
45 UNCLOS (n 40); Proelss (n 40).

fishing patterns, the interdependence of stocks and any generally recommended international minimum standards, whether subregional, regional or global.
4. In taking such measures the coastal State shall take into consideration the effects on species associated with or dependent upon harvested species with a view to maintaining or restoring populations of such associated or dependent species above levels at which their reproduction may become seriously threatened.
5. Available scientific information, catch and fishing effort statistics, and other data relevant to the conservation of fish stocks shall be contributed and exchanged on a regular basis through competent international organizations, whether subregional, regional or global, where appropriate and with participation by all States concerned, including States whose nationals are allowed to fish in the exclusive economic zone.[46]

Article 58 deals with the rights and duties of other States in the exclusive economic zone:

> 1. In the exclusive economic zone, all States, whether coastal or land-locked, enjoy, subject to the relevant provisions of this Convention, the freedoms referred to in article 87 of navigation and overflight and of the laying of submarine cables and pipelines, and other internationally lawful uses of the sea related to these freedoms, such as those associated with the operation of ships, aircraft and submarine cables and pipelines, and compatible with the other provisions of this Convention.
> 2. Articles 88 to 115 and other pertinent rules of international law apply to the exclusive economic zone in so far as they are not incompatible with this Part.
> 3. In exercising their rights and performing their duties under this Convention in the exclusive economic zone, States shall have due regard to the rights and duties of the coastal State and shall comply with the laws and regulations adopted by the coastal State in accordance with the provisions of this Convention and other rules of international law in so far as they are not incompatible with this Part.[47]

Again, there must be compliance of UNCLOS, coastal State's regulations, plus additional *ad hoc* normative. In addition, special treatment is provided to the Area, which:

> is the area outside the territorial jurisdiction of States. According to article 153 of UNCLOS, activities in the Area shall be carried out and controlled by the [International Seabed] Authority on behalf of mankind as a

46 UNCLOS (n 40); Proelss (n 40).
47 UNCLOS (n 40); Proelss (n 40).

> whole. The article contains rules about the manner in which these activities shall be carried out. In article 170 we find rules about the so-called Enterprise, which shall be the organ of the Authority which shall carry out activities in the Area.[48]

Lastly, to provide a common understanding of its international legal meaning, a general definition of *dumping* and *pollution* is provided in Article 1.[49]

According to UNCLOS' balance and summing up, it is evident that marine space division and regulation leads to a problematic treatment of deep seabed mining activities and elements: the relationship with biodiversity;[50] the relationship with environmental law, liabilities, and related areas; the relationship with other maritime industries.[51] Following these criteria, deep sea mining received special treatment in UNCLOS and its Annexes, upon which the International Tribunal on the Law of the Sea (ITLOS) delivered a Constative

48 Nordquist and Long (n 35).

49 '(4) "pollution of the marine environment" means the introduction by man, directly or indirectly, of substances or energy into the marine environment, including estuaries, which results or is likely to result in such deleterious effects as harm to living resources and marine life, hazards to human health, hindrance to marine activities, including fishing and other legitimate uses of the sea, impairment of quality for use of sea water and reduction of amenities.' UNCLOS (n 40); Proelss (n 40)..

50 'This specific concern has to be addressed because it leaves a gaping question with regard to the legal status on the biodiversity and genetic resources in those related areas with two different regimes. There must be a clarity in this particular situation whereby the biodiversity or genetic resources in the water column may have their life cycle starting from the subsoil and seabed or the other way around, as well as biodiversity or genetic resources that do have uninterrupted and well-connected biological linkage between the seabed and the water column in areas where overlap between extended continental shelf and BBNJ exist. The possible solution is to assign the water column above the seabed beyond 200 nautical miles that have been recommended by the Commission on Limits of the Continental Shelf to the coastal countries concerned. This may create fewer complications on the management of the biodiversity beyond national jurisdiction. The assignment may be followed with transparent monitoring by an international organ. However, some may consider this approach unfair because the coastal countries with extended continental shelves gain access to the resources in the water column. It is indeed an issue that has not been dealt with thoroughly at the intergovernmental conference (IGC) discussion in New York. A specific arrangement, a *sui generis*, may be considered in this unique overlapping situation of extended continental shelf and the water column above it which may be included as abnj. In this geographic situation of direct boundary with ABNJ, an issue of pollution or side effects coming from economic activities in the ABNJ going into the Exclusive Economic Zone or continental shelf of a coastal nation, especially small archipelagic and island developing States, must also be taken into consideration. As a matter of fact, many small island developing countries have become the epicentre of marine biodiversity in the world.' Nordquist and Long (n 35).

51 A. Boyle and D. Freestone (Eds), *International Law and Sustainable Development* (OUP 1999).

Opinion through its Deep Seabed Mining Chamber[52] and also the International Seabed Authority's production of reports, recommendations, and templates.[53]

3.2 Deep Seabed Mining in the United Nations Convention on the Law of the Sea and Annexes

The legal framework currently governing deep seabed mining is complex. UNCLOS' main provisions are included in: Part XI (The Area), Part XII (Protection and Preservation of the Marine Environment), Part XIV (Development and Transfer of Marine Technology), Part XVI (General Provisions), Annex III (Basic Conditions of Prospecting, Exploration and Exploitation), Annex IV (Statute of the Enterprise), Annex II (Commission on the Limits of the Continental Shelf); and, to a certain extent, in Annex I (Highly Migratory Species), and dispute settlement parts (XV) and Annexes (V through IX).[54]

3.2.1 Summary of UNCLOS Deep Seabed Mining Regime

Alberto Pecoraro recalls:

> Under article 136 of UNCLOS, the Area and the resources therein are the 'Common Heritage of Mankind'. This concept permeates the Area's legal regime but is not defined in UNCLOS. The provisions in Part XI of UNCLOS outline certain corollaries of that principle. Broadly speaking, they are: the prohibition on the acquisition of the deep seabed and its resources, the possibility of using the deep seabed for peaceful purposes only, the equitable sharing of benefits gained from deep seabed mining and an international management regime centered on the International Seabed Authority. Article 311(6) of UNCLOS, which prohibits any amendment to 'the basic principle relating to the common heritage of mankind set forth in article 136', makes it clear that this is one of the fundamental principles of the law of the sea.[55]

Then:

> Mankind is declared to be the one to dispose of the seabed and its resources while the International Seabed Authority acts as a trustee thereof. The meaning of the term 'mankind', which is employed by

52 R. Lopez and M. Toman, *Economic Development and Environmental Sustainability* (OUP 2006).

53 www.itlos.org/fileadmin/itlos/documents/cases/case_no_17/17_adv_op_010211_en.pdf, accessed 12 December 2021.

54 UNCLOS (n 40); Proelss (n 40).

55 A. Pecoraro, 'Law of the Sea and Investment Protection in Deep Seabed Mining' (2019) 20(2) *Melbourne Journal of International Law*.

> UNCLOS instead of the narrower 'states' or 'the international community', is clarified by article 140 of UNCLOS. This provision stipulates that activities in the Area shall be carried out for the benefit of mankind as a whole, irrespective of the geographical location of states, whether coastal or land-locked, and taking into particular consideration the interests and needs of developing states. Furthermore, the use of the term 'mankind' implies that the obligations contained in UNCLOS and the instruments adopted thereunder are owed to present and future generations too.[56]

And the following:

> The International Seabed Authority is tasked with the authorization, organization and control of activities in the Area. The Seabed Disputes Chamber of ITLOS has held that the notion of 'activities in the Area' includes the recovery of minerals from the seabed, their lifting to the water surface and the preliminary separation of materials of no commercial interest, but does not encompass transportation to points on land and onshore processing. The rules applied by the International Seabed Authority are principally set out in Annex III of UNCLOS as modified by the Part XI Implementation Agreement. These are supplemented by the 'Mining Code' a set of regulations issued by the Authority, binding by virtue of article 153 UNCLOS, which outline the procedure to apply for exploration rights and the standard terms of exploration contracts. To date, the Authority has issued Regulations on Prospecting and Exploration for Polymetallic, Regulations on Prospecting and Exploration for Polymetallic Sulphides in the Area and Regulations on Prospecting and Exploration for Cobalt-Rich Crusts. On the other hand, the Draft Regulations on Exploitation of Mineral Resources in the Area are yet to be completed and issued as a binding document.[57]

Finally:

> Article 153 of UNCLOS restricts seabed mining activities to specific categories of persons. These are states parties, state enterprises and natural or juridical persons possessing the nationality of states parties or effectively controlled by them or their nationals when sponsored by such states. For private persons, the sponsorship of one or more states is essential: the withdrawal of state sponsorship entails the end of the contractor's right to continue deep seabed mining in the Area. Applicants are qualified if they are entities possessing the nationality of a State party and effectively controlled by them or their nationals. According to the

56 Ibid.
57 Ibid.

wording of Annex III article 4(3) of UNCLOS, if effective control lies elsewhere the controlling state or the state of nationality of the controller must co-sponsor the applicant. However, in its past practice, the Authority has interpreted the effective control requirement as being equivalent to 'regulatory control', or incorporation in the sponsoring state. Consequently no contractor or potential contractor has ever been required to present a sponsorship certificate issued by the state of nationality of its controllers.[58]

3.2.2 UNCLOS and the Environment

UNCLOS Article 192 lays down the general obligation of states to protect and preserve the marine environment, while Article 193 stipulates that 'States have the right to exploit their natural resources pursuant to their duty to protect and preserve the marine environment'.[59]

In relation to the obligations of the sponsoring states, according to the Seabed Disputes Chamber, sponsoring states hold an obligation to adopt and enforce appropriate rules and measures *vis-à-vis* deep sea miners. The exercise of this duty will require sponsoring states to modify their legal framework in light of the rules adopted by the International Seabed Authority. But sponsoring states may also be tempted to modify the applicable legislation in order to better capture the economic benefits to be reaped from deep sea mining.[60]

3.3 Biodiversity

In relation to its definition, although not specifically referring to *ocean* or *marine*, Article 2 of the *Convention on Biological Diversity* states:

> 'Biological diversity' means the variability among living organisms from all sources including, inter alia, terrestrial, marine and other aquatic ecosystems and the ecological complexes of which they are part; this includes diversity within species, between species and of ecosystems.
>
> 'Biological resources' includes genetic resources, organisms or parts thereof, populations, or any other biotic component of ecosystems with actual or potential use or value for humanity.
>
> 'Biotechnology' means any technological application that uses biological systems, living organisms, or derivatives thereof, to make or modify products or processes for specific use.
>
> 'Country of origin of genetic resources' means the country which possesses those genetic resources in in-situ conditions.

58 Ibid.
59 Boyle and Freestone (n 51).
60 www.itlos.org/fileadmin/itlos/documents/cases/case_no_17/17_adv_op_010211_en.pdf, accessed 12 December 2021.

'Country providing genetic resources' means the country supplying genetic resources collected from in-situ sources, including populations of both wild and domesticated species, or taken from ex-situ sources, which may or may not have originated in that country.

'Domesticated or cultivated species' means species in which the evolutionary process has been influenced by humans to meet their needs.

'Ecosystem' means a dynamic complex of plant, animal and micro-organism communities and their non-living environment interacting as a functional unit.

'Ex-situ conservation' means the conservation of components of biological diversity outside their natural habitats.

'Genetic material' means any material of plant, animal, microbial or other origin containing functional units of heredity.

'Genetic resources' means genetic material of actual or potential value.

'Habitat' means the place or type of site where an organism or population naturally occurs.

'In-situ conditions' means conditions where genetic resources exist within ecosystems and natural habitats, and, in the case of domesticated or cultivated species, in the surroundings where they have developed their distinctive properties.

'In-situ conservation' means the conservation of ecosystems and natural habitats and the maintenance and recovery of viable populations of species in their natural surroundings and, in the case of domesticated or cultivated species, in the surroundings where they have developed their distinctive properties.

'Protected area' means a geographically defined area which is designated or regulated and managed to achieve specific conservation objectives.

'Regional economic integration organization' means an organization constituted by sovereign States of a given region, to which its member States have transferred competence in respect of matters governed by this Convention and which has been duly authorized, in accordance with its internal procedures, to sign, ratify, accept, approve or accede to it.

'Sustainable use' means the use of components of biological diversity in a way and at a rate that does not lead to the long-term decline of biological diversity, thereby maintaining its potential to meet the needs and aspirations of present and future generations.

'Technology' includes biotechnology.[61]

Most importantly, pursuant to Articles 1, 10, and 15 of the *Convention on Biological Diversity*, the sovereignty principle applies in relation to exploration, research, development, and exploitation of biodiversity, providing an agreed basis for future treaties pursuing biotech and biodata exchange.[62] This

61 Convention on Biological Diversity, www.cbd.int/convention/articles/?a=cbd-02, accessed 12 December 2021.

62 CBD, www.cbd.int/convention/text, accessed 12 December 2021.

basis was further developed in relation to access and benefit sharing by the Nagoya Protocol.[63]

In relation to its regulation, the key biodiversity agreements are over 11 relevant multilateral agreements.[64]

Since 2015 a preparatory committee has worked on the process of recommendations to UNGA in relation to BBNJ.[65] Several conferences have been held under UN mandate, dealing with sub-issues that constitute the core areas of development and conflict.[66]

Under the previously described legal framework scenario, it is possible to conclude that exploration and exploitation of deep seabed mining remains highly convenient for industrialized and financially developed states. In addition, it is probably also convenient for those states' corporations to promote foreign direct investments (FDI) in former colonial insular states in order to develop their deep seabed mining industry with lower costs. Why? Easy: lobbying availability, low rule of law and human rights compliance, and low transparency.[67]

4 Modelling Emerging Scenarios for Biodiversity Development: Blue Economy Planning

Blue Economy Plans (BEPs) can be provided pursuant to infinite business models. According to the extremely high levels of capital and technology requirements as well as the high risk of investment scenarios and strategically related clusters such as telecommunications, mining resources, and energy resources, BEPs will increase their feasibility by mixing private and public funding and management. These can be achieved, for example, through mixed participation projects (joint ventures; mixed corporations) or public–private partnerships (PPP), always trying to comply with United Nations 2030 Agenda.[68]

63 CBD, www.cbd.int/abs, accessed 12 December 2021.

64 Ardron, Rayfuse, Gjerde, and Warner (n 35); H. Govan, 'The Pacific Islands and Biodiversity Beyond National Jurisdiction.' Briefing note of the Council of Regional Organisations in the Pacific members of the Marine Sector Working Group (2014). http://dx.doi.org/10.13140/RG.2.1.1247.9527

65 'In 2015, UNGA Resolution 69/292 officially launched the two-year PrepCom process to make recommendations to the UNGA on elements of a draft text for an international legally binding instrument, implementing UNCLOS and open to non-UNCLOS parties as well.' C. Payne, 'Biodiversity in the High Seas: An Integrated Approach' (2017) 21(9) *American Society of International Law*. www.asil.org/insights/volume/21/issue/9/biodiversity-high-seas-areas-integrated-legal-approach, accessed 12 December 2021.

66 Nordquist and Long (n 35).

67 Jeff A. Ardron, 'Transparency in the operations of the International Seabed Authority: An initial assessment.' (2018) 95 Marine Policy. http://dx.doi.org/10.1016/j.marpol.2016.06.027, accessed 12 December 2021.

68 K. Mathur, *Public Private Partnerships and Public Accountability* (New Delhi: Centre for Democracy and Social Action 2021).

4.1 Public Policies and Regulations

BEP developments imply implementation by states of public policies and regulatory frameworks. Regarding public policy, it is possible to assert that the more developed the state, the more restrictive the environmental impact innovation activities will be. Therefore, developing countries and insular states show an incredible opportunity for BEPs.

BEPs' public policies normally determine fiscal burden (taxes, royalties, etc.), benefit or production sharing, the treatment of foreign investments (including special schemes for procurement and local content policies),[69] and any restrictions on the amount of production (for market or environmental reasons).

Regarding the regulatory framework, according to the principle of sovereignty, as observed, any state can wilfully adopt and implement international law. In addition, states can tailor normative pursuant to jurisdictional, legislative and constitutional criteria. Consequently, laws, decrees and regulations constitute the regulatory framework for most BEP industries (both nationally and sub-nationally), mainly including:

- The ownership of IP.
- The system of right to explore and research.
- The system of licences and authorizations.
- The system of contracts and operations (including risk sharing clauses, transfer clauses, operation, financing, confidentiality, corruption, termination, settlement of disputes, applicable law, etc.).
- The tax, financial management, and management of income (including taxes, fees, royalties, payment in kind option, currency and banking system, etc.) regime.
- The environmental regime (including risk analysis, insurance, etc.).
- The labour system (rights and obligations of workers, safety standards at work).
- The guarantee scheme (insurance, etc.) and responsibilities (contractual and tort).

4.2 A Top Hat Model

Probably, Singapore is the archetypical model of Blue Economy. This 718 square kilometres port-state has a population of 8,000,000 people with the second-highest annual per capita income; in addition, it has one of the lowest rates of unemployment and illiteracy. However, this was not the scenario in 1968, when Singapore obtained its independence from the United Kingdom. Back then, unemployment was nearing 80% and illiteracy 64%. Singapore's leader decided to apply his own BEP. He converted Singapore into a

69 For a detailed policy to promote local contracts, see M.Z. Ngoasong, 'How international oil and gas companies to respond local content policies in petroleum-producing developing countries: A narrative inquiry.' (2014) 73 *Energy Policy*.

totalitarian state with extreme rule of law, but a highly liberal economy – where a company may be incorporated in the early morning, hires and fires its employees during the day without labour charges or union inconveniences, and is dismantled by the end of the day. In relation to Singapore's trade matrix, it is highly techno-oriented: chemicals, pharma, cosmetology, circuit building, fuelling, repair shipyards, among others. This was the conclusion of 40 years of a BEP based on the understanding that political sacrifices, capital promotional measures, and intensive technological investment were necessary in order to provide economic high standards to 8 million people within 718km^2.

4.3 An Exponentially Growing Model

Italian scholars assert that '[s]ince 2014, China quickly embraced BE logics'.[70] The government's focus on BE:

> dates to the 11th 5-year Plan (2006–2011) and Beijing's strategic economic development plan, including data covering the marine economy (People's Republic of China, 2006). The performance during this period registered an annual growth of 13.5% with 33 million jobs in 2010 (China Briefing, 2013). As presented by the State Oceanic Administration (2014b), the country's BE consists of different economic sectors, ranging

70 'Recognizing that the high pollution and the overuse of natural resources were causing environmental degradation. Based on the most recent Marine Environment Bulletin of China's State Oceanic Administration, large parts of coastal areas and territorial seas are severely polluted. Furthermore, fish landings in China's coastal areas are decreasing due to the overuse of near-shore fisheries. As a second factor, China became aware of the opportunity offered by 9000 miles of coastline, linking the North Korean border to the Vietnamese border. Coastal regions are essential components for the Chinese "economic apparatus" since Deng Xiaoping's reform and opening-up of China's borders in 1979. Since then, China's coastal regions are considered one of the most economically developed and populated areas in the world, accounting for 20% of China's land area, 40% of the population and 60% of GDP. Firmly conceived as "gates" from which China should achieve a prosperous society, 40 years later the Chinese economic model has developed becoming the second largest global economic power in terms of GDP. Leveraging on these factors, in 2003 the State Council issued the first guideline document for the development of the national marine economy "An Outline of the Planning of National Marine Economy Development," which claimed a Gross Value-Added for the ocean economy, respectively, of 4% of GDP in 2005 and 5% in 2010.' Wei Song, Guang He, and Alistair Mcilgorm, A. 'From behind the Great Wall: The development of statistics on the marine economy in China.' (2013) Marine Policy 39 120–127. http://dx.doi.org/10.1016/j.marpol.2012.09.006. Then the National Marine Industrial Development Plan of 2008 discussed the strategic role that the Chinese authorities allocated to the marine industry as a crucial contributor to China's socialist modernization. In the following decades, China's marine economy reached these targets, and the value of Chinese marine activities was estimated at 9.87% of GDP in 2008 (Song et al.).

from fishing, shipping, oil and gas production to tourism. Additionally, the search for new organisms, known as marine bioprospecting, continued to provide insights to the pharmaceutical and cosmetic industries. Over 2012, the share of GDP from China's marine economy increased by 7.6% (Conathan & Moore, 2013) with tourism accounting for a great part of BE (35%) while activities for resource exploitation such as oil and gas grew only at 0.1%. On the contrary, ocean mining increased by 14% (Conathan & Moore, 2013) and marine bioprospecting grew by 21% (China Briefing, 2014). the 12th 5-Year Plan for National and Social Development (2011–2015), issued by the State Council, included a specific goal for BE to make up 10% of GDP by 2015 (China Briefing, 2013). Thus, 2011 represents an unprecedented step for China to release a '5-Year-Plan' in which 'developing the ocean economy' is presented as a major national strategy for economic development. The targets specified that the rate of R&D expenditure should reach 2% of the total output value for the marine economy (China Briefing, 2013). In this regard, the plan stated that scientific planning should be promoted for supporting the marine industries. As reported by Zhao et al. (2013), the plan focused on ensuring the sustainable usage of marine resources based on the improvement of marine management and marine environmental protection. Moreover, to further promote the development of seawater desalination and marine biomedicine, two national plans were released. Lastly, the 13th Five-Year Plan for Economic and Social Development restates the importance of ensuring the development of BE. For encouraging industrial development, the Plan proposes restoring the marine ecological environment and safeguarding the sustainable development of water fishery (Wenhai et al, 2019).'[71]

4.4 A Developing State Model

The most common and relevant variable for developing states such as Nigeria is the existence of vast natural resources along with the lack of capital and technological capacity.[72] In addition, the environmental risks of industrialization deter development initiatives, including sustainable BEPs.[73] In the end,

> [a]lthough the government may be required to provide the enabling environment to facilitate the diversification of resources into the development of the blue economy, the private sector also has a pivotal role to play in its success. There is the need for strategic local and private lump-sum investments in the development of the sectors of the blue economy

71 Sampaolo, Lepore, and Spigarelli (n 4).
72 Gbadegesin and Simisola (n 2).
73 Ibid.

> particularly in the maritime and fishing sub-sectors in order to improve their contributions to the economy. There is also the need to raise awareness to better educate stakeholders on what the blue economy is and why it matters. [...] It is further recommended that for the blue economy to be sustainable, the country needs to develop and/or strengthen national policies to better integrate blue economy considerations into national and sub-national policy and governance frameworks. [...] In a nutshell, a coordinated policy planning process for sustainable development of the country's ocean economy will require active participation and decisions by a wide range of public agencies, linked by common objectives and actively sharing information.[74]

4.5 A Simple and Feasible Model

Professional practice in developing countries shows that high-tech, high-cost private start-ups are uncommon. Programmes like these imply regional or global financing authorities (e.g. World Bank, Inter-American Development Bank, BICE). In those cases, a simple BEP model employed to reduce risks, increase stakeholders' interventions, and include social development is the following:

a Actions can be deployed by a mixed consortium or public–private partnerships (PPPs) (participation of companies and foreign and national research entities, on the one hand, and public and scientific research entities, on the other) through development agreements to carry out the general purposes which are divided into stages of work, topics of work, and objectives of work.

b It has three stages of work: (i) evaluation of the project, call for international tender, and the composition of the joint venture (JV); (ii) execution of the R&D programme (research and development of the marine area potential); and (iii) the implementation of the final BEP.[75]

c The corporations constituting the JV must be specialized in each of the mentioned 20 clusters.

d The corporations shall create the JV from a financial contribution and a technical contribution (know-how, qualified labour skills, or assets) in order to implement the project.

e The profit to be received by the corporations participating in the JV is the priority on the bidding rounds within each cluster (e.g. operation of ports, ports construction, housing construction); that is to say, when implementing the final development plan, the corporations that contributed

74 Ibid.

75 Carried out by the administrative branch with competence, according to the reports of the R&D stage based upon an international public tender in which priority will be given to companies that have participated in the R&D stage.

and worked for their creation shall have priority in the bidding process against their competitors.

The institutional role and the role of a BEP:

a The main purpose of the plan is to achieve sustainable economic and social development that has not yet existed in the country or area, achieving the training of specialists and technicians in sectors until now little developed or not developed at all, labour forces, small and medium-size enterprises and generators of public policies.
b The fulfilment of the first purpose entails the readjustment of the maritime coast in socio-economic, urban, infrastructural, etc. means.

5 Closing Remarks

According to the data analysed as well as the behaviours and governmental trends described in this work, the following conclusions regarding the performance and prospective development of deep seabed mining business can be identified:

Table 13.5 Conclusions and balances

State	*BE performance*	*Reasons*
Singapore	High	High Worldwide Governance Index (WGI) rating (especially anticorruption and rule of law) Low product diversification with high technology and high investment requirements High BE business incentives and regulation flexibility (low taxes, low union burden) High government interest in BE public policy improvement High liaison with marine colonial states
China	Increasing	High government interest in BE public policy improvement High business incentives and regulation flexibility (low taxes, low union burden) Low control of overseas BE business activities Aggressive overseas business policy
Nigeria	Low	Lack of government BE interest Low BE business incentives and legal framework Low WGI rate (especially anticorruption and rule of law) Low product diversification with low technology and low investment No overseas BE business activities

Traditional public policy analysis shows that any state must move cautiously and prudently when implementing or changing an offshore development model. Attempts to implement segregation of duties in countries where institutional prerequisites are absent can be counterproductive. These reforms can be useless and even harmful to the market and industry – for example, in the following ways:

- Initiatives that are overly broad and unworkable can displace more gradual reform efforts, which could actually be substantive and sustainable, and only serve the interests of those who benefit from the status quo.
- Reform efforts focused on the creation of new organs of government can better distribute the limited financial and human resources.
- Multiplying the points of engagement with government officials can increase opportunities for corruption.
- Corruption can lead to social cynicism and preclude future reforms.

The following considerations are also relevant:

- Separation of functions is rarely feasible in situations in states where there is either high political regime instability or low rule of law.
- Regimes with little support from public opinion simply have little inclination to allow a government agency to develop the independent regulatory authority.
- The ability of a country to implement the separation of functions meaningfully depends largely on the bureaucratic and administrative maturity of their state apparatus and its level of institutional development.
- The countries which lack institutional capacity (both in terms of their industrial sector and bureaucracy) can benefit from not establishing a model of separation of functions early.
- The existence of a strong system of checks and balances can provide fundamental stability and resistance against political or economic shocks, even if the separation of functions does not seem necessary, while political change remains low.

There is some debate on the merits of a gradual rearrangement improving many institutional factors. The latter approach, for example, might allow better coordination of reforms with political considerations, and may also yield benefits more quickly to provide credibility and maintain the momentum for reform. However, such larges deals are complicated by the lack of lasting institutions which can credibly negotiate and compromise with political concession-holders. There are simply no such credible institutions in countries like Nigeria, regardless of what the organizations look like on paper.

On the other hand, it is worth mentioning there are current scenarios where WGI conditions should affect performance negatively (e.g. China). Nevertheless, government centralization and strong political interests promote efforts to improve BE development with a heterogeneous package of measures that seems to be prospectively positive. Something similar happened in Singapore's transitional period towards its current condition, where measures had to be extremist due to the socio-economic and historical linkage of oil with the nation's economic evolution.

Lastly, countries with high local marine resource availability but extremely low WGI rates remain at the bottom of BE development. The lack of promotional public policies and very low private sector eagerness to be innovators increase this variable.

Concluding, with a newborn area of the economy and legal business practice, the potential of energy in the deep seabed mining cluster of Blue Economy is yet to be exercised, though its potential is evidently more than worth the try – particularly considering its underdeveloped context. Insular states with strong colonial liaisons already understand it. It is time for self-reliant states with technological and financial capacity such as India and Brazil to follow.

References

Ardron, J.A. 'Transparency in the operations of the International Seabed Authority: An initial assessment' (2018) 95*Marine Policy* 324–331. http://dx.doi.org/10.1016/j.marpol.2016.06.027, accessed 12 December 2021.

Ardron, J.A., Rayfuse, R., Gjerde, K., and Warner, R. 'The sustainable use and conservation of biodiversity in ABNJ: What can be achieved using existing international agreements?' (2014) 49*Marine Policy* 98–108. http://dx.doi.org/10.1016/j.marpol.2014.02.011, accessed 12 December 2021.

Boyle, A. and Freestone, D. (Eds) *International Law and Sustainable Development* (Oxford University Press 1999).

Convention on Biological Diversity, www.cbd.int/convention/articles/?a=cbd-02

Danzinger, P.N. '6 Trends Shaping the Future of the $532B Beauty Business.' Forbes, 1 September 2019. www.forbes.com/sites/pamdanziger/2019/09/01/6-trends-shaping-the-future-of-the-532b-beauty-business/?sh=36f7934d588d, accessed 12 December 2021.

Ellsmoor, J. 'RethinkingOurOceans:InvestinginTheBlueEconomy.' Forbes, 26 January 2019. www.forbes.com/sites/jamesellsmoor/2019/01/26/rethinking-our-oceans-investing-in-the-blue-economy/?sh=26de41423531, accessed 12 December 2021.

Ferguson, N. *The Ascent of Money: A Financial History of the World* (Penguin 2018).

Ferrara, P. *Don't Be Afraid, It's Nature: A Theory of Thermodynamic Pragmatism'* (Arraes 2020).

Gbadegesin, O. and Simisola, A. 'Charting the Course for a Blue Economy in Nigeria: A Legal Agenda' (2021) *1*(7) *Journal of Environmental Law and Policy*. https://doi.org/10.33002/jelp001.01, accessed 12 December 2021.

Goodin, R.E. *The Oxford Handbook of Political Science* (Oxford University Press 2018).

Govan, H. 'The Pacific Islands and Biodiversity Beyond National Jurisdiction.' Briefing note of the Council of Regional Organisations in the Pacific members of the Marine Sector Working Group (2014). http://dx.doi.org/10.13140/RG.2.1.1247.9527

Mathur, K. *Public Private Partnerships and Public Accountability* (New Delhi: Centre for Democracy and Social Action 2013).

Jadhav, A. *Minerals from the Sea* (Dakshin Foundation 2016).

Lopez, R. and Toman, M. *Economic Development and Environmental Sustainability* (Oxford University Press 2006).

Jones, M.C. and Cheung, W.W.L. 'Multi-model ensemble projections of climate change effects on global marine biodiversity' (2015) 72(3) *ICES Journal of Marine Science.* https://doi.org/10.1093/icesjms/fsu172, accessed 12 December 2021.

Modalities for Advancing Cross-Sectoral Cooperation in Managing Marine Areas Beyond National Jurisdiction. UNEP (DEPI)/RS.12/8.

Ngoasong, M.Z. 'How international oil and gas companies to respond local content policies in petroleum-producing developing countries: A narrative inquiry.' (2014) 73*Energy Policy* 471–479.

Nordquist, M.H. and Long, R. *Marine Biodiversity of Areas Beyond National Jurisdiction.* Leiden: Brill Nihjoff 2021).

Gbadegesin, O.A. and Akintola, S. 'Charting the Course for a Blue Economy in Nigeria: A Legal Agenda' (2021) 1*Journal of Environmental Law and Policy* 6–25. https://ssrn.com/abstract=3860675

Payne, C. 'Biodiversity in the High Seas: An Integrated Approach' (2017) 21(9) *American Society of International Law.* www.asil.org/insights/volume/21/issue/9/biodiversity-high-seas-areas-integrated-legal-approach

Pecoraro, A. 'Law of the Sea and Investment Protection in Deep Seabed Mining' (2019) 20(2) *Melbourne Journal of International Law* 530.

Placek, M. 'Protected size of the container shipping market worldwide 2020–2028.' Statista, 21 April 2022. www.statista.com/statistics/1097059/global-shipping-containers-market-size, accessed 12 December 2021.

Proelss, A. *UNCLOS: A Commentary* (Hart-NOMOS 2016).

Sampaolo, G., Lepore, D. and Spigarelli, F. 'Blue Economy and the quadruple helix model: The case of Qingdao' (2021) 23*Environment, Development and Sustainability* 16803–16818. https://doi.org/10.1007/s10668-021-01378-0

Shahbandendeh, M. 'Global seafood market value 2020–2025.' Statista, 16 December 2021. www.statista.com/statistics/821023/global-seafood-market-value

Song, W., He, G., and Mcilgorm, A. 'From behind the Great Wall: The development of statistics on the marine economy in China.' (2013) *Marine Policy* 39120–127. http://dx.doi.org/10.1016/j.marpol.2012.09.006.

Sustainable Development Goals, www.un.org/sustainabledevelopment/sustainable-development-goals, accessed 12 December 2021.

UNCLOS, www.un.org/Depts/los/convention_agreements/texts/unclos/closindx.htm, accessed 12 December 2021.

Warner, R. 'The sustainable use and conservation of biodiversity in ABNJ: What can be achieved using existing international agreements?' (2014) 49*Marine Policy* 98–108. http://dx.doi.org/10.1016/j.marpol.2014.02.011

Weaver, P.P.E., Benn, A., Arana, P.M., Ardron, J.A., Bailey, D.M., Baker, K., Billett, D.S. M., Clark, M.R., Davies, A.J., Durán Muñoz, P., Fuller, S.D., Gianni, M., Grehan, A. J., Guinotte, J., Kenny, A., Koslow, J.A., Morato, T., Penney, A.J., Perez, J.A.A., Priede, I.G., Rogers, A.D., Santos, R.S., and Watling, L. 'The impact of deep-sea fisheries and implementation of the UNGA Resolutions 61/105 and 64/72.' Report of an international scientific workshop, National Oceanography Centre, Southampton, 2011. http://hdl.handle.net/10013/epic.37995, accessed 12 December 2021.

14 Epilogue

Justice in Managing the Nexus between Mining, Energy Transitions, and Climate Change

Victoria R. Nalule

The mining sector's importance has been emphasized in the twenty-first century, given the increasing demand and role of critical minerals such as cobalt which are essential in the global transition to a low-carbon economy. The African continent is home to enormous reserves of oil, gas, energy, and solid minerals. These resources are crucial for these countries' economic development. Nevertheless, in recent years, while we have witnessed initiatives to address climate change, the 2022 global energy crisis spotlighted the continued role of extractives (both fossil fuels and minerals) in our economies. There is therefore a need to strike a balance in managing the extractive industries in order to adequately achieve the net zero targets and other climate change goals, as envisaged in the 2015 Paris Agreement and Goal 13 of the United Nations Sustainable Development Goals (SDGs). Scholars have advocated for just energy transitions,[1] while others, like Victoria Nalule, have insisted on the 'progressive' way of managing the extractive industries and climate change.[2] Consequently, the present and future laws, legislation, and policies of the African extractive industries, including oil, gas, and minerals, should take into consideration these environmental and sustainable global efforts.

The contributions to this book have highlighted several challenges and opportunities facing the mining sector, especially in this energy transition and climate change era. The book doesn't provide all the answers but rather, referring to different jurisdictions in Africa; it gives an analysis of different approaches and methodologies in regulating and managing the African mining sector. As the chapters in this book show, different African countries have different mining laws and policies. However, there are some common concerns shared by these countries, which can be addressed through regionalism or regional cooperation in the mining sector (Chapter 8). The book highlights the challenges and frustration of modernising the mining laws

1 McCauley D, Ramasar V, Heffron RJ, Sovacool BK, Mebratu D, Mundaca L. Energy justice in the transition to low carbon energy systems: Exploring key themes in interdisciplinary research. Applied Energy. 2019; 233: 916–921.

2 Nalule VR. How to Respond to Energy Transitions in Africa: Introducing the Energy Progression Dialogue. In *Energy Transitions and the Future of the African Energy Sector* (pp. 3–35). Cham: Palgrave Macmillan, 2021.

DOI: 10.4324/9781003284437-17

based on the categorisation of minerals including critical minerals, precious minerals, and construction minerals (Chapter 1). The increasing role of critical minerals in ensuring energy transitions and industrialisation is also espoused in Chapters 1 and 2. The fiscal regime is crucial in ensuring that African countries maximise the revenues from the mining sector, and this issue has been emphasised in Chapters 3 and 7. Although countries have made progress in formalising and legislating Artisanal and Small-Scale Mining, there are pertinent environmental, social, and governance issues that need more attention. Consequently, the book has authoritatively spotlighted the practical ways to address the challenges associated with ASM in Africa (Chapters 4, 5, and 6). Throughout the book, authors echo the importance of digitalisation and technological advancement in not only addressing the challenges in ASM, but also ensuring the achievement of ESGs in mining (Chapters 1 and 9). Moreover, the massive investments and capital needed in the development of the mining sector draws our attention to the influence of Chinese investments in the African mining sector (Chapter 10). Understanding how climate change impacts or is impacted by the mining sector is crucial in the twenty-first century, and, as such, authors have rightly analysed the nexus between climate change and mining (Chapters 1, 2, and 11). Lastly, the contributors have paid extra attention to the issues concerning deep sea mining (Chapters 12 and 13).

Although the book has emphasised the need to address the pertinent challenges in the African mining sector, we should emphasise the need to apply justice in addressing these challenges. The principles of energy justice[3] can be borrowed to understand how the mining sector can be governed and developed in a just way, not only to ensure economic development but also to address climate change and support the energy transition. For instance, the theory of justice is key when analysing the impact of land access on energy and mining projects in Africa. With land justice theory, the key aspects include:[4]

- procedure land justice: this focuses on the legal and regulatory processes for accessing land for energy and mining projects (including relocations of affected communities and fair compensation);
- distributive land justice: concerns the distribution of benefits and also the negatives that accrue from using land for extractive projects (including addressing the environmental damage; sharing the oil and mining revenues equally);
- recognition land justice: entails recognising and protecting the land rights of the different local communities (cultural land must be respected; community and customary land tenures must be recognised)

3 Jenkins K, McCauley D, Heffron R, Stephan H. Energy justice, a whole systems approach. *Queen's Political Review*. 2014; 2(2): 74–87.

4 Nalule VR. *Land Law and the Extractive Industries: Challenges and Opportunities in Africa*. Hart Publishers, 2021.

- restorative land justice: entails ensuring that the land is rehabilitated at the end of oil and mining activities (this includes mining closures and decommissioning);
- cosmopolitanism land justice: this stems from the notion that we are all citizens of the world, and, as such, we should consider the impact of extractives on land use beyond our borders and from a global perspective.

Besides land justice in the mining sector, the justice theory should also be applied when negotiating mining agreements. For instance, stabilisation clauses that are detrimental to the development of the mining sector should be avoided in the Global South, especially since most developed countries do not offer these stabilisation mechanisms.[5] In conclusion, therefore, protection of human rights and ensuring justice should be at the centre while addressing all the other challenges in the African mining sector as highlighted throughout this book.

5 Nalule VR. What is the Problem with Stabilization Clauses in Petroleum Agreements? *Journal of Sustainable Development Law and Policy.* 2022; 13(1): 85–102.

Index